COLD WATER OIL

Cold Water Oil: Offshore Petroleum Cultures is a collection of essays examining how societies conceive of fossil fuel extraction in the inhospitable but fragile waters of the North Atlantic and Arctic oceans.

What happens offshore matters. Currently, over a quarter of the world's oil and gas is produced from beneath the seas. The offshore petroleum industry is thus a crucial point of origin for global carbon emissions, and other environmental harms. *Cold Water Oil: Offshore Petroleum Cultures* illuminates ignored histories, influential contemporary narratives, and emerging energy and environmental futures. The volume centres on North Atlantic and Arctic regions; the continuing but often strongly contested pursuit of oil and gas in frigid, tumultuous, and environmentally sensitive seas enforces the lengths to which corporations and governments will go to maintain the centrality of fossil fuels. The book's contributors focus on the cultural, social, and ecological implications of oil and gas extraction in the oceanic territories of Canada, Norway, the UK, Russia, the US, and the Iñupiat of Alaska at a time of profound global uncertainty. In conversation with the energy and environmental humanities, and critical ocean studies, *Cold Water Oil* considers a region central to debates about climate change and the planet's future.

Cold Water Oil engages students and researchers interested in climate change, energy humanities, critical ocean studies, and North Atlantic and Arctic issues.

Fiona Polack is Associate Professor in the Department of English at Memorial University of Newfoundland and Labrador, and Academic Editor at Memorial University Press. She researches and publishes in the fields of energy and environmental humanities, island studies, and settler colonial studies. She currently leads the Social Sciences and Humanities Research Council of Canada Insight project

"Oil Rigs and Islands" (2020–25), which examines contemporary cultural figurations of place-making in the offshore world, and collaborates with Danine Farquharson on other energy humanities projects. Fiona Polack is co-author of *After Oil* ((2016), with the Petrocultures Research Group. Her edited collection *Tracing Ochre: Changing Perspectives on the Beothuk* was published in 2018.

Danine Farquharson is Associate Professor of English at Memorial University. Her early research focused on masculinity and violence in contemporary Irish literature and film. More recently, her interests are in energy humanities, specifically collaborating with Fiona Polack. Together, they examine how the North Atlantic offshore oil and gas industry is imagined in a wide range of high and popular contexts – everything from oil company websites, to visual art, to literary fiction. She is the co-editor of *Shadows of the Gunmen: Violence and Culture in Modern Ireland* and the co-founder with Julia Wright of Dalhousie University of SSHORE: Social Science and Humanities Ocean Research and Education.

Routledge Environmental Humanities

Series editors: **Scott Slovic** *(University of Idaho, USA)*, **Joni Adamson** *(Arizona State University, USA)* and **Yuki Masami** *(Aoyama Gakuin University, Japan)*

The *Routledge Environmental Humanities* series is an original and inspiring venture recognising that today's world agricultural and water crises, ocean pollution and resource depletion, global warming from greenhouse gases, urban sprawl, overpopulation, food insecurity and environmental justice are all *crises of culture.*

The reality of understanding and finding adaptive solutions to our present and future environmental challenges has shifted the epicenter of environmental studies away from an exclusively scientific and technological framework to one that depends on the human-focused disciplines and ideas of the humanities and allied social sciences.

We thus welcome book proposals from all humanities and social sciences disciplines for an inclusive and interdisciplinary series. We favour manuscripts aimed at an international readership and written in a lively and accessible style. The readership comprises scholars and students from the humanities and social sciences and thoughtful readers concerned about the human dimensions of environmental change.

For more information about this series, please visit: www.routledge.com/Routledge-Environmental-Humanities/book-series/REH

COLD WATER OIL

Offshore Petroleum Cultures

Edited by
Fiona Polack and
Danine Farquharson

Routledge
Taylor & Francis Group
LONDON AND NEW YORK

earthscan
from Routledge

Cover image: © Getty Images

First published 2022
by Routledge
2 Park Square, Milton Park, Abingdon, Oxon OX14 4RN

and by Routledge
605 Third Avenue, New York, NY 10158

Routledge is an imprint of the Taylor & Francis Group, an informa business

British Library Cataloguing-in-Publication Data
A catalog record for this book is available from the British Library

Library of Congress Cataloging-in-Publication Data
Names: Polack, Fiona, editor. | Farquharson, Danine, editor.
Title: Cold water oil : offshore petroleum cultures / edited by Fiona Polack and Danine Farquharson.
Description: Abingdon, Oxon ; New York, NY : Routledge, 2022. | Series: Routledge environmental humanities | Includes bibliographical references and index.
Identifiers: LCCN 2021034034 (print) | LCCN 2021034035 (ebook) | ISBN 9780367903930 (hardback) | ISBN 9780367903923 (paperback) | ISBN 9781003024125 (ebook)
Subjects: LCSH: Offshore oil industry—Arctic Regions. | Offshore oil industry—North Atlantic Region. | Arctic Regions--Environmental conditions. | North Atlantic Region—Environmental conditions.
Classification: LCC HD9578.A68 C65 2022 (print) | LCC HD9578.A68 (ebook) | DDC 333.8/23209162—dc23
LC record available at https://lccn.loc.gov/2021034034
LC ebook record available at https://lccn.loc.gov/2021034035

ISBN: 978-0-367-90393-0 (hbk)
ISBN: 978-0-367-90392-3 (pbk)
ISBN: 978-1-003-02412-5 (ebk)

DOI: 10.4324/9781003024125

Typeset in Bembo
by codeMantra

CONTENTS

ILLUSTRATIONS

Figures

Maps

Tables

ACKNOWLEDGEMENTS

Our first and most resounding thanks go to this volume's contributing authors. We're so grateful you agreed to participate in this project and for your help sketching *Cold Water Oil*'s parameters through your thought-provoking and methodologically diverse essays.

We first met some, but not all, of this book's contributors at *Petrocultures 2016: The Offshore*, an international conference we hosted at Memorial University of Newfoundland and Labrador. Our thanks to Imre Szeman and Sheena Wilson who initiated the Petrocultures Research Cluster at the University of Alberta in 2012 and to conference sponsors: the Social Sciences and Humanities Research Council of Canada, Memorial University's Conference Fund, and Memorial's Scholarship in the Arts program. Residual conference funds supported the graduate student assistant salaries associated with this volume. We have been incredibly fortunate to have the help of PhD students Katie Prowse and Morgen Mills; they've provided expert research, editing, and indexing assistance. Our gratitude also to Carol Sullivan in the School of Graduate Studies at Memorial who helped with financial administration.

Thanks to Brent Ryan Bellamy, Jason Haslam, and Rachel Webb Jekanowski for their insightful comments on preliminary drafts of "Heading Offshore" and to the anonymous peer reviewers and series editors who provided feedback on the project as a whole. Rosie Anderson and Rebecca Brennan at Routledge have been wonderful to work with.

Finally, thanks to our families: Danine's Jamie and Tess, and Fiona's Rob and Susannah. Our climate activist, Gen-Z daughters are an ongoing source of inspiration and admiration.

CONTRIBUTORS

Rosemary Ahtuangaruak, recipient of an Honorary Doctorate in Humanities from Oberlin College in 2017, is an Iñupiaq tribal leader and human rights advocate from Nuiqsut, Alaska. She is a founding member of Resisting Environmental Destruction on Indigenous Lands (REDOIL), an inter-tribal Alaska Native grassroots network that challenges the fossil fuel and mining industries. As a former mayor of the Native Village of Nuiqsut, Dr. Ahtuangaruak's congressional testimony to oppose oil and gas development in culturally and biologically significant places in Arctic Alaska contributed to President Obama's decision to ban oil drilling in large areas of the Arctic and Atlantic Oceans. In spoken and written words, as well as actions since the 1980s, Dr. Ahtuangaruak brings together the environmental, cultural, social, economic, and the political. Her activism is deeply rooted in traditional knowledge, cultural practices, social service, and environmental and social justice. She is an active member of the Nuiqsut City Council.

Gavin Bridge is Professor of Economic Geography at Durham University and a Fellow of the Durham Energy Institute. His research addresses questions of property, access, and control associated with emergent geographies of resource production and consumption, including the political ecologies of resource scarcity and security. Recent work includes research on raw material production networks associated with old and new carbon economies; and a continuing interest in how the materiality of resources shapes their appropriation and capitalization. He is the author of *Oil* (Polity 2017) with P. Le Billon; co-editor of the *Routledge Handbook of Political Ecology* (2015) with T. Perreault and J. McCarthy; and co-author of *Energy and Society: A Critical Perspective* (Routledge 2018) with S. Barr, S. Bouzarovski, M. Bradshaw, E. Brown, H. Bulkeley, and G. Walker.

Brigt Dale is Research Director of the Environment and Society Research Group at the Nordland Research Institute, Norway. He holds a PhD in Political Science and an MD in Visual Anthropology from the University of Tromsø, Norway, and focuses on theory development (particularly on societal transformation, ontological security, governmentality, biopolitics, and cultural theory) and research on the relation between politics, security, power, and resource management. Dale's empirical research interests include climate change and societal transformation, petroleum politics/petrocultures, and extractive industries' impact on local lives, tourism, fisheries and aquaculture, and cultural heritage. He has done fieldwork in Tobago (2001), the Lofoten Islands (from 2008 and up to the present), in Finnmark (2014, 2015), and in Greenland (2014, 2019 and forthcoming).

Amy Donovan is a PhD candidate in Anthropology at McGill University. Her research is on the ways humans know and listen to whales, and how we might expand our listening. She recently completed her masters in creative writing at Memorial University of Newfoundland and Labrador, where she wrote a novel, *The Ways of the Bird Rising*, that explores how a more-than-human community of narrators experiences offshore fracking. Her creative work, centred on multispecies storytelling, has been published in journals and collections including *Riddle Fence* magazine and Book*hug's *Write Across Canada*. She is from Cape Breton, Nova Scotia.

Danine Farquharson (co-editor of *Cold Water Oil*) is Associate Professor of English at Memorial University. Her early research focused on masculinity and violence in contemporary Irish literature and film. More recently, her interests are in energy humanities, specifically collaborating with Fiona Polack. Together, they examine how the North Atlantic offshore oil and gas industry is imagined in a wide range of high and popular contexts – everything from oil company websites, to visual art, to literary fiction. She is the co-editor of *Shadows of the Gunmen: Violence and Culture in Modern Ireland* and the co-founder with Julia Wright of Dalhousie University of SSHORE: Social Science and Humanities Ocean Research and Education.

Jason Haslam is Professor of English at Dalhousie University. His research includes work on prison studies, popular culture, science fiction, and the gothic. He is the author or editor of several books, including the monograph *Gender, Race, and American Science Fiction* (Routledge 2018), the essay collection *American Gothic Culture*, edited with Joel Faflak (Edinburgh University Press 2017), and a scholarly edition of Ian Fleming's novel *Casino Royale*, edited with Julia M. Wright (Broadview 2020).

Rachel Webb Jekanowski is an Assistant Professor of English Literature at Memorial University's Grenfell Campus and an interdisciplinary film and media scholar. Her research focuses on entanglements of visual culture, industry, settler colonialism, and environments in North America, historically and in the times-to-come. Her work has been published in *Canadian Journal of Film Studies*, *Canadian Journal of Communications*, *Middle East Journal of Culture and Communication*, and *The Goose*.

Berit Kristoffersen is a political geographer working on energy futures, environmental governance, climate change, and ocean management. She is primarily interested in the spatial outcomes of policy processes and notions or resources as part of space-making processes. She is currently researching renewable energy in northern Norway, global initiatives for supply-side climate policy, and political practices of ocean policy in dynamic ecologies in the Arctic. With co-authors Hannes Gerhardt and Kirsti Stuvoy, she published "Saving the Arctic: Greenpeace or Oil Riot?" in *The Politics of Sustainability in the Arctic: Reconfiguring Identity, Space, and Time*, edited by Ulrik Pram Gad and Jeppe Strandsberg (Routledge 2019). She also had a further chapter, co-written with Philip Steinberg, in the same collection titled "Building a Blue Economy in the Arctic Ocean: Sustaining the Sea or Sustaining the State?" Her article, with Phillippe Le Billon, "Just Cuts for Fossil Fuels? Supply-Side Carbon Constraints and Energy Transition," appeared in *Environment and Planning D*.

Graeme Macdonald is Professor in the Department of English and Comparative Literary Studies at the University of Warwick, UK. He is editor of the *Routledge Handbook of Energy Humanities* (2022) and journal collections *Powering the Future: Energy Resources in Science Fiction and Fantasy* (2019) and *Environment, Ecology and 'Nature' in 21st Century Scottish Literature* (2021). He also edited a recent edition of John McGrath's play *The Cheviot, the Stag and the Black, Black Oil* (2015). He is co-author, with the WReC collective, of *Combined and Uneven Development: Toward a New Theory of World Literature* (2015). A member of the Petrocultures Research Group, he is also co-author of *After Oil* (2016). He was Chief Investigator (CI) on the Royal Society of Edinburgh Humanities Research Network, *Connecting with a Low Carbon Scotland* and is presently CI on the FORMAS funded international research project *Climaginaries*: narrating socio-cultural transitions to a post-fossil society.

Fiona Polack (co-editor of *Cold Water Oil*) is an Associate Professor in the Department of English at Memorial University of Newfoundland and Labrador and Academic Editor at Memorial University Press. She researches and publishes in the fields of energy and environmental humanities, island studies, and settler colonial studies. She currently leads the Social Sciences and Humanities Research Council of Canada Insight project "Oil Rigs and Islands" (2020–25), which examines contemporary cultural figurations of place-making in the offshore world, and collaborates with Danine Farquharson on other energy humanities projects. Fiona Polack is co-author of *After Oil* (West Virginia University Press 2016), with the Petrocultures Research Group. Her edited collection *Tracing Ochre: Changing Perspectives on the Beothuk* was published by University of Toronto Press in 2018.

Nina Poussenkova is Senior Researcher of the Primakov Institute of World Economy and International Relations (IMEMO), Russian Academy of Sciences; and Researcher, ENERPO Center of the European University at Saint Petersburg. She graduated from the Department of Economics at Moscow State University and defended her PhD thesis in IMEMO. In the 1990s, she worked in the Center for Foreign Investment and Privatization and in Salomon Brothers and Lazard Freres investment banks. From 2006 to 2008, Dr. Poussenkova was director of the Energy Program of the Carnegie Moscow Center. She closely cooperated with WWF-Russia leading projects on corporate sustainability in Russia. Currently, she heads the IMEMO Oil and Gas Dialogue Forum. Her research interests are focused on Arctic petroleum development, national oil companies, and Russian energy policy. Dr. Poussenkova has authored over 100 publications on the Russian and international energy sector published both in Russia and internationally.

Helge Ryggvik is one of Norway's leading thinkers on the history and present-day realities of the petroleum industry. He is currently a Professor at the Centre for Technology, Innovation, and Culture at the University of Oslo. During recent years, Ryggvik has worked for Norsk Oljemuseum (The Norwegian Petroleum Museum) on a project initiated by the Norwegian Parliament about Norwegian offshore oil divers. He has also worked on safety projects financed by the Norwegian Research Council. From offshore divers in Norway, to offshore safety and supply systems, to railway histories, his research is based in both broad political economies and analyses of human relations to technology.

Chie Sakakibara is Associate Professor of Geography & the Environment and Native American & Indigenous Studies at Syracuse University. Prior to her current appointment she served as Associate Professor of Environmental Studies at Oberlin College. As a native of Japan, Sakakibara's academic training is in cultural geography, art history, and Native American and Indigenous studies. Her work explores the human dimensions of global environmental change among Indigenous peoples, and she is an adoptive member of three Iñupiaq whaling crews in Utqiaġvik, Alaska. She is the author of *Whale Snow: Iñupiat, Climate Change, and Multispecies Resilience in Arctic Alaska* (University of Arizona Press 2020).

Philip Steinberg is Professor of Political Geography and Director of IBRU: The Centre for Borders Research at Durham University, UK, where he also directs the Durham Arctic Research Centre for Training and Interdisciplinary Collaboration (DurhamARCTIC). His research focuses on the projection of social power to spaces whose geophysical and geographic characteristics make them resistant to state territorialization, including the ocean (*The Social Construction* *of the Ocean* 2001), the Arctic (*Contesting the Arctic* 2015), the universe of electronic communication (*Managing the Infosphere* 2008), the delta city (*What Is a City?* 2008), and a variety of other spaces beyond the limits of land (*Territory Beyond Terra* 2018). He researches these topics through analysing legal, artistic, cartographic, and literary modes of governance and representation.

Sue Jane Taylor studied fine art at Gray's School of Art, Aberdeen, the Slade School of Fine Art, London, and Konst Academie, Stockholm. She was born and brought up on the Black Isle in the eastern Highlands of Scotland, where her ancestors have lived for centuries. She is renowned for her work documenting the North Sea energy industry and continues to capture its transition: decommissioning and renewables. Over the years, she has gained access to extremely remote and publicly prohibited offshore installations, and in her work, she raises questions about the relationship between art, environment, and industry and develops ideas about labour, industry, and their place in the natural world. She actively engages in community public art and has exhibited nationally and internationally in private galleries, public galleries, and museums. Publications on her work include *Oilwork North Sea Diaries*, Birlinn 2005, *Beatrice Works* Aberdeen Art Gallery & Museums 2013, and *Age of Oil* National Museum Scotland 2017.

1

HEADING OFFSHORE

Introducing cold water oil

Fiona Polack and Danine Farquharson

As we write, in the autumn of 2021, the world remains in the grip of the COVID-19 pandemic. During the past year and a half of recurrent lockdowns and surging infection rates, the need for action to address the climate crisis became intertwined with a sharper awareness of the instability of global capitalism, the differing abilities of political leaders and global industries to respond to catastrophic threats, and the very real capacity of the nonhuman to overwhelm the human. Early hopes that travel bans, stay-at-home orders, and economic slowdowns might at least have the silver lining of reducing global carbon emissions in a meaningful way have proved premature (Tollefson 2021). Nonetheless, many have at least learned that it is possible to radically alter the way they live and work, particularly when it comes to circumscribing CO_2-producing travel.[1] Whether the will exists to make such changes permanent is another matter. Also still unclear is whether governments around the world will take the opportunity offered by COVID-related uncertainties to rebuild less carbon-intensive economies and societies. The Organisation for Economic Co-operation and Development (OECD)'s Green Recovery Database, which monitors COVID-19 recovery plans in 43 countries, noted that as of April 2021, environmentally positive spending measures were on par with environmentally detrimental ones (OECD 2021).

That the stakes are high, and the situation is urgent, is unquestionable. The United Nations (UN)'s 2020 *Emissions Gap Report* notes that "failure to significantly reduce global emissions by 2030 will make it impossible to keep global warming below 1.5°C" (United Nations Environment Program 2020, xi). Government and corporate decision-making in the wake of the pandemic will have significant ramifications for the long-term fate of the planet. It will also, inevitably, have consequences for energy industries. Oil and gas companies are under mounting pressure to circumscribe their activities, and their offshore

DOI: 10.4324/9781003024125-1

operations are in an especially tenuous position.[2] As Scott Carpenter notes, current economic challenges "could mean that a growing share of the world's oil production is pumped from cheaper onshore alternatives, rather than pricey offshore ones" (Carpenter 2020). Moreover, growing public demands to respond to the climate crisis have already seen some countries ban further offshore oil expansion altogether. Denmark recently cancelled all future licensing rounds for oil and gas exploration and production in its sector of the North Sea and resolved to end all offshore fossil fuel production by 2050 (Lepic 2020). Those decisions followed earlier similar moves by New Zealand (Graham-McLay 2018). In their *Net Zero by 2050: A Roadmap for the Global Energy Sector*, the International Energy Agency (IEA) clearly and forcefully directs policy makers that there "is no need for investment in new fossil fuel supply in our net zero pathway," and the roadmap describes that beyond "projects already committed as of 2021, there are *no new oil and gas fields approved* for development" (IEA 2021, 21, emphasis ours). Meanwhile, other nations that extract oil and gas from beneath the ocean, including Norway and Canada, scramble to rebrand their petroleum products as "greener" than those produced in other places by other methods (see Ryggvik, and Dale and Farquharson, this volume).

What happens offshore matters. Currently, more than a quarter of the world's oil and a rapidly growing proportion of gas are produced from beneath the seas (IEA 2018, 15). The offshore fossil fuel industry is, thus, a crucial point of origin for carbon emissions, as well as other environmental harms.[3] Notwithstanding this significance, humanities and social science scholars have so far made only sporadic efforts to comprehend the socio-cultural discourses that attend its operations.[4] As we have argued elsewhere, powerful imaginaries related to offshore petroleum extraction are invariably embroiled in long-standing conceptions of the sea (Polack and Farquharson 2017). Hence, this scholarly inattention, particularly by academics in the Global North, may be partially attributed to entrenched cultural narratives that position oceanic spaces as *aqua nullius*. As Tricia Cusack puts it, "Western conceptions of the ocean [have] continued to view it predominantly as 'empty space,' although it [has] remained a space available for exploration and appropriation" (Cusack 2014, 1).[5] This consideration is abetted by the fact that offshore petroleum extraction typically occurs well out of sight of land and within a context of deliberately cultivated corporate secrecy (see Macdonald this volume, and Polack 2019). The relative dearth of humanities and social science research about offshore oil extends to the interdisciplinary formations with which we most explicitly engage in this volume: the environmental and the energy humanities.

Cold Water Oil: Offshore Petroleum Cultures thus delves into overlooked histories, influential contemporary narratives, and emerging energy and environmental futures associated with offshore petroleum extraction. To focus our efforts, we pay special attention to industrial operations in the North Atlantic and Arctic, operations that are often entangled as nations such as Norway and Russia, with already well-established petroleum industries in the North Atlantic, seek to expand their operations in Arctic seas. As Julie Sze correctly notes, "as long as complex environmental problems are multiscalar across space and time, scale will remain a key conceptual tool in interdisciplinary environmental studies" (Sze 2016). Thinking about oil and its effects in geographically and environmentally delimited contexts makes it easier to perceive its complex influences and to better identify the narratives and mythologies that accompany its extraction and consumption. We see climate crisis-induced urgency in attending specifically to the region of cold water oil.

Cold water oil

Northern cold water oceans contain both well-established petroleum operations (in the Beaufort Sea, and on the Grand Banks, for instance) as well as fledgling and aspirational ones (such as around Svalbard and the Lofoten/Vesterålen/Senja Archipelago). The continuing pursuit of oil and gas in what we call the *cold water offshore* constitutes an extreme yet highly revealing example of the lengths to which corporations and governments have been prepared to go in order to maintain the centrality of fossil fuels. Cold water oil reserves in the North Atlantic and Arctic constitute a classic instance of what Stephanie LeMenager, borrowing from Michael T. Klare, terms "tough oil" (LeMenager 2014, 3). These deposits are typically located in frigid and tumultuous seas within ecologically sensitive and isolated environments. Ironically, the temptation to expand activities in "what may be the last large conventional oil frontier on earth" (Harriss 2016, 18)[6] grows as climate change's cryospheric and meteorological jolts increasingly impact the area. The Arctic has already seen climatic elevations two to three times greater than the global average (Abram et al. 2019; Barry and Hall-McKim 2018), and diminishing sea ice is now making navigation easier and extensive summer exploratory ventures more viable. However, the ramifications of rapid climate change in the cold water offshore extend in many nonhuman and human directions beyond the opening of newly navigable waters. These implications range from the endangerment of species such as ringed seals and polar bears, whose habitats are disappearing, to the growing "Atlantification" of Arctic oceans as the seas of the cold water offshore become increasingly indistinguishable (Csapó et al. 2021), to cultural disruption and ecological grief in Indigenous communities (Cunsolo and Ellis 2018; Watt-Cloutier 2015). Growing awareness of the fragility of Arctic ecosystems and the disastrous impacts associated with climate change means that garnering social license to exploit oil and gas reserves in the region is now much more difficult. The area has become absolutely central to debates about the planet's future.[7]

While its resources fall primarily under the jurisdiction of European and North American nations in the Global North,[8] exploration and extraction in the cold water offshore often have the most disruptive effects in socio-economically, and in the case of Indigenous communities also racially, marginalized locations within these countries. Close attention to the cold water offshore by necessity draws us into the task of complicating the term "North" and grappling with the ongoing legacies of colonialism and uneven development within wealthy, energy-hungry societies. We share Dolly Jørgensen and Sverker Sörlin's belief that North is "not a single, definable concept" and it is "difficult to say where North begins on a map" (Jørgensen and Sörlin 2014, 4, 12). We also agree with Ray and Maier that there is a long and troubled history of Western conceptions of the North marginalizing the perspectives of Indigenous peoples for whom the North is home, rather than peripheral to southern romantic and/or resource extraction fantasies (Ray and Maier 2017, 1). As Inuk visual artist and critic asinnajaq explains: "too much of the language around Inuit is about the 'isolation,' the 'remoteness,' the idea of being disconnected" (asinnajaq 2019). Perceptions of the North and Northern oceans as frontiers add "immensely to colonial narratives of indigeneity and nation" (Ray and Maier 2017, 1) and are often readily identifiable within the rhetoric of the contemporary fossil fuel industry.

Cold Water Oil considers offshore oil industrial development in and across the oceanic territories of Canada, Norway, the UK, Russia, the US, and the Iñupiat of Alaska at a time of profound global instability and uncertainty.[9] Our contributors also venture beyond the human to speculate about the perspectives of cetaceans whose habitats in cold oceans are disrupted by oil extraction. Substantial differences do, of course, exist between and within the respective Arctic and the North Atlantic peoples, regions, and nations under consideration here, a reality the genre of the edited collection with its successive distinctly focused essays can help respect. Indigenous and non-Indigenous communities, for instance, are very differently situated in relation to the cold water offshore because of long-standing and destructive histories of European colonialism and resource exploitation (see Sakakibara and Ahtuangaruak, this volume), and the continuing effects of Russia's Communist history and the Cold War linger in present-day tensions between Russia and North Atlantic Treaty Organization (NATO) members Norway, the UK, the US, and Canada.[10] Notwithstanding noteworthy distinctions such as these, we see great value in working with the ecologically inspired spatial designation of cold water oil. The concept is not purely localized or globalized, singular nation-focused, or, crucially, land-centred. A thalassic perspective is a major advantage given that, as Ursula Heise points out, "environmentalism needs to foster an understanding of how a wide variety of both natural and cultural places and processes are connected and shape each other around the world, and how human impact affects and changes this connectedness" (Heise 2008, 21). Moreover, the artificial boundaries of nation states can be especially obstructive to our approach. In 1982, the United Nations Convention on the Law of the Sea (UNCLOS) attributed coastal states' sovereign rights in exclusive economic zones (EEZs) extending 200 nautical miles offshore, thus updating jurisdictional considerations in the cold water offshore. However, while coastal countries' boundaries technically extend into the ocean, as Brian Russell Roberts (Roberts 2021) points out, land rather than sea-based epistemologies are powerfully responsible for shaping national imaginaries. Furthermore, as ecocritic Katie Ritson emphasizes, it is crucial to think: "across the borders of nation-states to the underlying environmental factors that shape the development of areas and regions" (Ritson 2018, 6). Just as importantly, *Cold Water Oil* counters approaches that see the oceans as undifferentiated expanses valuable only for the exploitable resources within and beneath them.

Scales and scapes

To think in the geo-environmentally centred terms that a focus on cold water oil demands also counteracts approaches to offshore extractivism that elide differences between places. The offshore oil industry is heavily reliant on what Hannah Appel terms "modularity," a commercial capacity to deploy elements including a peripatetic labour force, complex financialization, and mobile infrastructure in ways that "enable offshore work in Equatorial Guinea to function 'just like' offshore work in Ghana, Brazil, or the North Sea" (Appel 2015, 259). The supposedly "smooth space" (Deleuze and Guattari 1988, 479) of the ocean operates to the industry's advantage here, providing a seemingly history-less and territory-less expanse on which to operate. Indeed, Christopher Connery describes the sea as "Western capitalism's primary myth element" (Connery 2010, 686). Chasing resources and potential profits wherever they can be found, large petroleum companies and suppliers do not confine their extractive operations to specific geographies or nations. To give but two examples, Norway's Equinor (formerly Statoil) pursues oil in onshore and offshore locations in more than 30 countries ranging from Angola, to Brazil, to the US (Equinor 2021), and US-founded Transocean, former owner of the ill-fated *Deepwater Horizon*, moved its base to land-locked Switzerland to lower taxes, and contracts its rigs and crews to companies operating in offshore drilling provinces around the world (Jordans 2010; Transocean 2021). The deterritorializing predilections of petroleum producers are complemented by the diffuse distribution and consumption patterns of the oil and gas they extract. As John Urry notes, from the mid-20th century, most societies replaced localized and decentralized traditions of energy production and consumption with a reliance on power from far afield (Urry 2014, 102). It is virtually impossible to determine the origins of the petrol we pump into our cars or the furnace oil we burn to heat our houses. At a time when accountability for climate change is of paramount concern, this phenomenon also helps enable those who extract petroleum offshore to abdicate some measure of responsibility for the greenhouse gases their products generate when burned.

Notwithstanding these powerful homogenizing forces, distinct implications *do* attach to particular locations and modes of oil production and consumption, and to experiences of and responses to the planetary climate crisis. The petroleum industry's offshore activities are ultimately determined by the geology of specific sea beds and influenced by the oceanic and climatic conditions in their vicinity. Drilling in tumultuous, ice-laden seas, which can be exceedingly difficult to remediate in the event of spills, poses unique engineering challenges and heightened risks to fragile environments, and oil industry workers' lives, as well as considerable increased expense. Indeed, cold water physical environments are so uniquely challenging that the International Organization for Standardization (ISO) issues specific guidelines to inform operations within them.[11] Companies must dedicate resources to developing the capacity to operate in the cold water offshore in order to mitigate the elevated financial and insurance-related risks of doing so.[12] As SNC-Lavalin engineer Mark Manton notes, "the control of energy consumption by a facility will determine its profitability. There is little point extracting these hard to reach hydrocarbons if the majority of them are going to be burnt in the process" (Manton 2014). Failing to acquire the expertise necessary to operate in cold water environments can have consequences far more serious than financial ones. The joint Canada-Newfoundland and Labrador Royal Commission into the sinking of the *Ocean Ranger*, with the loss of 84 lives in a winter storm on the Grand Banks in 1982, found that the exploration rig's American managers failed to properly account for the extremities of local sea and weather conditions, and operated the Ranger as if it was located in the much warmer Gulf of Mexico (Royal Commission on the *Ocean Ranger* Marine Disaster 1984).[13] In fact, countries such as Norway and Canada, which already produce oil from the North Atlantic, cite their long-standing experience in this ocean to justify their expansion into the even more risk-laden waters of their Arctic territories.

The deterritorializing tendencies of corporations who seek and extract hydrocarbons from beneath the seas meet other kinds of impediments, too. Most notably, offshore petroleum exploration and extraction by oil companies invariably occurs in partnership or tension with particular national and provincial governmental regimes, local contractors and workers, community members, and environmental groups. Activities at sea have tentacular infrastructural, economic, legal, social, and environmental networks with myriad implications for adjacent and more distant onshore locations. *Cold Water Oil*'s attention to place-based research is no coincidence; most of our contributors live in onshore communities with close connections to the offshore oil industry.

Ideas across oceans

Our decision to focus this collection on the cold water offshore originates in our positioning as scholars who live and work in Newfoundland and Labrador, on the homelands of the Beothuk, Mi'kmaq, Inuit, and Innu Nations. The search for oil began here on the Grand Banks in the 1960s, with commercially viable quantities discovered in 1979. However, jurisdictional wrangles between the provincial and federal governments and other challenges meant production did not commence until 1997 (Higgins 2009). Since then, the industry has cycled through a series of dramatic economic up and downturns. At present, the fossil fuel economy contributes roughly one quarter of Newfoundland and Labrador's GDP and, despite increasing public resistance but with government support, is pursuing an aggressively expansionist program (Government of Newfoundland and Labrador 2020). Over the past two decades, we have experienced oil's transformation of Newfoundland and Labrador in both overt and subtle ways. Indeed, our very employment with a provincially-funded university is entirely bound up in Newfoundland and Labrador's fiscal, petroleum-price dependent, realities.[14] We have also become increasingly aware that development here is imbricated with that of other cold water oil jurisdictions. Norwegian expertise, for instance, was critical to the construction of *Hibernia*, Newfoundland and Labrador's first production platform (Fusco 2007). More recently, the knowledge gained about how to effectively extract oil in harsh, ice-laden environments on the Grand Banks has been touted as evidence that the skills exist to open up similar, currently undeveloped, areas. Vladimir Putin, contemplating oil extraction in Russia's Arctic offshore, has taken inspiration from the fact that production platforms like *Hibernia* can withstand icebergs (Poussenkova, this volume). Relatedly, Newfoundland and Labrador has endeavoured to position itself as a key launching pad for new Arctic development (Sandlos 2015).

The value of sharing ideas and strengthening alliances across cold ocean geographies and societies is not only important to the oil industry and its boosters. Organizations such as the Inuit Circumpolar Council (ICC) further communal and environmental goals in these regions and with an awareness of the artificiality of their non-Indigenous national borders. The first ICC, held in Utqiagvik (formerly Barrow), Alaska in 1977, for instance,

> issued ICC Resolution 77-06, a resolution concerning environmental policy which called for Denmark, Canada, the United States, and the USSR to agree on a common set of rules for all onshore and offshore Arctic oil and gas development; and to involve our Inuit community in this rule making (Hopson 1978).

The ICC is also a permanent participant in the Arctic Council: an intergovernmental forum intended to promote cooperation between nations with Arctic territories including Russia, the US, Canada, Norway, Finland, Sweden, Denmark, and Iceland. Meanwhile, academic entities such as the University of the Arctic network and the North Atlantic Forum facilitate research exchange and learning across the cold water offshore.

Other scholars (sometimes working with and sometimes against the petroleum industry) precede us in studying explicitly petro-oriented connections between North Atlantic and Arctic locations.[15] For instance, in August 1967, earlier even than the OPEC upheavals of the 1970s that accelerated interest in oil extraction from non-Middle Eastern reserves, Columbia University brought together geologists from 16 countries to consider the stratigraphic and structural origins of the North Atlantic and Arctic.[16] In the following decade, an extensive socio-economic-focused research program at Memorial University examined how an offshore petroleum boom was impacting communities in Norway and Scotland with close ties to the industry (see Jekanowski, this volume). Like Newfoundland and Labrador, these Northern societies possessed relatively small, rural-based populations, traditionally strong community cohesiveness but extensive out-migration, overwhelmingly white racial identities, a long-standing reliance on the fishing industry, a history of economic challenges, and a peripheral importance to the nations to which they belonged. More recently, sociologists Mark Stoddart, Alice Mattoni, and John McLevey have examined how tourism and oil development coexist, or fail to, in Norway, Newfoundland and Labrador, Scotland, Denmark, and Iceland (Stoddart et al. 2020), and geographers Philip Steinberg, Jeremy Tasch, and Hannes Gerhardt have addressed tensions between climate change and resource extraction across the Arctic in their *Contesting the Arctic: Politics and Imaginaries in the Circumpolar North* (Steinberg et al. 2015). *Cold Water Oil* builds on previous efforts to draw experiences of petroleum development and its consequences in Northern locations on both sides of the Atlantic together but addresses somewhat different national and geographic contexts, applies a more broadly multidisciplinary lens, and focuses more extensively on human and nonhuman cultural and ecological concerns.

Energy and environmental humanities at sea

Cold Water Oil: Offshore Petroleum Cultures approaches the question of how societies conceive of fossil fuel extraction in the inhospitable but fragile waters of the North Atlantic and Arctic oceans from a variety of disciplinary and interdisciplinary perspectives. However, the majority of contributors are preoccupied with the socio-cultural imaginaries associated with petroleum exploration and extraction and their lived, political ramifications. We agree with Ross Barrett and Daniel Worden that the:

> cultural analysis of oil aims to reconstruct those symbolic forms and practices that enabled the emergence, development, and entrenchment of oil capitalism. [In addition] attending to those moments in which oil's ceaseless colonizing expansion has engendered uncertainty, ambivalence, or resistance [...] can uncover symbolic materials that may be useful to the construction of alternative perspectives on fossil fuels and the energy economy (Barrett and Worden 2014, xxvi).

Our concerns, and the methods by which we address them, situate *Cold Water Oil* within the rapidly growing, multidisciplinary and interdisciplinary fields of the energy humanities and the environmental humanities.[17] Both these interrelated domains are, along with other new humanities formations such as the Arctic humanities and the Blue humanities, fostering the development of research with wider scope than has traditionally been the case; this scholarship is less "esoteric" and more "*of* the world and *for* the world" (Sörlin 2017, 434). The shift reflects the monumental scale of the problems currently facing the earth and its denizens and awareness of the unique viewpoints that humanities and social sciences scholars contribute. Questions about our energy futures cannot, for instance, be adequately addressed by scientists and technologists alone. Participating in the energy humanities and the environmental humanities prompts many of us to venture well-beyond the conventional boundaries of the disciplines in which we were originally trained. Literary scholars, for instance, might be found grappling with technical reports from governmental organizations, redeploying their long-honed skills in close reading and discourse analysis in the service of drawing more far-reaching conclusions than they would venture to put forward in the past. Concomitantly, anthropologists experiment with creative writing to confront the energy and environmental-related issues they study from fresh perspectives and in ways that their home discipline might not have previously countenanced. That said, the energy and the environmental humanities are multidisciplinary and interdisciplinary endeavours. While they share the same thematic foci, and a preoccupation with contributing to social and energy and environmental justice, not all scholars in the fields depart radically from the approaches and methodologies acquired in their original training. Within this volume, as a quick glance at the biographies of our contributors in tandem with their chapters reveals, both approaches are in evidence.

In their introduction to *Energy Humanities: An Anthology*, Imre Szeman and Dominic Boyer state that the energy humanities "[retains] a deep kinship and intimate conversation with environmental humanities" (Szeman and Boyer 2017, 9).[18] This statement reflects the fact that the primary impetus for launching energy humanities research over the last decade or so has been global warming and other deleterious environmental effects associated with extensive use of fossil fuels within hegemonic capitalist systems. However, in distinction to many working in the environmental humanities, energy humanists "position oil and energy as the fulcrum around which many of today's most pressing social, economic, and political issues must be analysed and understood" (Wilson, Carlson and Szeman 2017, 4), unpack the often invisible ways in which different energy sources facilitate specific kinds of "social structures, lived and material infrastructures, and even cultural practices" (Szeman and Boyer 2017, 3), and speculate about how best to prompt energy justice, and its interlinked social and environmental transitions. Energy humanists are typically environmental humanists by default. In fact, their work, despite the distinct institutional, publication, and collegial infrastructures developing around it, is viewed by some as simply part of a "burgeoning subfield of environmental humanities" (Goodbody and Smith 2019, 11).[19] We are not especially interested in attempting to definitively carve out the respective turfs of these two fields. For one thing, many scholars work seamlessly across both scholarly formations (Stephanie LeMenager and Lawrence Buell are especially good examples). Far more importantly, as Nicolas Howe frankly insists, at a moment in which we are running out of time to confront the ecological problems besetting the planet, such preoccupations should not be our main concern (Howe 2017, 114).

To date, no extended work on the cultural, social, and environmental implications of offshore petroleum production has appeared in either the energy humanities or the environmental humanities. This lacuna is especially surprising given research on oil and petrocultures more generally has certainly been the dominant focus of energy humanities scholars; Christopher Jones goes so far as to suggest that energy humanism has suffered from "petromyopia" (Jones 2016). This research gap also persists despite the fact environmental humanities scholars are striving hard to address a supposed "ocean deficit" (Dobrin 2021, 35) in their domain. Important work within critical ocean studies is challenging the land ethic that has long preoccupied researchers.[20] That said, Craig Santos Perez (2020) contends that critical ocean studies' emphasis on the supposedly "new" turn toward the sea does not sufficiently acknowledge the long-standing thalassic-focused work of Pacific Islander scholars, such as Albert Wendt and Teresia Teaiwa. His argument can be extended to encompass Caribbean writers and thinkers, such as Kamau Brathwaite and Derek Walcott, and their especially close interest in the legacies of the Transatlantic slave trade as well as landmark texts such as Paul Gilroy's *The Black Atlantic: Modernity and Double Consciousness* (1993). Indeed, in *The Black Shoals: Offshore Formations of Black and Native Studies*, Tiffany Lethabo King comments that within Black and Indigenous studies, "the ocean and its legacy has crested again and again" (King 2019, 5). There is still, however, a dearth of socio-culturally focused research on offshore oil. Elizabeth DeLoughrey's recent call for increased attention within critical ocean studies to the contemporary exercise of militaristic and industrial power offshore is justified (DeLoughrey 2017). *Cold Water Oil* seeks to contribute to this undertaking at a moment of acute environmental and social instability.[21]

Wave patterns

Kimberley Peters and Philip Steinberg write that more-than-wet-ontologies are a valuable antidote to terrestrial-inspired approaches when thinking about the nonhuman world. They suggest we take inspiration from the ocean's materiality, its "flows, connections, liquidities and becomings" as well as the fact it can take non-liquid forms such as "solid (ice) and air (mist)" and generate winds felt far inland (Peters and Steinberg 2019, 293). We have kept their comments in mind in thinking through how best to structure this volume. As a result, we eschew strictly bounded subgroupings of chapters in favour of a more thalassic structure.

This collection's opening chapter is by Chie Sakakibara and Rosemary Ahtuangaruak. Rooted in the performances of Indigenous artist Aku-Matu and their powerful message that animals make humans visible, "'We are torn about our future': Big Oil and Iñupiaq community health in Arctic Alaska," combines community field work, economic analysis, human health studies, and nonhuman considerations to reveal how political and economic turbulence inaugurated by Big Oil is manifest in rural Arctic Alaska at the expense of Indigenous health and community integrity. Sakakibara and Ahtuangaruak contextualize Iñupiaq modernity by exploring how the bowhead whale, a cultural and physical cornerstone species for Iñupiat, is entangled in the making of human health and well-being and, by extension, facilitates the "complex matrix of sovereignty in relation to extractive industries and the fear for future environmental destruction." In chronicling the history of oil industry development in the Beaufort Sea and Chukchi Sea, Chie Sakakibara and Rosemary Ahtuangaruak illustrate how the cold water offshore contributes to contemporary Indigenous identities in Arctic Alaska and compellingly argue with Aku-Matu that it's "time for us to borrow animal voices."

"We are torn" is followed by five chapters with a common methodological undercurrent: each focuses on cultural figurations of offshore oil. However, the specific materials and discourses on which they dwell range from documentary films to social media.

Fiona Polack's chapter "Encountering the nonhuman in North Atlantic oil catastrophes" analyses the rhetorical and discursive structures of commissioned government reports and autobiographical recollections of three offshore petroleum catastrophes: the *Alexander Kielland*, the *Ocean Ranger*, and the *Piper Alpha* disasters. Polack shows how close attention to accounts of offshore disasters sheds light on the "conceptual and narrative machinations that underpin false assumptions about the extent of human control" over the North Atlantic Ocean. The "rigwrecks" she examines can be read as "ecological parables" within the context of current environmental predicaments. More specifically, accounts of petroleum industry catastrophes in the North Atlantic highlight human willingness to cling to comforting routines and assumptions even in moments of crisis, the troubling implications of personifying the nonhuman, and the acute discomfort experienced by witnesses to processes of watery embodiment. Polack's chapter seeks to identify and widen fissures in entrenched petronarratives about the instrumentality of the nonhuman world.

Graeme Macdonald's chapter, "Dynamic positioning: North Sea petroculture's backwash," takes the stranding of Transocean's decommissioned *Winner* rig on the Scottish coast in 2016 as its conceptual touchstone. It does so because the incident prompted widespread reflections on "existential questions that projected into larger, already existing debates about the future of oil itself." Macdonald contends that concerns about petroleum's future have beset discussions about North Sea Oil since its inception and have only become more pronounced in the era of climate crisis. In the course of proving his point, he offers a comprehensive consideration of recurrent political and cultural preoccupations associated with North Sea Oil over the past 50 years. Drawing on literature, film, drama, visual art, and documentary, Macdonald demonstrates how various genres, figurative devices, and other effects have been deployed "by vested parties from multiple sides engaged in a dynamic war of position over the past, present and future form and meaning of oil – not only in the North Sea, but around the fossil-world-system in general."

In resonance with "Dynamic positioning," Jason Haslam's essay, "The call of Cthoilu: the weird subject of cold water oil" presents a call to renarrativize offshore oil disasters. Starting with the question, "what does it mean when offshore oil disasters become spectacle, rendered as objects to be observed for pleasure rather than terrors that demand physical and immediate reaction?" Haslam goes on to assert that, rather than "accidents" or "inevitabilities," offshore oil disasters are active agents too often misrecognized as passive effects. Based in the history of the North Atlantic, where dominant epistemologies are framed and maintained through colonialism, slavery, capitalism, and the exploitations embedded therein, Haslam draws on China Miéville's short fiction to illustrate how reserving the category of "humanity" for only a select few risks that term "being used to bait-and-switch the 'cause' of global warming to the transcendent subject of 'humanity,' which in turn makes necessary local actions within particular contexts seem inadequate or impossible."

In Rachel Webb Jekanowski's chapter, we shift focus from the literary to the filmic while maintaining attention to the documentary impulse. "Contested and emergent futures: film and energy regimes of the Newfoundland offshore" examines historical, and highly geographically specific, associations between offshore oil and future imaginaries through a collection of Canadian films produced during the early days of exploratory drilling in Newfoundland and Labrador. Jekanowski analyses how these media forms act as "sites of future-making," examines how the gendered and "racial dynamics of these imagined oil futures frame social relations with energy, oceans, and more-than-human life," and discusses the ways the films help produce the oil fields in the Grand Banks as an *offshore*, an extractive zone and new form of nature. From these specific analyses, "Contested and emergent futures" moves beyond the local and poses the wider cold water oil-related question: "what future are we planning for?"

Sharing an interest in what comes next for offshore oil cultures, Brigt Dale and Danine Farquharson analyse conventional and social media in their "Dispatches from two cold water oil cultures: Norway and Newfoundland and Labrador." Describing the quickly changing conversations about offshore oil development in the COVID-19 pre-pandemic and pandemic years, Dale and Farquharson present the historical connections across the North Atlantic between Norway and Newfoundland and Labrador, together with profound differences in how the two places are currently managing offshore oil production and futures. Offshore oil matters in these two locations but in diverse ways for their political leaders and communities. Dale and Farquharson analyse social media, government documents, and traditional media sources within the context of their own experiences to provide snapshots of two cold water oil cultures.

Sue Jane Taylor's illustrated chapter "Art and the offshore" constitutes an interregnum of sorts, a place where the tide shifts. Taylor provides an autoethnographic account of her work as an artist chronicling offshore petroleum production in the North Sea. Taylor, who grew up on the Black Isle in the midst of Scotland's 1970s oil boom, tells stories of her visits to offshore ships, rigs, and production platforms in the North Sea and related infrastructural sites onshore. Her visual art emphasizes the importance of oil and gas industry labourers and the environments in which they work. "Art and the offshore" charts profound changes in new methods of oil and gas extraction, attitudes toward natural and social environments, and emerging large-scale offshore renewables. Sue Jane Taylor's chapter is part history, part memoir, and part political-social reflection on 35 years of making art about the North Sea oil and gas industry.

Taylor's essay is followed by another tidal current of chapters: the next three essays work primarily, although not exclusively, across social science disciplines. They also share a common interest in the respective factors pushing for and against the expansion of oil and gas activities in previously little-developed areas of the cold water offshore. In "Time for oil: competing petrotemporalities in Norway's Lofoten/Vesterålen/Senja Archipelago," co-authors Berit Kristoffersen, Gavin Bridge, and Philip Steinberg use anthropological, philosophical, and historical methodologies to analyse how the petroleum industry operates across multiple temporal frames. They then go on to illustrate how, when communities debate their petroleum futures, the "anticipatory temporalities" of petroleum abundance, technology, price, and demand are contrasted with alternate temporalities that, in many cases, fail to align with those articulated by the oil industry. Through a reflection on historic and ongoing debates over offshore oil production in Norway's Lofoten/Vesterålen/Senja (LoVeSe) region, they propose that debates over petroleum futures are not simply struggles over shared space, exemplified in oil vs community, oil vs fisheries, or oil vs environment narratives; they are also debates about the intersections and alignments of different temporalities and the meaning and significance of time.

Nina Poussenkova's chapter, "From land to polar sea: Russia's ill-starred quest for a new petroleum province in the Arctic offshore" unpacks Russia's recent oil exploration and development ambitions, particularly in the Arctic. Through the lens of economic history and analysis, Poussenkova paints a portrait of Russia's continuing drive to exploit oil and gas reserves, despite adverse macroeconomic and political factors working against such risky ambitions. Her case study serves as a chilling reminder of both the tenacity of national and industrial efforts to drill more and further afield as well as the precarious state of Arctic and Northern energy futures.

Historian Helge Ryggvik's chapter "Svalbard and oil: dangerous games in the Arctic?" considers the historical and political contexts for Norway's current efforts to expand its oil industry into a region of the Arctic subject to an ongoing dispute with, most notably, Russia. Ryggvik argues that the Norwegian government has taken advantage of the distraction of the COVID-19 pandemic to introduce measures aimed at facilitating expansion in the Barents Sea, including into the controversial Svalbard Zone. He attributes this aggressive stance to the fact Norway finds itself with a growing offshore supply industry tied to ever-diminishing oil resources in the North Sea. Ryggvik highlights that while international oil companies often draw the most ire, oil producing states, even those such as Norway which promote themselves as environmentally conscious, also play a key role in hampering efforts to transition away from fossil fuels.

Cold Water Oil: Offshore Petroleum Cultures concludes with an essay that circles back in important respects to the collection's opening chapter. Amy Donovan's "Raw, dense, and loud: a whale's perspective on cold water energy" is deeply attentive to multispecies relationships between humans and whales, but from a non-Indigenous perspective. Donovan's chapter is informed by the question: "what is energy for a whale?" Her work presents seismic exploration and oil and gas extraction in the North Atlantic as energy depletion on cultural, psychological, and physical levels for both particular animal species and entire ecosystems. Following Anna Lowenhaupt Tsing's imperative to write stories with nonhuman protagonists, anthropologist and creative writer Donovan seeks a perspective on the cold water offshore that sidesteps motives of modern progress, linear time, and human exceptionalism. Her chapter deploys literary analysis and creative writing to recalibrate our narratives about petroleum exploration and extraction in the North Atlantic and Arctic toward nonhuman protagonists and multispecies engagement.

Gazing out at the cold water offshore from locations such as Aberdeen City Beach at the North Sea, one now sees wind turbines, as well as ship traffic travelling to and from petroleum installations further out of sight. As scholars including Laura Watts (2018) and Alexandra Campbell (2019) articulate, new ocean energy imaginaries are beginning to emerge, ones capable of presaging potentially more equitable, sustainable, and multi-layered relationships between and within human and nonhuman communities. As is increasingly clear, however, moves toward renewable energy, despite the work they might do in offsetting carbon emissions, do not inevitably guarantee such positive outcomes.[22] It is vital we better understand historical and current destructive petroleum imaginaries, and the best-articulated responses to them, so that we can recognize when they recur in the context of new energy regimes.

Notes

1 A further consideration here is that the information and communications technology sector that facilitated the move to replace physical with virtual encounters also has a significant carbon footprint (Jones 2018).

2 In late May 2021, at least two major developments in the oil sector made global headline news: first, a Dutch court ordered Royal Dutch Shell to reduce its CO_2 emissions by 45% by 2030 and exposed Shell's publicized climate policy as "full of conditions" and "not enough" (Buosso 2021), and second, Exxon Mobil lost a battle with its own shareholders – despite Exxon's $35 million campaign – which saw a small investors group, Engine No.1, install two of its own nominees, both of whom advocate for Exxon's business strategy to account for climate change realities ("Powerful Signal" 2021).

3 Beyond spectacular, well-publicized environmental catastrophes such as the 2010 *Deepwater Horizon* blowout, the offshore oil industry is also implicated in slower forms of violence including, among other examples, chronic small-scale spills of oil and drilling fluids (Avery 2020) and disruption of marine and bird life through activities

such as seismic testing and flaring (see, for instance, Donovan in this volume, and Fraser and Racine 2016).

4 Notable among these efforts is work by LeMenager (2014), Urry (2014), Appel (2015), Chalfin (2015), Phillips (2018), Gladwin (2018), Ritson (2018), Polack (2019), Campbell (2019), Pitt Scott (2020), Stoddart, Mattoni and McLevey (2020).

5 Over the past century, these conceptions have been bolstered by governmental managerialism and related scientific and actuarial attempts to make the ocean as predictable as possible (Bavington 2010).

6 Harriss cites US research on the Arctic Circle indicating that around 84% of estimated petroleum reserves in the region are located offshore (Harriss 2016, 21).

7 Questions and controversies about oil and gas development in the Arctic National Wildlife Refuge have existed since 1977 (Shogren 2005). The area enters in and out of general awareness as activist groups such as Greenpeace launch more or less successful campaigns to stop drilling; their "Everything is Not Awesome" video intervention being particularly potent (Perera et al. 2017). Indigenous communities such as the Iñupiat of Kaktovik pass resolutions calling companies such as Shell Oil a "hostile and dangerous force" ("Native Village" 2006). Concurrently, successive US governments allow exploration and drilling (Bush and Trump administrations) then reverse those decisions (Obama and Biden administrations).

8 The only exception considered in this volume is Russia, which is not typically classified as belonging to the Global North.

9 While activity occurs in Chile, Argentina, and around the Falkland Islands, cold water offshore petroleum exploration and production efforts are currently far less extensive in southern regions of the planet than northern ones. The Protocol on Environmental Protection to the Antarctic Treaty, the 1991 Madrid Protocol banned all forms of extraction south of 60 degrees south (Secretariat 2021).

10 John Kampfner (2021) writes in *The New European* that a "very cold war" is under way in the Arctic and "the warming climate has changed everything, militarily and economically, in the northernmost reaches of the world."

11 ISO 19906 addresses "design, construction, transportation, installation and decommissioning of offshore structures related to the activities of the petroleum and natural gas industries in arctic and cold regions" (ISO 2019).

12 For example, within the Scottish-based multinational Wood Group, business units Wood Group Kenny, Wood Group Mustang, and Wood Group PSN concentrate on "extreme environment technology" (Wood Group 2013), and Transocean's fleet of 38 mobile offshore drilling units contains "11 harsh water environment floaters" designed "for any manner of cold, windy, high seas environments" (Transocean 2021).

13 Ironically, BP's Oil Spill Response Plan for what became the disastrous Deepwater Horizon project made an antithetical error: the company cut and pasted material from environmental impact statements developed for prospective Arctic operations into its Gulf of Mexico-related document. For instance, the Deepwater Horizon Plan referred to prospective impacts on walruses and other species with northern habitats (National Commission 2011, 84).

14 Memorial University is closely imbricated with the offshore oil industry in more direct ways, too. For instance, the Scottish Wood Group promotes its local image and abets the operation of the oil industry within Newfoundland and Labrador's offshore, and other proximate cold water environments, by funding The Wood Group Chair in Harsh Environment Engineering. Similarly, US-founded, Hong-Kong financed, Calgary-based Husky recently contributed 1 million dollars to Memorial's new (primarily federal and provincial government-funded) Harsh Environment Research Facility (Green 2019).

15 Our own previous efforts to expand scholarship on oil development in the North Atlantic and Arctic have included orienting the third biannual international Petrocultures conference we hosted in St. John's in 2016 around the offshore. A number of the contributors to this volume presented their research at that event.

16 The event and the subsequent publication it spawned were also sponsored by British-Newfoundland Exploration Ltd, Mobil Oil, Shell, and the American Association of Petroleum Geologists Foundation (Kay 1969).
17 "Humanities" here is often taken to encompass the social sciences, too. This is reflected, for instance, in the title of the Center for Energy and Environmental Research in the Human Sciences at Rice University. The Center has been an important institutional home for energy humanities research.
18 To be clear, Szeman and Boyer use "humanities" in a way that encompasses those working on energy in traditional social sciences *and* humanities disciplines.
19 Biannual international Petrocultures conferences, annual Cultures of Energy symposia at Rice University, and dedicated energy humanities-related book series at the University of Alberta Press, the University of West Virginia Press, and Johns Hopkins University Press are such examples.
20 See, for instance, Blum (2015), Shewry (2015), Mentz (2015), Alaimo (2017), DeLoughrey (2019), Campbell and Paye (2020), and Jue (2020).
21 DeLoughrey's call also resonates with Allan Sekula and Nöel Burch's 2010 documentary *The Forgotten Space*, and Sekula's *Fish Story*, which highlight the necessity of more closely examining labour and industry at sea to better comprehend the ocean's role as a highway for the world's commodities and the exploitative labour relations central to the global shipping industry.
22 In Newfoundland and Labrador, for instance, oil revenues have been ploughed into the disastrous Muskrat Falls mega-dam project (see Crocker and Moore, forthcoming with Memorial University Press, and Dale and Farquharson in this volume).

References

Abram, Nerilie, et al. 2019. *IPCC Special Report on the Ocean and Cryosphere in a Changing Climate*. https://www.ipcc.ch/srocc/.

Alaimo, Stacy. 2017. "The Anthropocene at Sea." In *The Routledge Companion to the Environmental Humanities*, edited by Ursula K. Heise, Jon Christensen, and Michelle Niemann, 153–161. London and New York: Routledge.

Appel, Hannah. 2015. "Offshore Work: Infrastructure and Hydrocarbon Capitalism in Equatorial Guinea." In *Subterranean Estates: Life Worlds of Oil and Gas*, edited by Hannah Appel, Arthur Mason, and Michael Watts, 257–273. Ithaca: Cornell University Press.

asinnajaq. 2019. "Isuma Is a Cumulative Effort." *Canadian Art*. https://canadianart.ca/features/isuma-is-a-cumulative-effort/.

Avery, Dwayne. 2020. "Death by a Thousand Spills: How Corporate Branding and Media Strategies Downplay the Risk of Offshore Oil Spills in Canada." *Journal of Environmental Media* 1(2): 225–244.

Barrett, Ross, and Daniel Worden, editors. 2014. *Oil Culture*. Minneapolis: University of Minnesota Press.

Barry, Roger, and Eileen Hall-McKim. 2018. *Polar Environments and Global Change*. Cambridge: Cambridge University Press.

Bavington, Dean. 2010. *Managed Annihilation: An Unnatural History of the Newfoundland Cod Collapse*. Vancouver: University of British Columbia Press.

Blum, Hester. 2015. "Terraqueous Planet: The Case for Oceanic Studies." In *The Planetary Turn: Relationality and Geoaesthetics in the Twenty-First Century*, edited by Amy J. Elias and Christian Moraru, 25–36. Evanston: Northwestern University Press.

Bousso, Ron, et al. 2021. "Shell Ordered to Deepen Carbon Cuts in Landmark Dutch Climate Case." *Reuters*. https://www.reuters.com/business/sustainable-business/dutch-court-orders-shell-set-tougher-climate-targets-2021-05-26/.

Campbell, Alexandra. 2019. "Extractive Poetics: Marine Energies in Scottish Literature." *Humanities* 8 (1): https://doi.org/10.3390/h8010016.

Campbell, Alexandra, and Michael Paye. 2020. "Water Enclosure and World-Literature: New Perspectives on Hydro-Power and World Ecology." *Blue Humanities* 9 (3). doi: 10.3390/h9030106.

Carpenter, Scott. 2020. "'On Pause' – Offshore Oil Industry Faces Biggest Setback in Years." *Forbes*, 26 May 2020. https://www.forbes.com/sites/scottcarpenter/2020/05/26/on-pause-offshore-oil-industry-faces-its-biggest-setback-in-years/?sh=c2a4f39327b4.

Chalfin, Brenda. 2015. "Governing Offshore Oil: Mapping Maritime Political Space in Ghana and the Western Gulf of Guinea." *South Atlantic Quarterly* 114 (1): 101–118. https://doi.org/10.1215/00382876-2831312.

Connery, Christopher. 2010. "Sea Power." *PMLA* 125 (3): 685–692.

Csapó, Hedvig Kriszta, Michał Grabowski, and Jan Marcin Węsławski. 2021. "Coming Home – Boreal Ecosystem Claims Atlantic Sector of the Arctic." *Science of the Total Environment* 771. https://doi.org/10.1016/j.scitotenv.2020.144817.

Cunsolo, Ashlee, and Neville R. Ellis. 2018. "Ecological Grief as a Mental Health Response to Climate Change-Related Loss." *Nature Climate Change* 8: 275–281.

Cusack, Tricia. 2014. *Framing the Ocean, 1700 to the Present: Envisaging the Sea as Social Space*. Abingdon: Routledge.

Deleuze, Gilles, and Félix Guattari. 1988. *A Thousand Plateaus: Capitalism and Schizophrenia*, trans. Brian Massumi. London: Athlone.

DeLoughrey, Elizabeth. 2017. "Submarine Futures of the Anthropocene." *Comparative Literature* 69 (1): 32–44.

———. 2019. *Allegories of the Anthropocene*. Durham: Duke University Press.

Dobrin, Sidney I. 2021. *Blue Ecocriticism and the Oceanic Imperative*. Abingdon: Routledge.

Equinor. 2021. "Where We Are." *Equinor*. https://www.equinor.com/en/where-we-are.html.

Fraser, Gail and Vincent Racine. 2016. "An Evaluation of Oil Spill Responses for Offshore Oil Production Projects in Newfoundland and Labrador, Canada: Implications for Seabird Conservation." *Marine Pollution Bulletin* 107 (1): 36–45.

Fusco, Leah. 2007. "Offshore Oil: An Overview of Development in Newfoundland and Labrador." *Oil, Power and Dependency: Publications and Reports*. http://www.ucs.mun.ca/~oilpower/pages/papers.html.

Gilroy, Paul. 1993. *The Black Atlantic: Modernity and Double-Consciousness*. Harvard: Harvard University Press.

Gladwin, Derek. 2018. *Ecological Exile: Spatial Injustice and Environmental Humanities*. Abingdon: Routledge.

Goodbody, Axel, and Bradon Smith. 2019. "Stories of Energy: Narrative in the Energy Humanities." *Resilience: A Journal of the Environmental Humanities* 6 (2–3). https://doi.org/10.5250/resilience.6.2-3.0001.

Government of Newfoundland and Labrador. 2020. *Budget 2020: Today. Tomorrow. Together*. https://www.gov.nl.ca/budget/2020/wp-content/uploads/sites/3/2020/09/The-Economy-2020.pdf.

Graham-McLay, Charlotte. 2018. "New Zealand Cites Climate Change in Banning Offshore Drilling." *New York Times*, 12 April 2018. https://www.nytimes.com/2018/04/12/world/asia/new-zealand-ban-oil-drilling.html.

Green, Jeff. 2019. "Innovative Breakthroughs: Significant Investment for Harsh Environment Research Facility at Memorial." *Memorial University Gazette*, 27 August 2019. https://gazette.mun.ca/research/innovative-breakthroughs/.

Harriss, Robert. 2016. "Arctic Offshore Oil: Great Risks in an Evolving Ocean." *Environment: Science and Policy for Sustainable Development* 58 (3): 18–29.

Heise, Ursula. 2008. *Sense of Place and Sense of Planet: The Environmental Imagination of the Global*. Oxford: Oxford University Press.

Higgins, Jenny. 2009. "Oil Industry and the Economy." *Heritage: Newfoundland and Labrador*. http://www.heritage.nf.ca/articles/economy/oil-economy.php.

Hopson, Eben. 1978. "Hopson's Address to the London Press Corps: 23 June 1978." *Eben Hopson: An American Story*. http://old.ebenhopson.com/papers/1978/London.html.

Howe, Nicolas. 2017. "Place and Pluralism in the Environmental Humanities." *English Language Notes* 55 (1–2): 113–120.

IEA. 2018. *Offshore Energy Outlook*. https://www.iea.org/reports/offshore-energy-outlook-2018.

IEA. 2021. *Net Zero by 2050: A Road Map for the Global Energy Sector*. IEA Paris. https://www.iea.org/reports/net-zero-by-2050.

ISO. 2019. "ISO 19906: 2019: Petroleum and Natural Gas Industries – Arctic Offshore Structures." https://www.iso.org/standard/65477.html.

Jones, Christopher. 2016. "Petromyopia: Oil and the Energy Humanities." *Humanities* 5 (20): 36. doi: 10.3390/h5020036.

Jones, Nicola. 2018. "How to Stop Data Centres from Gobbling up the World's Electricity." *Nature*, 13 September 2018. https://www.nature.com/articles/d41586-018-06610-y.

Jordans, Frank. 2010. "Swiss Retreat Helps Transocean Save Millions." *Boston.com*, 13 May 2010. http://archive.boston.com/business/articles/2010/05/13/swiss_retreat_helps_transocean_save_millions/.

Jørgensen, Dolly, and Sverker Sörlin, editors. 2014. *Northscapes: History, Technology, and the Making of Northern Environments*. Vancouver: University of British Columbia Press.

Jue, Melody. 2020. *Wild Blue Media: Thinking Through Seawater*. Durham: Duke.

Kampfner, John. 2021. "The Very Cold War under Way in the Arctic." *The New European*, 29 May 2021. https://www.theneweuropean.co.uk/brexit-news/europe-news/political-tensions-in-the-arctic-7995376.

Kay, Marshall. 1969. "Introduction." In *North Atlantic–Geology and Continental Drift: A Symposium*, edited by Marshall Kay. Tulsa: The American Association of Petroleum Geologists.

LeMenager, Stephanie. 2014. *Living Oil: Petroleum Culture in the American Century*. New York: Oxford University Press.

Lepic, Bojan. 2020. "Denmark to End Oil and Gas Era by 2050." *Offshore Energy*, 4 December 2020. https://www.offshore-energy.biz/denmark-to-end-oil-and-gas-era-by-2050/.

Lethabo King, Tiffany. 2019. *The Black Shoals: Offshore Formations of Black and Native Studies*. Durham: Duke University Press.

Manton, Mark. 2014. "Engineering for a Cold Climate." *Offshore Engineer*, 3 November 2014. https://www.oedigital.com/component/k2/item/7372-engineering-for-a-cold-climate.

Mentz, Steve. 2015. *Shipwreck Modernity: Ecologies of Globalization 1550–1719*. Minneapolis: University of Minnesota Press.

National Commission on the BP Deepwater Horizon Oil Spill and Offshore Drilling. 2011. *Deep Water: The Gulf Oil Disaster and the Future of Offshore Drilling: Report to the President*. https://www.govinfo.gov/app/details/GPO-OILCOMMISSION.

"Native Village Battles Big Oil Company over Drilling" 2006. *Idianz.com*, 25 May 2006. https://www.indianz.com/News/2006/014138.asp.

OECD. 2021. "OECD Policy Responses to Coronavirus (COVID-19): The OECD Green Recovery Database: Examining the Environmental Implications of COVID-19 Recovery Policies." *OECD.org*, 19 April 2021. https://www.oecd.org/coronavirus/policy-responses/the-oecd-green-recovery-database-47ae0f0d/.

Perera, B. Yasanthi, et al. 2017. "When Corporate Partnerships are NOT Awesome: Leveraging Corporate Missteps and Activist Sentiment in Social Media." In *The Dark Side of Social Media*, edited by Angeline Close Scheinbaum, 125–146. New York: Routledge.

Peters, Kimberley, and Philip Steinberg. 2019. "The Ocean in Excess: Towards a More-Than-Wet Ontology." *Dialogues in Human Geography* 9 (3): 293–307.

Phillips, Jon. 2018. "Order and the Offshore: The Territories of Deep-Water Oil Production." In *Territories Beyond Terra*, edited by Kimberley Peters, Philip Steinberg and Elaine Stratford, 51–67. London: Rowman and Littlefield.

Pitt Scott, Harry. 2020. "Offshore Mysteries, Narrative Infrastructure: Oil, Noir, and the World Ocean." *Humanities* 9 (3). https://doi.org/10.3390/h9030071.

Polack, Fiona. 2019. "Corporate and Worker Photographs of the Offshore Oil Industry: The Case of the *Ocean Ranger*." *Journal of Canadian Studies* 53 (1). DOI: 10.3138/jcs.2018-0009.

Polack, Fiona, and Danine Farquharson. 2017. "Offshore Rig." In *Fueling Culture: 101 Words for Energy and Environment*, edited by Imre Szeman, Jennifer Wenzel and Patricia Yaeger, 252–254. New York: Fordham University Press.

"Powerful Signal: In a Single Day Big Oil Suffers Historic Blows on Climate." *Politico*, 31 May 2021. https://www.politico.com/news/2021/05/26/big-oil-exxon-climate-491104.

Ray, Sarah J. and Kevin Maier, editors. 2017. *Critical Norths: Space, Nature, Theory*. Chicago: University of Chicago Press.

Ritson, Katie. 2018. *The Shifting Sands of the North Sea Lowlands: Literary and Historical Imaginaries*. London: Routledge.

Roberts, Brian Russell. 2021. *Borderwaters: Amid the Archipelagic States of America*. Durham: Duke University Press.

Royal Commission on the *Ocean Ranger* Marine Disaster. 1984. *Report One: The Loss of the Semisubmersible Drill Rig Ocean Ranger and its Crew*. Ottawa: Canadian Government Publishing Centre.

Sandlos, John. 2015. "MUN and the Arctic Oil Frontier." *The Independent*, 1 April 2015. https://theindependent.ca/2015/04/01/mun-and-the-arctic-oil-frontier/.

Santos Perez, Craig. 2020. "'The Ocean in Us': Navigating the Blue Humanities and Diasporic Chamoru Poetry." *Humanities* 9 (66). doi:10.3390/h9030066.

Secretariat of the Antarctic Treaty. 2021. *The Protocol on Environmental Protection to the Antarctic Treaty*. https://www.ats.aq/e/protocol.html.

Sekula, Allan. 2003. *Fish Story*. Dusseldorf: Richter Verlag. Revised edition.

Sekula, Allan and Nöel Burch. 2010. *The Forgotten Space*. Amsterdam: Doc.Eye Film.

Shewry, Teresa. 2015. *Hope at Sea: Possible Ecologies in Oceanic Literature*. Minneapolis: University of Minnesota Press.

Shogren, Elizabeth. 2005. "For 30 Years, a Political Battle over Oil and ANWR." *All Things Considered*. NPR, 10 November 2005. https://www.npr.org/templates/story/story.php?storyId=5007819.

Sörlin, Sverker. 2017. "Scaling the Planetary Humanities: Environmental Globalization and the Arctic." In *The Routledge Companion to the Environmental Humanities*, edited by Ursula K. Heise, Jon Christensen, and Michelle Niemann, 433–442. London and New York: Routledge.

Steinberg, Philip, and Jeremy Tasch and Hannes Gerhardt. 2015. *Contesting the Arctic: Politics and Imaginaries in the Circumpolar North*. London: I.B. Tauris.

Stoddart, Mark, Alice Mattoni and John McLevey. 2020. *Industrial Development and Eco-Tourisms: Can Oil Extraction and Nature Conservation Co-Exist?* London: Palgrave.

Sze, Julie. 2016. "Scale." In *Keywords for Environmental Studies*, edited by Joni Adamson, William A. Gleason and David Pellow, New York: New York University Press. Ebook. Credo Reference.

Szeman, Imre, and Dominic Boyer, editors. 2017. *Energy Humanities: An Anthology.* Baltimore: Johns Hopkins University Press.
Tollefson, Jeff. 2021. "COVID Curbed Carbon Emissions in 2020 – But Not by Much." *Nature,* 15 January 2021. https://www.nature.com/articles/d41586-021-00090-3.
Transocean. 2021. "Our Fleet." *Transocean.* https://www.deepwater.com/our-fleet.
United Nations Environment Program. 2020. *Emissions Gap Report 2020.* https://www.unep.org/emissions-gap-report-2020.
Urry, John. 2014. *Offshoring.* Cambridge: Polity.
Watt-Cloutier, Sheila. 2015. *The Right to be Cold: One Woman's Story of Protecting her Culture, the Arctic and the Whole Planet.* Toronto: Penguin.
Watts, Laura. 2018. *Energy at the End of the World: An Orkney Islands Saga.* Cambridge, MA: MIT Press.
Wilson, Sheena, Adam Carlson and Imre Szeman, editors. 2017. *Petrocultures: Oil, Politics, Culture.* Montreal and Kingston: McGill-Queens University Press.
Wood Group. 2013. "Wood Group Continues Investment in Arctic Technology Development," Memorial University, 9 December 2013. https://www.mun.ca/engineering/news.php?id=5663&type=news.

2

"WE ARE TORN ABOUT OUR FUTURE"

Big Oil and Iñupiaq community health in Arctic Alaska

Chie Sakakibara and Rosemary Ahtuangaruak

Concerns about accelerating climate change, resource extraction, and cultural survival have pushed the Arctic to the centre of global conversations about Indigenous sovereignty. In her one-person performance *Multispecies Solidarity*, Iñupiaq performance artist Allison Akootchook Warden (Aku-Matu) freely turns herself into the bowhead whale, polar bear, and porcupine caribou to highlight the idea of permeability and shape-shifting between humans and nonhumans. Aku-Matu is an Anchorage-based artist who is also an environmental activist. She is originally from Kaktovik, an Iñupiaq settlement on the Arctic National Wildlife Reserve (ANWR). The Iñupiat are the Indigenous people who call the northernmost rim of the United States home: Arctic Alaska. During her performance, Aku-Matu explains to the audience about the attention that Arctic wildlife species manage to receive from the public (Williams et al. 2018). She also emphasized the point during a personal interview: "People love animals. They care when they see animals suffer, and often say things like oh, poor polar bears!" Talking about how global climate change and fossil fuel extraction in the offshore cold water collectively affect her people and their Arctic homeland, she continues: "We, Iñupiat, are invisible in our own home. Poor animals get more thoughts and attention. Our health is not anyone's priority. But never forget, there are people who live where the polar bears live" (Sakakibara 2010). As Aku-Matu's remark clearly demonstrates, the Arctic ecology and social landscape are meticulously intertwined because where the polar bears and humans live, there is also the black gold – petroleum – calmly laying under the icy cold water.

"Yep, humans are invisible in the Arctic," repeats Aku-Matu. "So I gave voice to polar bear, connecting ourselves with polar bear through dreams, speaking as a polar bear about our home. Our Elders have many stories about polar bears. We [Iñupiat and bears] are related." She grinned. "Because when they [animals]

DOI: 10.4324/9781003024125-2

take off their parkas, they are people inside." Aku-Matu's performance sends a powerful message that *animals make humans visible.* The animals make humans real in an environment where the cold water seasonally turns into the ice and crude oil ominously flares from the land and the sea. "It's time for us to borrow animal voices," says Aku-Matu (Sakakibara 2010).

This chapter reveals how political and economic turbulences inaugurated by Big Oil – the world's largest multinational publicly traded oil and gas companies – have often manifested in rural Arctic Alaska at the expense of Indigenous health rooted in community integrity. Despite the impression of the Arctic as an isolated realm covered by ice and snow, the ever-accelerating pace of climate change and its consequences have been making the region and its resources accessible to extractive industry more than ever. This increasing accessibility means that Arctic Alaska is grappling with environmental issues similar to other North Atlantic regions resulting from commercial offshore oil and gas development. More specifically, our work contextualizes Iñupiaq modernity by exploring how the bowhead whale, a cultural and physical cornerstone species for Iñupiat, is entangled in the making of human health and well-being and, by extension, facilitates the complex matrix of sovereignty in relation to extractive industry and the fear for future environmental destruction. We chronicle the history of the Beaufort Sea and Chukchi Sea that has been connected with the history of Iñupiaq modernity through commercial development. In so doing, we discuss how "cold water oil" has contributed to the formation of a contemporary Indigenous identity in Arctic Alaska.

While some Iñupiat hold the view that resource exploration can be done with great care, others are often motivated to resist in subsistence communities (Ahtuangaruak 2005, 2010; Edwardson 2017; Kingik 2016). The Iñupiaq way of life, steeped in whaling, would be effectively terminated by any oil spill or industrial accident. In this fragile extractive landscape that intricately intersects with the global economy, deteriorating Indigenous community health and loss of social cohesion underscore a violent process through which the Indigenous people become outsiders in their own lands. Colonization and settler colonialism have been frequently linked to the poor health and lack of well-being of marginalized communities, creating intergenerational waves of trauma that destabilize emotional and physical health.

This chapter is co-authored by Chie Sakakibara, cultural geographer, and Rosemary Ahtuangaruak, former mayor of Nuiqsut, Iñupiaq environmentalist, educator, and Nuiqsut City Council member. In what follows, Sakakibara first contextualizes the integration between contemporary Iñupiaq subsistence and Big Oil, then Ahtuangaruak presents her narrative about Iñupiaq traditions as the foundation of human health and well-being in the face of global climate change and extractivism. In our final section, we collectively discuss the contemporary crises of Iñupiaq community health to ponder its future. Our chapter contributes to the aim of this book by revealing Indigenous voices that show their deep and inevitable entanglement with petrocultures. In so doing, it is our goal to provide a narrative on the intimacy that brings together the three central elements of this book: cold, water, and oil.

Oil and Iñupiat of the North Slope Borough

Nature and culture are never mutually exclusive, and, it is certainly the case in Arctic Alaska. In other words, despite the romanticism that is frequently associated with Indigenous ways of life, carbon-intensive capitalism is now part of Indigenous reality as much as a deep respect for nonhuman others that still penetrates Indigenous cosmologies. Calling the northernmost rim of Alaska home, Iñupiat of the North Slope Borough live a complex modernity that is built on many paradoxes and challenges associated with the process of being intertwined with energy production and global economy. To the people, the ocean and land are the source of life. An oil spill in the Arctic Ocean would have a detrimental impact on their society. That said, Iñupiat are also embedded in the oil economy, and the petroleum industry is part of their future (see Figure 2.1).

The profits generated from oil and gas extraction have been an integral part of Iñupiaq governance at least since the passing of the Alaska Native Claims Settlement Act (ANCSA) in 1971. In 1972, the villages north of the Brooks Range of northern Alaska were bound together in a state-chartered Iñupiaq governing institution called the North Slope Borough. This borough would be funded by

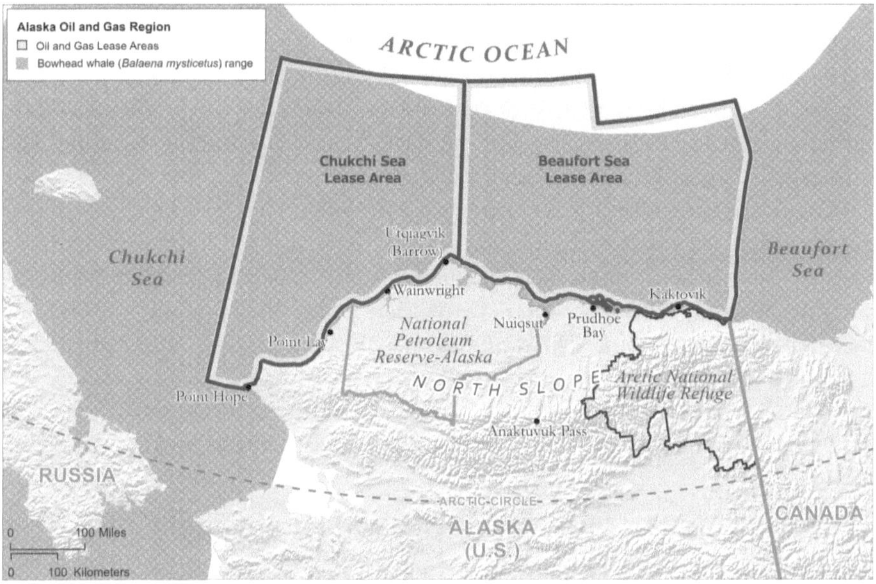

FIGURE 2.1 Alaska oil and gas region and bowhead whale range

Source: Courtesy of Todd Fagin, created using sources from: Alaska Department of Fish and Game; Bureau of Ocean Energy Management; North Slope Science Initiative; U.S. Department of Fish and Game taxing oil company leases, thus providing funds for the new government from Prudhoe Bay wealth. Fast-forward about half-a-century, the North Slope Borough's current economy is still driven by Prudhoe Bay, and the wealth the oil has brought to the community still enables myriad elements of cultural revitalization. In this context, Iñupiaq sovereignty is cultivated on their access to fossil fuel. The North Slope communities, however, are divided over how to deal with many dilemmas that the petroleum economy brings to them. At this time of global social anxiety and environmental uncertainty, political and economic turbulence often manifest at the community level by pitting family members against one another among Iñupiat. Despite the long history of opposition to the offshore development, journalists often report that the Iñupiaq community is warming up to fossil fuel extraction as global climate change accelerates its pace.[1] True, it is impossible to separate Iñupiaq subsistence from the wealth oil brings.

In the Western Arctic, regardless of extractive activities on the sea, oil production on land has declined in recent years. Nevertheless, in March 2017, some 1.2 billion barrels of oil were discovered on the North Slope, marking the largest U.S. onshore discovery in three decades (Egan 2017). The discovery is near the north-eastern edge of the National Petroleum Reserve-Alaska: leased state land that had been in decline and close to the Iñupiaq village of Nuiqsut. The Spanish oil company Repsol and its U.S. partner Armstrong Energy announced the find, predicting production could begin as soon as 2021 and lead to 120,000 barrels of output per day as President Trump signed the opening of the ANWR to oil and gas exploration into law at the end of the year. However, fossil fuel extraction in the Arctic Ocean came to a sudden halt in May 2020. Oil prices plummeted while the industry faced the lowest demand seen in decades due to the global recession incurred by the COVID-19 pandemic, with the newly introduced requirements of social distancing and the overall impact of business closures, layoffs, reduced work hours, and other restrictions.

The Arctic Slope Regional Corporation (ASRC), the largest regional Iñupiaq corporations on the North Slope, has announced that it will relinquish 21 leases in the Beaufort Sea, which were originally sold to them by Shell in 2015 after their departure from Arctic waters (McArthur 2020). Undoubtedly, this is good news for the ecosystem of the Arctic Ocean and is a victory for environmentalists and many subsistence hunters. That said, this good news for the environment has turned the future of the North Slope murky due to the intricate interconnection of the people and oil economy that has sustained the community since the 1970s.

Solastalgia, extractivism, and global climate change[2]

On the global scale, recent environmental degradation, the consequential disruption of subsistence, and systematic oppression against marginalized populations are closely linked with critical social problems experienced among Indigenous peoples.[3] Such problems often result in destruction of human health and the status of well-being. Yup'ik anthropologist Angayuqaq Oscar Kawagley states that the imbalance of the self among Alaska Native experience is rooted in Western contact as "the ruptures [caused by cultural collisions] allowed some aspects of Native characteristics to flow out or become modified by allowing new fragmented ideas, ways of being, thinking, behaving and doing to seep in. This has caused much confusion among the Native people" (Kawagley 2008).

Despite countless turbulences that modernity has brought to the Arctic, the bowhead whale remains a cornerstone species for Indigenous cultural and physical survival (Sakakibara 2020). In "Shell's Arctic Drilling Will Destroy our Homeland and Culture" co-author Rosemary Ahtuangaruak explains: "Whales give us food and bind our communities. Iñupiat communities across the Arctic coast of Alaska primarily depend on Bowhead whales for subsistence food and our culture is tied intimately with the whales and the sea" (Ahtuangaruak 2010). The villagers of Point Hope, Alaska, tell a creation story in which the whale is the land and the land is the whale (Lowenstein 1993). But now, Arctic shorelines are eroding and, as a consequence, human kinship with the whale is conceivably imperilled. Ironically, however, the initial social and economic frictions that Iñupiat experienced were brought to them by the bowhead whale, which attracted the commercial whalers who brought with them Christianity, diseases, and alcohol, all of which ultimately caused the loss of countless lives, local customs, and community integrity. In northern Alaska, especially since the 1980s, physical isolation, access to firearms, and abuse of alcohol among youth put the risk of suicide much higher than in other American communities,[4] and this timeline coincides with the aggressive encroachment of fossil fuel industry to the Arctic homeland.

Extractivism presents another form of colonial endeavour that is encroaching on Indigenous lands, and Iñupiat are caught in the middle. As a premise, Indigenous ideas of health and well-being are steeped in their relations within the community, between communities, and community and its environment. Coined by Glenn Albrecht, the idea of *solastalgia* – a combination of the Latin word *solacium* (comfort) and the Greek root *algia* (pain) – underscores the Iñupiaq experience of global climate change (2005, 2019). Unlike nostalgia, distress, or sadness experienced when one is separated from a home, solastalgia is caused by environmental change affecting people while they are still directly connected to their home (environment). The ramifications of climate change cause deep distress and mental and physical problems, which are directly and indirectly triggered by natural resource extractions.

Human health is intimately tied to the health of the earth. This point is also elaborated by many villagers of Point Hope who express concerns about the loss of their homeland to erosion. Today, the shorefast ice of the Arctic Ocean retreats too soon and exposes more open water to the action of the wind and brings waves toward the land (Milman 2017). Normally, the ice provides a buffer for the coastline and helps prevent erosion, but now the coastline is vulnerable without sufficient sea ice to protect it. The resulting erosion increases the impacts of rising sea levels. In Point Hope, when the villagers lost their primordial homeland to the sea in the 1970s, the Village Council decided to move their settlement of Tiġara (today known as Old Town) to its present location, New Town. Old Town used to contain sod houses made of whale jawbones. As the remainder of Old Town disappears into the sea, Iñupiat relationship with the land, the foundation of their social health and well-being, increases uncertainties (Sakakibara 2008). As for coastal erosion, Point Hope is not the only example of the community that is affected by the exacerbating problem.

About 12 years ago, in February 2008, a small Iñupiaq village of Kivalina – it is now just a matter of time before the village goes under water – filed suit against 24 fossil fuel companies for contributing to the village's erosion through large amount of greenhouse gas emissions (Shearer 2011). Today, Newtok, a Yup'ik village near the coast of the Bering Sea with 21 households and 137 residents (and long threatened by erosion and flooding) has begun its relocation process to Mertarvik, the new village on Nelson Island, nine miles away from the original settlement, after a wait of two decades.[5] Reflecting on this strong correlation between climate change, community health, and commercial resource extraction, what does it mean for Iñupiat to live within the economic system sustained by oil?

As noted in the previous section, Iñupiat did not interact with the fossil fuel industry in any meaningful way until the mid-20th century. With the passage of the Alaska Native Claims Settlement Act (ANCSA) of 1971, Alaska Natives including Iñupiat exchanged their ancestral lands for 44 million acres of land, $962.5 million, and a royalty of $500 million on mineral rights in compensation. To this day, Alaska Native responses to the ANCSA remain controversial, as it ushered in the extinction of aboriginal rights and replaced such rights with legal definition of villages and the distribution of community rights to individual shareholders through the system of corporations.[6]

As powerful as whaling is to empower the society, this corporatization of tribal communities has given Alaska Natives previously unwielded political influence (McPherson 1982). In this way, the ANCSA reversed the usual historical experience of Indigenous peoples with resource extractors by landing the wealth that Big Oil bought (Williams 2017). With it, and despite the apparent conflict between capitalism and subsistence, many Alaska Native leaders and corporations learned to expand and prosper as capitalists and entrepreneurs as much as whaling captains, which undoubtedly contributed to the making of a modern Alaska Native sovereignty. That said, it is important to note that Iñupiat are never one unified voice. While petroleum has become an integral element of Alaska Native lives today, Republican Alaska Senator Lisa Murkowski's recent visit to Kaktovik was met with powerful signs, and "Whale Oil, Not Crude Oil" remains a strong voice among Iñupiat who now call the oil fields home.

Life in the Arctic oil fields

I, Rosemary Ahtuangaruak, co-author of this chapter, am an Iñupiaq environmentalist from Arctic Alaska. I have lived and raised my family in Nuiqsut, an Iñupiaq community on the Beaufort Sea coast since 1986. In 2002, I helped to found REDOIL (Resisting Environmental Destruction on Indigenous Lands), a nonprofit Native grassroots network to protect and promote Indigenous lives and environmental well-being. REDOIL exemplifies a movement of Alaska Natives of the Iñupiat, Yupik, Yup'ik, Aleut, Tlingit, Eyak, Gwich'in, and Denaiana Athabascan Tribes who came together in June 2002 in Cordova, Alaska, to form a powerful entity to challenge the fossil fuel and mining industries and demand our rights to a safe and healthy environment conducive to subsistence. REDOIL aims to address the human and ecological health impacts brought on by unsustainable development practices of the fossil fuel and mineral industries and the ensuing effect of catastrophic climate change. We strongly support the right of self-determination for tribes in Alaska and a just transition from fossil fuel and mineral development to sustainable economies and sustainable development. The three core focus areas of REDOIL are: sovereignty and subsistence rights, human and ecological health, and climate change and climate justice.

Nuiqsut is a small remote village of 400 people in the middle of this Arctic oil field. It is surrounded by land leased for development by oil companies among a large patchwork of active and planned wells. I completed the Physician Assistant Program at the University of Washington in 1991. I worked as a health aide in Nuiqsut for 14 years. I also served as mayor and community health and environmental advocate in Nuiqsut. In the village, I live a very traditional lifestyle – hunting, fishing, whaling, gathering, and teaching our family and community members the traditional and cultural activities as my Elders taught me. Hunting and consuming various wildlife species in collaboration have been the foundation of human health and well-being in my community, including birds such as ptarmigan, ducks, and geese; fish including char, salmon, whitefish, dolly varden trout, grayling, pike, trout, and cisco; land mammals including caribou, moose, and muskox; and marine mammals such as bearded seals, walrus, beluga, and bowhead whales. We harvest berries, plant roots, and herbs. Our subsistence is characterized by the extensive sharing traditions that unite families and communities, which sustain both individual and community health by enhancing a variety of relations. We acknowledge family ties that bind us to Nuiqsut and other communities across the Arctic, and exposure to hunting and gathering is an important part of these ties. The presence of petroleum industries directly affects my community and our traditional whaling practice.

The bowhead whale is a central existence for us to carry on subsistence, which allows us to promote our cultural and physical health. Nuiqsut whalers hunt for bowhead whales in the Beaufort Sea during the fall. We await the migration of the whales from Camden Bay, 144 miles east of Nuiqsut, for this hunt. The crews go to Cross Island, about 100 miles from Nuiqsut, in August. My sons help the whaling crews with their preparation. My ex-husband has gone whaling from Cross Island with his uncle a few times. There is a feast in the village after a successful whale hunt. Everyone is invited to the captain's home to eat whale meat and other food, tea, and coffee. These feasts are a time of celebration, when stories are told about the hunt and our changing environment.

We have a unique sharing system to divide the harvested whale among the crew striking the whale and the crews helping to land and butcher the whale. The whalers also set aside a portion of the whale meat to share with those in need. Part of the catch is saved for feasting. We share the feast three times during a year at the blanket toss, a traditional celebration during the summer, Thanksgiving, and Christmas. Sharing of the whale binds our communities. Whale meat feeds families throughout the long, dark winter and provides nourishment, warmth, and fuel for our daily activities during the Arctic winter.

Looking back to the 1990s, we faced the first threat of offshore drilling in our whaling grounds. If the whale hunts are less successful, the community suffers from food shortage. To give an example, earlier in the decade, there were seismic activities and exploration drilling in Camden Bay, which severely affected our whaling. The following winter, I heard of an unusually high rate of domestic violence in Nuiqsut and increases in suicide attempts and suicides. I heard of and witnessed increases in drug and alcohol use as well. As a community health aide, I listened to people's stories of how difficult it was to hunt without success. That winter was the worst I spent, and the experience prompted me to speak out at meetings about oil development. Our Elders told us how whaling would be made harder by aggressive industrial extraction, which would result in increasing hunting distances and, ultimately, loss of harvest.

Despite extensive testimony against drilling in our village, the oil and gas industry has made it to the Arctic land and ocean, which we call our garden. Through my career, I have cultivated a special relationship with many people of the Nuiqsut community. Having people come every day to the clinic for help, needing food, and offering me stories of whale hunts of hundreds of miles and coming home with no harvest, having no whale or other foods to share, of fish numbers dwindling, was an awakening moment for me. As an Iñupiaq woman who has faith in human health and well-being, and a deep entanglement with our environment, I learned to ask questions and learned to speak up. Our Elders had my back. I never wanted to face what happened with increased social ills incurred by the encroachment of extractive industry, domestic violence, alcohol and drug abuse, suicides, and many other problems that used to be rare in our tight-knit community.

All communities need to be committed to public health to protect our environments and our health. For generations, Indigenous and tribal communities have had a fundamental and close relationship with our lands. Our lands and waters have sustained and nourished us. Traditional knowledge, shared through storytelling, is based on generations of Iñupiaq relations with our ecology, weather, species and their interactions, historical events, and more. Our traditional knowledge supports our life, health, and safety. The impacts of resource extraction in Alaska on my community's health and well-being are immeasurable. The integration of traditional knowledge of our ancestors, along with the skills and knowledge of current community members and new tools and technology, can guide the solutions to protect our lands and ensure the health of future generations. It is critical that research and policy processes engage with and learn from the knowledge and experience of tribal communities.

Just as the river and ocean shores are eroding, however, so is our traditional knowledge as those values compete with dollars per barrel and acres to sustain the families and villages. The erosion causes the loss of the nests of many birds, including the bald eagle (the U.S. national symbol), at the cost of the national energy policy; the irony is that such policy includes words for health and other budgetary buzz words.

In May 2010, immediately after the Deepwater Horizon oil spill, I toured the Gulf of Mexico. It was after I participated in the bowhead whale census to assess the health of the whale population in the Arctic Ocean, Beaufort, and Chukchi Seas to protect our subsistence rights, and thus, the temperature difference was more than a hundred degrees. I saw the migratory birds on the coast spinning as the spill reached the shores, soaking the vegetation with reeds drawing oil up into the plants, oil in the mangroves, and the sick sweet crude smell increased as we left the shore and later got closer to the spill site. The blue plume visible miles away darkened as we got closer. The droplets of oil splashed on my skin, and the stench grew gradually. The painful views continued as boats and fishing gear were being taken out and put away, an attempt to prevent further contamination. We saw porpoises in the oil slick. The slick continued for miles, with workers lacking any respirators. The sight reminded me of the Exxon Valdez oil spill that occurred in Prince William Sound, Alaska, on March 24, 1989; oil dispersants being sprayed by planes, stocked in yards, the coast guard telling people everything is okay, and the fishermen attending countless meetings as we did in Alaska and often being misled and misinformed. After the Exxon Valdez disaster, an acquaintance told me how his father waited 20 years for his settlement for a mere thousand dollars. There are many lessons we must learn, or must have learned, from these unnatural disasters.

In Nuiqsut, I saw the number of people treated for respiratory illness during that time rise from 1 to 75 between 1986 and 2000, an increase that far outpaced population growth, at the same time the oil wells crept closer to town. In Nuiqsut and other Arctic villages with no year-round road access, the situation is grim, as the closest hospitals are hundreds of miles away. But even as I tried to raise awareness about it, I struggled to get any help. My family and I also faced many harassments and obstacles in our own community due to the intricate relationship that our modernity produces with Big Oil. Meanwhile, climate change multiplies adverse effects on people living in areas vulnerable to sea level rise, sea ice reduction, and environmental contaminants because of the lack of access to health-care services and the inability to relocate or rebuild after a disaster. We are torn.

Among our people, such community dysfunction often results in the rise of depression, alcoholism, domestic violence, and suicide in addition to physical manifestations of adverse effects caused by environmental destruction including asthma and other respiratory diseases.

Meanwhile, the cancer rates in the community and birth defects among newborns are going up. As an elixir for this trauma today, illicit drugs, from marijuana and cocaine to heroin and hallucinogens, take away the pain temporarily, but further deepen the colonial wound. To our youth, such pains and social problems bound up in substance abuse are equivalent to, or an even more immediate fear than, life-threatening issues posed by global climate change. Let me give you an example from my personal life. My son shot his first caribou near Nuiqsut in March this year, but the harvest was lost due to the overflight of a helicopter, which drove the frightened animal into a man-made lake on the tundra. My son has been fighting an uphill battle with social illness. In this divided community, my environmentalist career that advocates for environmental justice has often taken a personal toll, occasionally affecting the financial and personal health of my family and kin. This struggle and polarized values represent the imbalance between the industrial extractivism and our physical, emotional, and community health.

Furthermore, the bowhead whales' health is directly connected to human health and well-being because the animals we eat are important to our lives and health. Thankfully, the bowhead population has successfully recovered since the end of the commercial whaling in the early 20th century that had devastated the fate of the whales and, by extension, the social vibrancy of Iñupiat. Since then, the people and whales have come a long way together. Today, the International Union for Conservation of Nature (IUCN) lists the global population of the bowhead as a whale of "least concern" as its populations are on a steady increase and its health in the ecosystem is sustained.[7] Yet the bowhead whale is still designated an endangered species by the 1975 Convention on International Trade in Endangered Species of Wild Fauna and Flora (CITES) and by the National Marine Fisheries Services. As a result, the bowhead whale is not allowed to enter the commercial market. Iñupiat and Siberian Yupik on St. Lawrence Island had been sharing the whale for over a thousand years before such outsider-imposed regulations.[8] Through an international treaty signed by the U.S. government, Iñupiat and Yupik hunters (through the NOAA-AEWC Cooperative Agreement) must prove the health of the stock every ten years to the International Whaling Commission to continue the traditional subsistence hunt. This aspect demonstrates the perseverance of the subsistence community and adaptability forced on them by regulation and laws determined by those in Washington, D.C., who clearly do not understand life and communities in the Arctic.

We have activities associated with our whale harvesting throughout the year, such as seal skin sewing to create a whaling boat, caribou sinew preparation, and craft making. The teaching of the activities and stories continues throughout the year, with each generation sharing family hunting stories. Even though Shell's proposals to drill for oil and gas in the Chukchi and Beaufort Sea were permanently dropped in 2015, they were seen as detrimental to the traditional and cultural activities of our family and village. We depend on traditional foods that migrate through both Beaufort and Chukchi Seas. Any harm from the oil industry's activities to our resources, including bowhead whales, seals, fish, and caribou, threatens our food and our health.

I was concerned that industrial drilling, icebreaking, aircraft and helicopter flights, and other noisy activities in the Beaufort and Chukchi Seas would keep whales from their feeding areas or otherwise harm them. Any changes in the whale populations could affect our hunting. We educate our families that a whale hunt must be respectful and quiet. If the noise from drilling and icebreaking causes whales to exist further offshore, it would place more wear on our equipment and put the whalers at greater risk. We have been taught not to put anything in the water that may cause the whales to turn away from the hunting grounds. Water discharges from drill ships or the presence of drilling muds in the water where oil companies conduct their activities could cause whales to avoid our hunting grounds.

In November 2010, there were federal government hearings in several Iñupiat communities including Utqiaġvik, Kotzebue, Point Hope, Point Lay, and Wainwright to hear our concerns about Shell's Arctic drilling. I attended the hearing in Utqiaġvik, where I used to live. We had about 60 people attend the hearing, and overwhelmingly, the statements were in opposition to Shell's drilling plan. The hearing showed the continued concerns of the lack of ability to respond to a spill, the lack of taking the concerns of the people into meaningful consultation, unwillingness to protect our traditional and cultural activities, and the continued stress and strain this is causing to our people. We want to be Iñupiat (the real and completed people) into the future, not just residents in an industrialized area destroyed by Shell's offshore oil drilling. Our animals, land, and seas in the Arctic are already severely stressed by climate change. We did not want Shell's dangerous offshore drilling to add to our difficulties. I urged President Obama to hear our concerns and deny Shell the permit to drill for oil in the Beaufort and Chukchi Seas, which resulted in his decision to ban oil drilling in large areas of the Arctic and Atlantic Oceans.

"We Are Torn About Our Future"

Petroculture, modernity deeply shaped by fossil fuel, remains the core of Iñupiaq governance. The contradictions of oil are confronted by the young Iñupiaq artist, Crystal Naataq Martin. Originally from Utqiaġvik, and now a resident of Anchorage, she pronounced that "Some of us are a little torn about our future" (Figure 2.2):

> The oil industry is a touchy topic amongst the Alaska Natives. But working in the oilfield myself, along with many other Iñupiats, I know I'm not the only one who feels torn about the growing industry and the influences it has on our Elders, classmates, and our children in the communities all across the North Slope. And there's the question [...] can there be a reasonable compromise that doesn't include subsistence or cultural sacrifice and bribery from big oil? (Martin 2018)

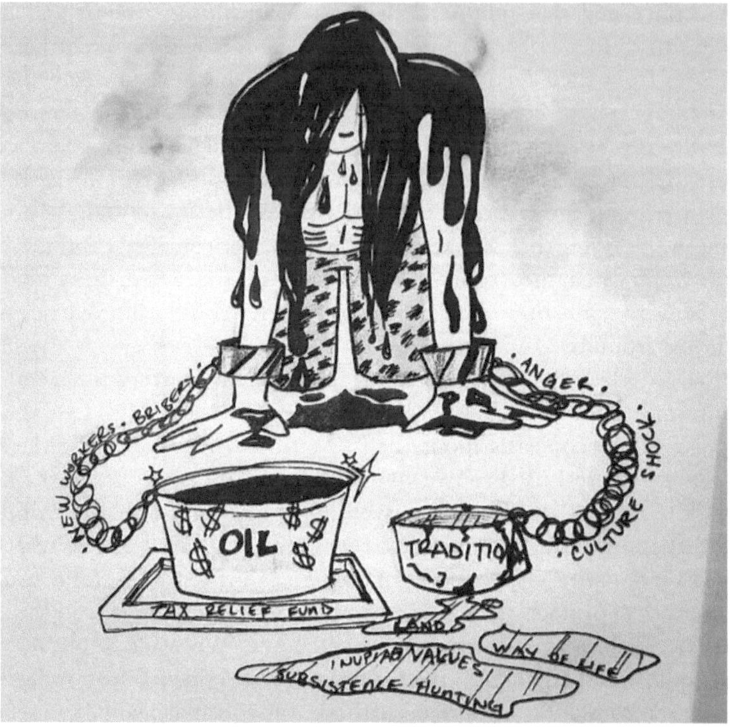

FIGURE 2.2 "Some of us are torn about our future"
Source: Crystal Naataq Martin, 2018

The man is on all fours looking down. Oil covers his head, shoulders, and arms. The black, shiny, and sleek texture of the liquid slowly covers his body and drips down from his shoulders and arms and stains his seal-skin pants. His wrists are chained to two buckets; the larger one is filled with "oil" and the other bears "tradition." The oil bucket with the dollar signs is shiny and well-protected with the words "tax relief fund," "new workers," and "bribery." The other bucket – tradition – is clearly old and exploited, and transparent liquid leaks out of it with the words "land," "way of life," "Iñupiaq values," and "subsistence hunting." These words are linked with "anger" and "culture shock." The man is caught in the middle of oil and tradition. Clearly, oil and tradition are tied in a very complex way, creating a new identity politics and reality.

As one of the Alaska Native corporations, the Arctic Slope Regional Corporation (ASRC) currently enrols approximately 12,000 shareholders. In 2014, the ASRC and six North Slope village corporations joined to create a new company named Arctic Iñupiat Offshore, LLC (AIO). The AIO and Shell Gulf of Mexico Inc. (Shell) signed an agreement that allows the AIO the option to acquire an interest in Shell's acreage and activities in its Chukchi Sea leases.[9] In September 2015, Shell's President and AIO Executive Vice President visited Utqiaġvik to have a meeting with leaders representing the borough and the ASRC. Despite fear of an oil spill and the concern that the noise of drilling would disrupt the feeding and migration patterns of the bowhead whales and other marine animals, it was reported that many local stakeholders, citing economic necessity, supported Shell in the debate over drilling, local impacts, and climate change (Mufson 2015). Some leaders commented that climate change *is* a reason to push ahead with developing new oil resources, not a reason for holding back. As Crawford Patkotak, Chair of the ASRC, put it, "We are not going to be victims of change, climate change or social change [...]. We wanted to plan for the future of our people" (Patkotak, in Newton 2015). In Patkotak's view, oil secures an Iñupiaq future through the resources it brings to the community. His perspective shows the embedded nature of Iñupiat to capitalistic economy without being the miner's canary. Later in the month, however, to the disappointment of many, Shell abandoned its plans to drill in the Chukchi Sea for the foreseeable future due to a "clearly disappointing exploration outcome" (Chappell 2015). Meanwhile, the ASRC currently supports Kivalina's initiative to build an evacuation road from the rising sea. This situation may appear contradictory as the ASRC contracted with every one of those extractive companies Kivalina had a suit with. Nevertheless, the ASRC responds to climate crisis promptly. This example shows the complex reality that Iñupiaq people, their ideologies, and local political institutions face today.

The whalers had been broadly opposed to offshore drilling for decades (Kishigami 2010). But the reality is that even before the partnership between the ASRC and fossil fuel companies came to fruition, the oil companies had become major employers in the North Slope. The risks of oil exploration prevail, but many see oil as a means to strengthen their culture and to, ultimately, displace colonial narratives. Richard Glenn testified to the U.S. Senate Committee on Energy and Natural Resources at a 2017 hearing on ANWR that Iñupiat cherish land for both food and resource development. To Iñupiat, Glenn stated, the land can be sacred and produce oil (Ruskin 2019).

I (Rosemary Ahtuangaruak) attended the air quality hearing and testified at Washington, D.C., with Richard Glenn, ASRC representative, and Daniel Scott Sullivan, member of the Republican Party serving as the junior U.S. Senator from Alaska in 2017. I managed to get the last word in: "The cost of oil and gas development should not cost the health of the people who live where oil and gas development will occur." The words fell on deaf ears.

The North Slope communities are divided over how to deal with the petroleum juggernaut. The past several North Slope Borough mayors have openly endorsed ocean drilling. Although their supporters agree with the need for impact assessments and holding oil companies to high standards, those against oil are equally strong in their opposition. The 2010 North Slope Borough mayoral election displayed the results of this polarization. Among three candidates, the most pro-oil candidate (601 votes) and the most anti-oil candidate (583 votes) lost to the middle-ground candidate (639 votes) who garnered 29% of the vote by advocating the development of resources with caution.[10]

Political and economic turbulences often manifest at the community level by pitting family members against one another. While Ahtuangaruak kept herself busy with attending various meetings to vote against Arctic drilling, I (Chie Sakakibara) was in Point Hope one summer when seismic testing on the outer continental shelf for a Chukchi Sea oil survey was announced. Such surveys take place when the sea ice is most diminished and have a significant impact on the fall whaling season. While all residents stood to gain from oil revenues, the village was split between the anti-oil Native Village of Point Hope and the pro-development Tikiġaq Corporation. The villagers were caught in the midst of divided loyalties based on kinship relations. Anti-development leaders organized meetings, and the heated discussions carried on until late. Despite their resistance led by Caroline Cannon, then-mayor of the village who later received the Goldman Environmental Prize in 2013, the testing was done, and several marine mammals were sighted stranded on the beaches later in the year.[11] While the cause of such incidents was not officially validated, community consensus blamed seismic testing.[12] The whalers feared that these sounds will drive the whales further north, out of hunting range.[13] This development also intensified relationships within the small community, fracturing its integrity.

The Village of Point Hope, the lead plaintiff in a lawsuit seeking to overturn federal approval of Shell's leases, dropped its legal challenge in March 2015. Nevertheless, many residents continue to oppose the drilling. Point Hope, a village of about 700 people bound by kinship, supports *and* opposes oil drilling. This divide affects the community-based solidarity that has historically constituted the social ability to survive. Mae Hank of Point Hope, an active and vocal opponent of offshore development for over 40 years, raises concern that the harsh Arctic weather will get in the way of any clean-up process when a spill occurs. If the spill went under the ice, the ice would make it impossible to remove.[14] Pointing at how Shell responded to a rig fire in Kulluk, near Sitkalidak Island in 2012, Earl Kingik argues that oil companies are not ready to operate in the challenging conditions of the Arctic Ocean (2016).[15]

Oil may well be a strong agent to form and influence human perspectives. The divided views among the people on oil encapsulate not only the diverse values and perspectives among community members but also how complicated human lives with and around oil can be. Since its discovery and entrance into Alaska Native modernity, oil's significance cannot be explained clearly by a compartmentalization of values labelled as good or evil, convenient or inconvenient, and traditional or hegemonic. As much as oil has the capacity to serve as a source of community sustainability and resilience, it can certainly ruin Iñupiaq sovereignty with mismanagement or an accident.[16]

Can oil be a new whale that the ANCSA, the new harpoon, has landed? Or is it at best a whale that was harpooned but sinks before retrieval? There is no straightforward answer to these questions. Meanwhile, the Alaska Federation of Natives (AFN), the largest state-wide Native organization in Alaska representing 178 villages, 12 regional Native corporations, and 12 regional non-profit tribal consortiums, declared a state of emergency for climate change on October 16,

2019, after an emotional split over climate change and resource extraction dividing the convention floor in Fairbanks for well over an hour. The Anchorage Daily News coverage by Alex DeMarban indicates the disagreement was based on "generational differences, as mostly young people concerned about thawing permafrost and eroding villages faced off against older counterparts fearful of losing access to oil and mineral resources, and even the whales and seals they hunt" (DeMarban 2019).

An Iñupiaq origin story from Point Hope has it that the giant creature (likely the bowhead whale) harpooned by trickster Raven became the land for Iñupiat. The new harpoon struck the oil out of the land-whale through a deep entanglement in the ocean, and now the tundra is thawing, releasing long-dormant spores of the highly infectious germs came back to life as well as causing a series of violent explosions in a blowout of highly flammable methane gas. Meanwhile, the oil economy is about to rupture in the time of the contemporary pandemic. The whale and oil, the two entities residing deeply in the cold water that are central to the future of Iñupiaq health and well-being, collectively tie the humans to the land and ocean, only to increase the intricacy of the embeddedness.

Acknowledgments: The authors are thankful to many mentors and colleagues whose encouragements and wisdom have informed and inspired our collaboration (in alphabetical order): Subhankar Banerjee, Lucy Brown, Caroline Cannon, George and Debby Dahl Edwardson, Todd Fagin, Aaron A. Fox, Mae Hank, Earl Kingik, Crystal Naataq Martin, Mae Masuleak, Jack Schaefer, Sean Asiqłuk Topkok, Allison Akootchook Warden (Aku-Matu), Maria Shaa Tláa Williams, and our colleagues and students at Oberlin College and Syracuse University. Several parts of this chapter have appeared in *Whale Snow* by Chie Sakakibara © 2020 The Arizona Board of Regents. Reprinted by permission of the University of Arizona Press.

Notes

1 See Rosen 2019; Welch 2015.
2 Parts of this section are modified versions of Chie Sakakibara's *Whale Snow: Iñupiat, Climate Change, and Multispecies Resilience in Arctic Alaska* (2020) © The Arizona Board of Regents. Reprinted with permission from the University of Arizona Press.
3 See Isaac, Finn, Joe, Hoover, Gone, Lefthand-Begay, and Hill 2018; Kral, Idlout, Minore, Dyck, and Kirmayer 2011; Sakakibara 2020; Topkok and Green 2016.
4 See Napoleon 1996; Reimer 1999; Wexler 2009.
5 See D'Oro 2008; Lester 2019.
6 Douglas Jones, assistant to Senator Mike Gravel of Alaska, testified that the ANCSA was a form of "social engineering" (Berger 1985, 21). Interpreting this statement, Tlingit scholar Maria Shaa Tláa Williams says the ANCSA has "elements of genius and elements of destruction," the latter of which sought to better integrate subsistence societies into the capitalistic economy (Williams 2017).
7 IUCN 2020; Also see Cooke and Reeves 2018.
8 See Huntington 1992; Jolles 2002.
9 Arctic Slope Regional Corporation, "North Slope Village Corporations and Shell Announce Historic Venture." Last modified August 5, 2014. Accessed August 9, 2017. https://www.asrc.com/PressReleases/Pages/Arctic-Iñupiat-Offshore.aspx.

10 Since 40% is the required percentage to win the election, there was a runoff election between the top two candidates in the following month. The winner was Charlotte Brower, a whaling captain's wife from Utqiaġvik, the first female mayor of the borough.

11 There is a marine mammal stranding record kept by the NOAA in Alaska; however, if no one reports, no record is made. There are not regularly scheduled systematic beach surveys flown for a specific purpose of monitoring strandings on the North Slope or Western Alaska. Most information that I acquired while in Alaska was anecdotal, mentioning many dead whales floating near Point Hope and also near Prudhoe Bay. None of those were reported to NOAA. Rosa 2009 presents a summary of strandings and floating carcasses since 2006.

12 It is likely that seismic surveys would disrupt migration patterns, expose marine mammals to the risk of hearing impairment, interfere with feeding, resting, and influence the population levels of the Western Arctic stock of the bowhead whale (Robertson et al. 2013). Oil exploration involves sonar and explosives, and whales are known to be very sound-sensitive. It has been internationally reported that the increasing number of whale beaching worldwide is partially due to the noise pollution in the oceans, and the bowhead whales are not exceptions (Bernstein 2013; Calloway 2013).

13 Jack Schaefer of Point Hope said: "There was also seismic work that was being done without our knowledge over a ten-year period, and that may have wiped out a lot of the food chain. We had no idea until last year that took place. Intense underwater sound can have a range of deleterious effects on marine mammals and other marine life, the animals who make us who we are, Iñupiat." (Sakakibara 2006).

14 Hank says, "We hunt off our ocean and that provides us with the majority of our food resources. Mitigation of such crisis appears helplessly impossible. If they [Shell] have a spill we no longer have food security. Our security for our food resources [is] gone" (Hank in Ottery 2015).

15 Referring to the politics that involves the whale and ocean, Iñupiaq author and artist Nasuġraq Rainey Hopson deplores how colonization and the boarding school curriculum instill propaganda into Iñupiat minds. She finds the curriculum at odds with "Iñupiat values and how we care for and treat the natural world should never change [because they are] at the core of our personal and social identity" (personal communication, October 20, 2018). Her concerns are echoed by ICAS Vice-Chair, George Edwardson. George remembers when he, as a young student in mineral and petroleum technology, was told by the Elders "to keep the industry off of the ocean for the whales," which he interpreted as for the future Iñupiat (Edwardson 2017).

16 The North Slope Borough Economic Profile and Census Report for 2015 (2020) reveals that the major sources of income for the North Slope Borough households are wage work, corporation dividends, the permanent fund, and income from other sources. The unemployment rate is still very high for North Slope households at the rates of 20% or more. Discussing the controversy that the oil business has brought to northern Alaska and how it directly and indirectly causes illness among animals that constitute the Iñupiaq way of life, Siqiniq Maupin, art and youth organizer at Native Movement, a Fairbanks-based nonprofit activism group, writes:

> The cost of living in rural Alaska is significantly higher than the national average. Despite the idea of wealth gained from development, a majority of North Slope residents live below the poverty line. My family and other Inupiat depend on our traditional foods for sustenance and survival. Our traditional foods are direct link to our ancestors—they are a part of our culture that I hope our future generations will get to experience (Maupin 2019).

References

Ahtuangaruak, Rosemary. 2005. Resisting Environmental Destruction on Indigenous Lands (REDOIL). Press Release.

Ahtuangaruak, Rosemary. 2010. "Shell's Arctic Drilling Will Destroy Our Homeland and Culture." *ClimateStoryTellers: A Gathering Place for Stories on All Things*

Global Warming, edited by Subhankar Banerjee. Accessed May 4, 2020. http://www.climatestorytellers.org/stories/rosemary-ahtuangaruak-shell-arctic-drilling/.

Albrecht, Glenn. 2005. "Solastalgia: A New Concept in Human Health and Identity." *Philosophy, Activism, Nature* 3: 41–55.

Albrecht. Glenn A. 2019. *Earth Emotions: New Words for a New World*. Ithaca, NY: Cornell University Press.

Arctic Slope Regional Corporation. "ASRC, North Slope Village Corporations and Shell Announce Historic Venture." Last modified August 5, 2014. Accessed August 9, 2017. https://www.asrc.com/PressReleases/Pages/Arctic-Iñupiat-Offshore.aspx.

Berger, Thomas R. 1985. *Village Journey: The Report of the Alaska Native Review Commission*. Vancouver, BC: Douglas and McIntyre.

Bernstein, Lenny. "Panel Links Underwater Mapping Sonar to Whale Stranding for First Time." *The Washington Post*. Last modified October 6, 2013. Accessed August 14, 2017. http://www.washingtonpost.com/national/health-science/panel-links-underwater-mapping-sonar-to-whale-stranding-for-first-time/2013/10/06/52510204-2e8e-11e3-bbed-a8a60c601153_story.html.

Calloway Whiting, C. "Mass Whale Stranding Proven to Be Caused by Sonar." *Huffpost Hawaii*. Last modified September 26, 2013. Accessed July 3, 2017. http://www.huffingtonpost.com/candace-calloway-whiting/whale-stranding-sonar_b_3997569.html.

Chappell, Bill. "Shell Oil Abandons Controversial Drilling Off Alaska's Shore." Last modified September 28, 2015. Accessed August 1, 2017.

Cooke, Justin G., and Reeves, Randal. 2018. *Balaena Mysticetus. The IUCN Red List of Threatened Species* 2018: e.T2467A50347659. https://dx.doi.org/10.2305/IUCN.UK.2018-1.RLTS.T2467A50347659.en. Downloaded on 21 April 2020.

D'Oro, Rochael. 2008. "State Gives $3 Million to Fight Newtok Erosion." *Alaska Dispatch News*, October 1.

DeMarban, Alex. "AFN Declares 'State of Emergency' for Climate Change." *Anchorage Daily News*. Last modified October 20, 2019. Accessed October 21, 2019. https://www.adn.com/alaska-news/2019/10/20/afn-declares-state-of-emergency-for-climate-change/.

Edwardson, George. "OPINION: We've Won an Important Battle." *Arctic Sounder*. Last modified February 28, 2017. Accessed August 14, 2017. http://www.thearcticsounder.com/article/1409weve_won_an_important_battle.

Egan, Matt. "Massive Oil Discovery in Alaska is Biggest Onshore Find in 30 Years." *CNN Business*. Last modified March 10, 2017. https://money.cnn.com/2017/03/10/investing/alaska-oil-discovery-repsol-spain/.

Huntington, Henry P. 1992. *Wildlife Management and Subsistence Hunting in Alaska*. Seattle: University of Washington Press.

Isaac, Gwyneira, Symma Finn, Jennie R. Joe, Elizabeth Hoover, Joseph P. Gone, Clarita Lefthand-Begay, and Stewart Hill. 2018. "Native American Perspectives on Health and Traditional Ecological Knowledge." *Environmental Health Perspectives* 126 (12): 1–10.

IUCN Red List. 2020. "Bowhead Whale: *Balaena mysticetus*." Accessed April 22, 2020. https://www.iucnredlist.org/species/2467/50347659.

Jolles, Carol Zane. 2002. *Faith, Food, and Family in a Yupik Whaling Community*. Seattle: University of Washington Press.

Kawagley, Angayuqaq Oscar. 2008. "An Alaska Native's Theory on Events Leading to Loss of Spiritual Life." Alaska Native Knowledge Network. Last modified September 30, 2008. Accessed May 9, 2020. http://www.ankn.uaf.edu/Curriculum/Articles/OscarKawagley/AKNativeTheory.html.

Kingik, Earl. "As a Member of the Iñupiaq People of Alaska, My Biggest Concern Is Offshore Drilling." *AlterNet*. Last modified June 13, 2016. Accessed August 13, 2017.

Kishigami, Nobuhiro. 2010. "Climate Change, Oil and Gas Development, and Iñupiat Whaling in Northwest Alaska." *Études Inuit Studies* 34, no. 1: 91–107.

Kral, Michael J., Lori Idlout, J. Bruce Minore, Ronald J. Dyck, and Laurence J. Kirmayer. 2011. "Unikkaartuit: Meanings of Well-Being, Unhappiness, Health, and Community Change among Inuit in Nunavut, Canada." *American Journal of Community Psychology* 48: 426–438.

Lester, Marc. "A Western Alaska Village, Long Threatened by Erosion and Flooding, Begins to Relocate." *Anchorage Daily News*. Last updated October 19, 2019. Accessed October 21, 2019. https://www.adn.com/alaska-news/rural-alaska/2019/10/19/a-western-alaska-village-long-threatened-by-erosion-and-flooding-begins-to-relocate/.

Lowenstein, Tom. 1993. *Ancient Land: Sacred Whale: The Inuit Hunt and Its Rituals*. New York: Farrar, Straus and Giroux.

Martin, Crystal Naataq. "Some of us Are Torn about the Future." Last updated June 7, 2018. Accessed October 31, 2020. www.facebook.com

Maupin, Siqiniq. "North Slope Iñupiat Who Oppose on Development Aren't Being Heard." *Anchorage Daily News*. Last modified May 30, 2019. Accessed July 5, 2019. https://www.adn.com/opinions/2019/05/30/north-slope-Iñupiat-who-oppose-oil-development-arent-being-heard/.

McArthur, Kaden. "Forfeit of Offshore Leases Is Good News for the Arctic Ocean." Alaska Wilderness League. Last modified May 18, 2020. Accessed June 23, 2020. https://www.alaskawild.org/blog/return-of-offshore-leases-is-good-news-for-the-arctic-ocean/#:~:text=The%20Arctic%20Slope%20Regional%20Corporation,leases%20in%20the%20Beaufort%20Sea.&text=The%20largest%20company%20to%20leave,the%20Chukchi%20Sea%20in%202016.

McPherson, Karen Michel. *The Corporate Whale: ANCSA, the First 10 Years of Program*. Last modified 1982. Accessed October 18, 2019. http://www.ankn.uaf.edu/curriculum/ANCSA/TheCorporateWhale/whale10.html.

Milman, Oliver. "Alaska Towns at Risk from Rising Seas Sound Alarm as Trump Pulls Federal Help." *Guardian*. Last modified August 10, 2017. Accessed August 14, 2017. https://www.theguardian.com/us-news/2017/aug/10/alaska-coastal-towns-sea-level-rise-climate-change?CMP=Share_iOSApp_Other.

Mufson, Steve. 2015. "Some Alaska Native Companies Invest in Arctic Oil." *Washington Post*, September 12.

Napoleon, Harold. 1996. *Yuuyaraq: The Way of the Human Being*. Fairbanks, AK: Alaska Native Knowledge Network.

Newton, Jonathan. "Some Alaska Native Companies Invest in Arctic Oil." *Herald Net*. Last modified September 12, 2015. Accessed August 14, 2017. http://www.heraldnet.com/news/some-alaska-native-companies-invest-in-arctic-oil/.

North Slope Borough. 2020. "NSB 2015 Economic Profile and Census Report." Accessed April 18, 2020. http://www.north-slope.org/your-government/nsb-2015-economic-profile-census-report.

Ottery, Christine. "Interview: 'There Is No Way they could Ever Clean It' Say Indigenous Leaders on Arctic Oil Spill." Last modified May 28, 2015. Accessed August 14, 2017. http://energydesk.greenpeace.org/2015/05/28/interview-native-alaskans-warn-of-offshore-oil-exploration-and-climate-change-consequences/.

Reimer, Catherine S. 1999. *Consulting the Iñupiat Eskimo*. Westport, CT: Greenwood Press.

Robertson, Frances C. et al. 2013. "Seismic Operations Have Variable Effects on Dive-Cycle Behavior of Bowhead Whales." *The Beaufort Sea Endangered Species Research* 21: 143–160.

Rosa, Cheryl. June 2009. "A Summary of Dead, Stranded Bowhead Whales Reported in the Chukchi and Beaufort Seas over the Last Twenty-five Years." *International Whaling Commission Scientific Committee* (SC/61/E12). Madeira, Portugal: Paper presented to the IWC Scientific Committee.

Rosen, Yereth. "Oil-rich Inupiat Corporation to Quit Alaska Federation of Natives." *Arctictoday*. Last modified December 17, 2019. Accessed July 15, 2020. https://www.arctictoday.com/oil-rich-inupiat-corporation-to-quit-alaska-federation-of-natives/.

Ruskin, Liz. "Two Worlds that Overlap: Richard Glenn Sees ANWR Drilling as a Boon to Iñupiaq Communities." *Alaska Public Media*. Last modified July 19, 2019. Accessed October 18, 2019. https://www.alaskapublic.org/2019/07/19/two-worlds-that-overlap-richard-glenn-sees-anwr-drilling-as-a-boon-to-Iñupiaq-communities/.

Sakakibara, Chie. 2006. Personal Interview with Jack Schaefer. September 5, 2006.

Sakakibara, Chie. 2008. "'Our Home Is Drowning': Climate Change and Iñupiat Storytelling in Point Hope, Alaska." *Geographical Review* 98, no. 4: 456–475.

Sakakibara, Chie. 2010. Personal interview with Allison Akootchook (Aku-Matu). September 30.

Sakakibara, Chie. 2020. *Whale Snow: Iñupiat, Climate Change, and Multispecies Resilience in Arctic Alaska*. First Peoples: New Directions in Indigenous Studies Series. Tucson, AZ: University of Arizona Press.

Shearer, Christine. 2011. *Kivalina: A Climate Change Story*. Chicago, IL: Haymarket Books.

Topkok, Sean Asiqłuq, and Carie Green. 2016. "Following the Pathways of the Ancestors: Well-Being Through Iñupiaq Dance." In *Indigenous Perspectives on Education for Well-Being in Canada*, edited by Frank Deer and Thomas Falkenberg, 173–186. Winnipeg: ESWB Press.

Welch, Craig. "Why Alaska's Inupiat Are Warming to Offshore Oil Drilling." *National Geographic*. Last modified May 22, 2015. Accessed July 15, 2020. https://www.nationalgeographic.com/news/2015/05/150522-Inupiat-Shell-offshore-oil-Arctic-Alaska-ocean-whale-sea/.

Wexler, Lisa. 2009. "The Importance of Identity, History, and Culture in the Well-being of Indigenous Youth." *Journal of the History of Childhood and Youth* 2, no. 2: 267–276.

Williams, Maria Shaa Tláa. 2017. "Indigenous Expressive Culture and Environment in the Twenty-First Century: The Role of the Artist in Alaska." Unpublished presentation. Bettman Family Fellowship Public Lecture, Oberlin College. February 9, 2017.

Williams, Maria, David Solomon, Allison Akootchook Warden, and Kimberly Pinder. 2018. "Multispecies Solidarity." The Last Oil: A Multispecies Justice Symposium on Arctic Alaska and Beyond. Albuquerque: University of New Mexico. February 21–23, 2018. Accessed July 31, 2020. https://thelastoil.unm.edu/watch-online/.

3

ENCOUNTERING THE NONHUMAN IN NORTH ATLANTIC OIL CATASTROPHES

Fiona Polack

Three catastrophic petroleum industry accidents in the North Atlantic in the 1980s caused significant loss of human life and destroyed offshore infrastructure. On March 27, 1980, the *Alexander Kielland*, a pentagonal rig deployed as an accommodation platform for workers in Norway's Edda Field, overturned when one of its faultily constructed legs collapsed. Of the 212 men on board 123 were killed (Inquiry Commission 1981, 1). Less than two years later, in the early hours of February 15, 1982, 84 lives were lost on Newfoundland's Grand Banks when the *Ocean Ranger* capsized in an horrendous winter storm after a wave smashed a porthole and water saturated the electronic panel controlling the rig's stability. There were no survivors (Royal Commission on the *Ocean Ranger* Marine Disaster 1984, iii). Then, on July 6, 1988, what historian Alex Kemp terms "by far the most serious and dramatic accident in the history of North Sea Oil" occurred when a large-scale blowout on the United Kingdom's *Piper Alpha* killed 165 of the 226 people on board, and two rescue personnel (Kemp 2012, 605).

These three accidents were by no means the only ones to occur in the North Atlantic oil industry in the latter part of the 20th century.[1] However, they remain among the most significant and deadly offshore petroleum disasters in history.[2] Each had a profound effect both on the local onshore communities that lost men as well as the global industry more generally. They also prompted high profile government-commissioned inquiries into their causes within Norway, Canada and the United Kingdom, respectively.[3] Moreover, the stories of the *Alexander Kielland*, *Ocean Ranger,* and *Piper Alpha* catastrophes have continued to be retold and reimagined in a range of contexts. In keeping with Stephanie LeMenager's observation that trauma-inflected testimony is a key oil industry-related genre (LeMenager 2014, 16) survivors, rescuers, and those closely connected to victims have personally recorded their experiences in writing and/or provided oral accounts to others.[4] Literature, music, art, documentary films and public

DOI: 10.4324/9781003024125-3

monuments help keep memories of the catastrophes alive, as do museum exhibits and remembrance ceremonies.[5] There was even an ill-conceived experiment to create a *Piper Alpha*-themed version of Monopoly in order to raise money for the foundation formed in Scotland after the accident (Duckett 2018).

Significantly, except for those still grappling with long-term trauma sparked by direct or familial involvement in them,[6] the three North Atlantic disasters are now seen as of primarily historical rather than contemporary import. The lessons the catastrophes taught about safety and risk are presumed to have been learned. Norwegian union leader Hilde-Marit Rysst recently contended that "'Kielland' doesn't work anymore, the major accident which struck us is no longer alive in people's consciousness. We must bring this story to the forefront again, so that another major accident will not be needed in order to give us perspective" (Tungland et al. 2020, 241).[7] Mark Stoddart and B. Quinn Burt observe in relation to contemporary Newfoundland and Labrador that there is "a shared sense that the oil industry operates to high worker safety and environmental standards" (Stoddart and Burt 2020, 397). Regarding Scotland, they note that focus has primarily now turned to winding down the industry and "decommissioning oil rigs in ways that manage local environmental risks" (Stoddart and Burt 2020, 397). In short, there is a sense that future major catastrophes are highly unlikely in all three of these offshore jurisdictions. While it is true that changes to safety procedures recommended by inquiries into the *Alexander Kielland, Ocean Ranger,* and *Piper Alpha* were largely, although not completely, implemented by corporations and governments (Higgins 2018; Sutton 2013, 64–6; Thomas 2020), it is also the case that petroleum catastrophes tend to "spill across history to happen again as if they are happening for the first time" (LeMenager 2014, 65). Offshore petroleum exploration and production in the North Atlantic remain inherently risky.

Even if the industry does manage to avoid future, large-scale, spectacular events, it already persistently threatens human and nonhuman lives in less obvious ways. Cumulative small-scale spills, ongoing stress to bird and marine life, increasing plastics pollution, as well as direct and indirect contributions to escalating global heating, are all examples of the slowly violent (Nixon 2011, 2) and deeply harmful consequences of continuing petroleum-related activity in the cold water offshore. Writing about the media strategies Husky Energy used to downplay a recent spill of 250,000 litres of oil at its *SeaRose* production facility on the Grand Banks, Dwayne Avery persuasively argues that such spills belong to a "category of chronic disasters which, despite their gradual and piecemeal temporality, rival the sensationalistic catastrophes featured in the mass media. Small recurring oil spills are *creeping events* that slowly and ceaselessly expose ecosystems to environmental degradation" (Avery 2020 3, emphasis in original). While there have not been precipitate disasters on the scale of the *Alexander Kielland, Ocean Ranger,* and *Piper Alpha* catastrophes since 1989, the North Atlantic offshore has today arguably subsided into "crisis ordinariness" (Berlant 2011, 9). At the root of this situation is the petroleum industry's ultimate disinterest in the environment. While oil companies might go through the motions of preparing government-mandated environmental impact statements and the like,[8] as Timothy Morton puts it, industrial capitalism has decimated Earth because "Nature is the featureless remainder at either end of the process of production" (Morton 2013, 112). As the advent of the Anthropocene, and especially the climate crisis, makes clear, this stance is causing increasingly alarming consequences not just for the nonhuman world but for the human one as well.[9]

Notwithstanding that they occurred more than 30 years ago, the *Alexander Kielland, Ocean Ranger,* and *Piper Alpha* catastrophes remain exceedingly timely case studies for better comprehending our current environmental predicaments. The North Atlantic accidents on which I focus were shocking events which – for a while, at least – undermined conceptions of the nonhuman world as a "featureless remainder" that could be safely manipulated and exploited for profit via sophisticated technology. In this chapter, I examine accounts by direct participants and witnesses as well as those charged with determining the causes of the disasters, of moments at which hubristic and expedient assumptions about the more-than-human world were thoroughly challenged. Notably, these attestants are overwhelmingly male and white; the companies that administered the rigs, and the inquiries into the accidents, were all overseen by men. In addition, no women were present on the *Alexander Kielland, Ocean Ranger,* or *Piper Alpha,* when catastrophe struck, and most of the workers on the rigs were from local, predominantly white, onshore, Northern communities. As Sheena Wilson notes, women and minority citizens, especially those who adopt environmentalist stances, are often "perceived not only as potential obstacles to oil extraction but also as threats to the proliferation of capitalism itself, since oil and capitalism are imagined as symbiotic" (Wilson 2014, 245). Hence, it was those most likely to be committed to petro-capitalism and its accompanying attitudes to the nonhuman world that were forced to confront their assumptions in the *Alexander Kielland, Ocean Ranger,* and *Piper Alpha* disasters. It is also important to acknowledge, however, that the many rig workers who lost their lives were also clear victims of oil capitalism; inquiries into each of the three respective accidents found evidence of reckless corporate and government behaviour (Cullen 1990; Inquiry Commission 1981; Royal Commission on the *Ocean Ranger* Marine Disaster 1984).[10] The aspects of the nonhuman world by which participants and witnesses to catastrophe were overwhelmed were primarily watery ones, although non-aquatic elements – wind and weather, as well as volatile gases – also played important roles. Even human-constructed material objects, most notably, the rigs themselves, became threats that consequently had to be perceived anew. As I demonstrate, closely attending to accounts of offshore petroleum catastrophes can shed light on the conceptual and narrative machinations which underpin false assumptions about the extent of human control. The *Alexander Kielland, Ocean Ranger,* and *Piper Alpha* disasters provide invaluable "ecological parables" (Mentz 2015, xix).

Rigwreck in the Naufragocene

Before turning to the accounts themselves, a word on my theoretical approach is necessary. Central to the conceptual framing of my chapter is a repurposed version of what Steve Mentz calls "shipwreck modernity." In his book of the same title, Mentz, a literary and critical ocean studies scholar, focuses primarily on shipwreck narratives written between the mid-16th and the early 18th century. However, he argues that, during this first period of globalization, shipwreck became an "ecological parable" (Mentz 2015, xix), which proves resilient as a way of describing the disorder into which human experience is thrown by violent encounters with the nonhuman environment. Shipwreck is an historically mobile trope that "makes palpable the stark contradiction between human desires and mortal limits" (Mentz 2015, xxxvi). Relatedly, Mentz proposes the term "Naufragocene" in order to supplement and complicate alternatives including the Anthropocene; for the last 500 years or so, he argues, we have inhabited the Age of Shipwreck: "Tangible dislocation punctuates both the early modern globalizing process that wiped out so many human and nonhuman populations in the early modern period and our own era of warming oceans and killing storms" (Mentz 2015, xii). Mentz's argument is highly persuasive. There is a case to be made, however, for introducing "rigwreck" into conversation with his theory, too.

Offshore rigs and platforms are not straightforwardly analogous to ships. In addition to usually being quite different in design and appearance, they may be secured in the same ocean location within a nation's offshore exclusive economic zone for years. Moreover, in contrast to ship-based industrial fishing, for example, rig and platform-based activity focuses primarily on extracting what lies beneath the sea floor; a pursuit that requires quite different skills to fish harvesting. Concomitantly, with the exception of divers, rig workers have less direct contact with the ocean than fishers. Indeed, their work stations are literally metres above the surface of the sea. While sailors on modern cargo vessels, cruise ships and other gargantuan contemporary craft may also rarely need to get wet, they still have a more vested interest in ocean and weather conditions as they attempt to navigate from one location to another. They are unlikely, for example, to be helicoptered to and from their offshore place of work.

Category confusion between ships and rigs can have dangerous consequences. Paradoxically deploying nautical terms such as port and starboard (Inquiry Commission 1981, 6) in their descriptions of the *Alexander Kielland*, the final report of the Norwegian commission of inquiry also notes that, "The safety control of mobile platforms exercised by Norwegian authorities has been based somewhat uncritically on experience from ships design [sic] and operation without fully recognizing the particulars of mobile units. Now, there are greater differences between various platform concepts and between platforms and ships than between various ship types" (Inquiry Commission 1981, 231). However, ignoring affinities between rigs and ships is risky, too. The commissioners of inquiry into the *Ocean Ranger* disaster ultimately concluded that safety was seriously compromised by the lack of seafaring experience among those controlling the rig's day-to-day operations. They observed that land-based oilfield techniques were deployed offshore with little regard for the fact that "despite this industrial focus the activity takes place at sea" (Royal Commission on the *Ocean Ranger* Marine Disaster 1984, ix–x). Their description of offshore drilling as a hybrid industrial-marine undertaking acknowledges that rigs resist simple categorization within existing taxonomies.

Despite the lack of clear equivalence between rigs and ships, and the quite different relationships offshore oil workers and sailors typically have with the sea, Mentz's notion of shipwreck modernity remains useful in the context of the rigwrecks of the *Alexander Kielland, Ocean Ranger,* and *Piper Alpha.* In fact, these events are arguably even more revelatory ecological parables than the shipwrecks Mentz addresses. First, because those who work on offshore oil rigs are more disconnected from their ocean locations than the shipbound and, in terms of their work with volatile gases, engaged in a dangerously new kind of strikingly environmentally hubristic activity. Second, while Mentz's argument that we currently live in an era whose contours were first delineated in the early modern period is convincing, and resonates, too, with Jeffrey Bolster's assertions in *The Mortal Sea* that the North Atlantic has been ecologically compromised by human activity for centuries (Bolster 2012, 9), Katie Ritson is correct that the unsustainable relationship between humans and their environments currently has "a new pitch of urgency" (Ritson 2019, 22). The oceans upon which Mentz's sailors were wrecked are in a more parlous state now than at any other time in human history, as the burgeoning environmental humanities sub-discipline of critical ocean studies argues and examines.[11] Moreover, the contemporary petroleum industry in the service of which the *Alexander Kielland, Ocean Ranger,* and *Piper Alpha* were deployed bears significant direct and indirect blame for this situation. Third, and finally, the tendency towards comforting resolution that Mentz identifies in conclusions to early modern shipwreck narratives persists in accounts of contemporary rigwrecks but now rests on invocations of the ultimate power of human ingenuity rather than the grace of the more-than-human figure of God.

Mentz identifies three recurrent elements in early modern manifestations of the literary microgenre of the shipwreck: "The story begins by performing the radical break, deepens the crisis through physical and conceptual instability, and then often concludes with a tantalizing promise of Providential recuperation" (Mentz 2015). The framework Mentz identifies still largely holds in the context of accounts of the *Alexander Kielland, Ocean Ranger,* and *Piper Alpha* disasters, although, again, some qualifications need to be made. By the late 20th century, what Ulrich Beck (1999) famously termed risk society had largely superceded a religious-centred one. Consequently, a (fragile and delusory) return to stability is conjured via government-commissioned investigations into the causes of the catastrophes rather than via belief in divine intervention.[12] As Andrew Brown explains, public inquiries can "depoliticize disaster events, legitimate social institutions, and extend the hegemonic influence of dominant groups" (Brown

2003, 95). Notwithstanding that they can offer opportunities for communities to grieve together in the wake of disaster, facilitate the sharing of knowledge across jurisdictions, and lead to practical improvements in operational practices, public inquiries ultimately tend to elaborate "fantasies of omnipotence and control" (Brown 2003, 107) and act as vehicles for reestablishing the status quo after the rupture a disaster causes. In the cases under examination here, they reinforce beliefs that humans can ultimately control the nonhuman world.

Storying disaster

In what follows, I draw on first-person accounts of the overturning of human hubris at sea provided by direct participants in and witnesses to rigwreck: those who worked on the *Alexander Kielland, Ocean Ranger,* or *Piper Alpha* during the disasters and/or were involved in rescue and recovery operations. Much of this material is harrowing and disturbing. However, it also remains irredeemably incomplete. Missing are the insights of the nearly 400 people who lost their lives. The absence of their voices creates a profound silence at the heart of the historical record.

In the context of the *Alexander Kielland* capsizing, I rely particularly heavily on testimonies contained in Else Tungland, Marie Smith-Solbakken and Ellen Kongsnes' recently published *Kielland: Based on the Stories of Those Who Were There.*[13] With the exception of brief introductory and concluding essays, *Kielland* consists entirely of direct, attributed, quotations of varying lengths, from interviews with those associated with the accident. While considerable editorial intervention underlies the book's structure, and the choices of excerpts, the quotes themselves appear in direct succession and are presented without overt editorial commentary. This medley of voices makes *Kielland: Based on the Stories of Those Who Were There* quite different from the *Kielland* inquiry report, in which witnesses are not named, rarely quoted directly, and have their evidence compressed with that of others. *Kielland: Based on the Stories of Those Who Were There* also allows for conflicts between perspectives in a way inquiry reports, keen to synthesize findings, can tend to restrict. The archives of the *Ocean Ranger* disaster are markedly distinct from those of the Norwegian accident. Accounts of the loss of this rig have at their heart a complete and deafening silence: not one of the men working on the night of February 14, 1982 survived. I thus draw most heavily on the testimony of those involved in the rescue operation (particularly sailors on the *Seaforth Highlander*), workers who were on adjacent rigs the night the *Ranger* sank and divers who investigated the wreck. Their accounts are contained in transcripts of the public hearings associated with the Royal Commission into the *Ocean Ranger*'s sinking and in Mike Heffernan's *Rig: An Oral History of the Ocean Ranger Disaster.* Heffernan's book contains 34 thematically organized "monologues" drawn from interviews with people ranging from former offshore rig workers, to media representatives, to individuals who lost family members when the rig sank. As in *Kielland: Based on the Stories of Those Who Were There*, the questions posed to respondents are excised, so that the speakers' voices dominate. Finally, in discussing the *Piper Alpha* accident, I draw on witness testimonies contained in the final reports of public inquiry commissioner Lord Cullen (1990). I also rely extensively on memoirs written by three men who escaped the platform and/or were involved in the rescue: Diver Ed Punchard's *Piper Alpha: A Survivor's Story* (1989); paramedic Jon May's *Divers: from Piper Alpha to the Gulf War* (1995); and control room operator Geoff Bollands' *Baptism of Fire: Life, Death and Piper Alpha* (2018).

Notable and recurrent thematic preoccupations emerge across these accounts. This is despite the fact the narratives I draw on vary in terms of the circumstances of their creation (some were composed quite soon after the events they discuss, others at a greater temporal distance; some were first related in public, others composed in private), the means by which they were first communicated (oral versus written; with a great deal or no editorial intervention) and the genres to which they belong (from official witness testimony, to autobiographical trauma narrative, to social history). Like Mentz's shipwreck narratives, the more extended stories of rigwreck trace a movement from "crisis to immersion to salvage" (xxxii). That said, the process of salvage – as in Punchard's and Bollands' accounts of the lengthy ordeal of recovering from post-traumatic stress – can take many years to conclude. In what follows, though, I hone in on what I see as three especially striking ways in which the narratives illuminate human perceptions of the nonhuman world in moments of extremity. The first concerns the desensitizing effects of quotidian human routines and practices; the second conceptualizations of the vital materiality of the nonhuman; and the third the transmogrification of the human into the once human. Stories of oil-related disaster in the North Atlantic highlight blind-spots in human interactions with the nonhuman world within and beyond this offshore region and hint at possible solutions to this myopia.

The power of the quotidian

Frederick Buell (2014) contends that cultures of oil production are grounded in the dramatic twin poles of exuberance and catastrophe. The industrial-related element of offshore oil drilling and production in the North Atlantic partakes in such extremes, but in different ways to land-based operations. Rigs are perhaps most obviously "exuberant" in the grandiosity of their status as industrial sites. Geoff Bollands describes his first tour offshore and initial sight from the helicopter of the *Ninian Southern* platform by saying "My initial reactions were surprise and admiration; I was amazed at the sheer size of the rig. It was a clear day, and the platform was massive on the horizon, standing some 300 feet above sea level, with the production, accommodation and drilling derrick all stacked on top of each other" (Bollands 2018, 32). Ed Punchard describes his initial impression of the *Frigg* complex in similar terms as a "staggering sight" one which "looked like something out of a Mad Max movie. Everything about it was completely alien. It was noisy and dirty with bizarre bits sticking out and flames shooting up in the air" (Punchard 1989, 26). Edmund Burke famously wrote that "astonishment [...]

is the effect of the sublime in the highest degree" (Burke 1792, 80). Bollands and Punchard are both so awestruck by the colossal structures upon which they are to be stationed that they resort to the Burkean sublime in order to conjure them for their readers.[14]

Yet at the same time as they partake in the monstrous ambitions of oil culture, offshore rigs are also, unlike land-based ones, quotidian, domestic spaces in which off-shift workers eat, sleep, socialize and participate in recreational activities. The combination of the domestic and the industrial, and their presence on platforms for continuous blocks of time, helps workers become habituated to the weirdness of the endeavours in which they are engaged; it also makes them better able to repress feelings of vulnerability associated with being marooned in tumultuous oceans while dealing with highly volatile substances. When Jon May goes to work on the *Ocean Victory*, located close by the *Piper Alpha*, in 1988, he is especially disturbed by the fact that the platform is a sister rig to the lost *Ocean Ranger* and situated on a notoriously noxious gas field: "at first the atmosphere of sitting on a bomb got to everyone. But after a few days when nothing lethal had transpired, we all began to relax into a routine" (May 1995, 41). Relatedly, striking in all accounts of the *Alexander Kielland, Ocean Ranger,* and *Piper Alpha* disasters are the incongruous nature of the collisions between the quotidian, daily, reassuring routines of being on and off-shift and the catastrophic. These are particularly obvious in two recurrent contexts: first, in the descriptions of what the men on the rigs were doing when disaster struck; and, second, in accounts of rescuers encountering displaced signs and artefacts of domesticity during their efforts to locate survivors.

Theis Salvesen, a caterer on the *Alexander Kielland*, remembers how the first instincts of those in the mess hall were to ignore signs of trouble:

> 40 men have sat down to eat. Cutlery and plates clink and clatter. There is a thump, a crash, and a loud bang. Everything gets quiet, everyone silently agrees this is quite the weather. They resume eating. A couple minutes later comes another big thump and the room grows quiet again, wolf, wolf, how long is this going to go on? A moment later the platform spills over on its side and everything is swept down, cups and plates, pots – the whole lot fell. That's when I thought I just need to get out. (Tungland et al. 2020, 19)

Helge Georg Knutsen, elsewhere on the *Kielland*, recalls a similar reluctance to accept that something is seriously wrong: "Heard a god-awful bang. We laughed, and kept playing cards" (Tungland et al. 2020, 28). As the platform began to capsize, however, nonthreatening objects transformed into lethal missiles – "Typewriter, radio, TV and all other loose objects, fell towards us" (Tungland et al. 2020, 30); even more frighteningly, "the hallway was now transformed into a slide into the abyss" (Tungland et al. 2020, 30). With extraordinary swiftness, the workers' apparently stable surroundings become unrecognizable and terrifying.

One of the most poignant details from the *Ocean Ranger* report is that men's bodies were recovered from the freezing, tumultuous North Atlantic clad only in pyjamas. The sense of human vulnerability conjured by these images is profound. The inference that the quotidian can morph into catastrophe at lightning speed, and before people can prepare themselves, is also unavoidable. The shock caused by the quotidian being sundered from its appropriate context and exposed to the elements is echoed by rescuers in Norway who are moved that "Everywhere, there were rubber boots floating in the ocean" (Tungland et al. 2020, 132). Their shock is grounded in the fact that, as Tim Edensor observes, "Social order is partly maintained by the predictable and regular distribution of objects in space" (Edensor 2005, 311). Moreover, the boots function metonymically, drawing attention to the absence of the people to whom they once belonged. This first ecological parable, then, warns of the devastating consequences of becoming habituated to danger and placing false trust in the seemingly protective quality of human routines and procedures. The quotidian lulls the rig workers into a misguided sense of security to the point at which they are compelled to ignore pressing signs of imminent catastrophe. In this, they resemble populations that continue to dwell in what Mark Simpson and Imre Szeman (2018) call an "impasse" about climate crisis: we know we are in trouble, but we do not take sufficient action based on our knowledge. Adherence to everyday patterns and routines aids and abets this situation.

"There were enormous forces in play"

When offshore infrastructure and systems fail, human bodies become intensely vulnerable to impacts from the physical world. In consequence, attestations to the agency of the nonhuman recur in accounts of the *Alexander Kielland, Ocean Ranger,* and *Piper Alpha* disasters. Implied or overt declarations that things have the capacity to act as quasi agents or forces with trajectories, propensities or tendencies of their own (Bennett 2010, viii) occur not just in relation to domestic objects transforming into lethal missiles but also in evocations of weather, the ocean, and oil and gas under pressure. As radio operator Harry Vike says of the *Alexander Kielland* accident, "[t]here were enormous forces in play" (Tungland 2020, 58). Moreover, these forces exceeded human control and comprehension: "people were [...] well, it was terrible to look at" (Tungland et al. 2020, 58, ellipses in original). Crucially, witnesses to the offshore catastrophes do not perceive manifestations of the nonhuman as neutral, but rather attribute hostile, even evil, intentions to the forces themselves.

In testimony to the *Ocean Ranger* public inquiry, Jerry Higdon, Second Officer on the standby vessel the *Seaforth Highlander*, categorizes the storm on the night of February 14–15 as the worst he had encountered in his ten years at sea (Royal Commission-Public Inquiry 1983, vol. 36, 6135). When he came on shift, "It was quite bad, 60 foot seas, snow, freezing spray, the wind was 80 knots or better and the temperature was quite low" (Royal Commission-Public Hearing

1983, vol. 36, 6063). Outside of the formal confines of the public inquiry (in which factual information such as the precise times at which radio messages were logged, and where exactly boats were located, were obsessively and distractingly dissected), the men who were on the Grand Banks that mid-February choose powerfully evocative language to describe their experiences. Strikingly, they repeatedly personify the forces with which they had to contend, and in derogatory ways. Thomas Kane, on the *Boltentor*, for instance, describes the weather as "nasty" (Kane 2009, 68); Wayde Butler on the *Sir Humphrey Gilbert* characterizes the night on which the *Ocean Ranger* was lost as "the dirtiest of nights" (Butler 2009, 84); and Carl Staubitzer on the *Java Sea* recalls the North Atlantic as "savage" (Staubitzer 2009, 90). The sailors perceive the North Atlantic environment as wilfully destructive and harmful. The language by which the ocean is othered here is reminiscent of the disparaging, but also fear-induced, terms colonial-era settlers from Europe deployed to describe Indigenous peoples in the lands they invaded; the North Atlantic is perceived as possessing malevolent, uncivilized, violent intent.[15] Underlying these assumptions is the unexamined conviction that members of the oil industry have a right to be doing the work they do upon the ocean; the sea's assaults are unjustified and disproportionate.[16]

While not as extreme as the evening two years later on the Grand Banks, conditions were also adverse the night the *Alexander Kielland* capsized. This time we have survivors' as well as rescuers' accounts of what it feels like when the North Atlantic takes control. While one comments "When I leapt into the water, the ocean was neither cold nor wet – only a means to make my escape, toward rescue" (Tungland et al. 2020, 46),[17] others perceived the sea as a powerfully destructive force. One survivor comments: "something black was pushing me down [under the water]. All hope was lost" (Tungland et al. 2020, 69). Another recalls how "I had to fight to get back up [to] the surface […] I got dragged under once again by the waves" (Tungland et al. 2020, 59). Surveying the catastrophic scene offshore from the supply vessel the *Normand Skipper*, Didrik Stonghaugen articulates his sense that malevolent powers have been at work: "It was as if the devil himself had been let loose" (Tungland et al. 2020, 61). Again, then, the North Atlantic is granted malign agency.

On board the *Piper Alpha*, it was predominantly explosive gases and the heat they generated once ignited, rather than the combined powers of water and weather, that were perceived as constituting the most disturbing evidence of a malicious vital materiality. The evening on which the accident occurred was, as Ed Punchard relates, "a most pleasant summer evening" (Punchard 1989, 122). The first cues for most workers that there were problems on the platform were auditory ones. In a sub-section of his report devoted to "Noises associated with initial explosion," Lord Cullen provides a panoramic perspective, moving between the simultaneous viewpoints of differently located personnel at the time of the initial serious gas leak (Cullen 1990, 49–50). The imagery used by witnesses to describe this leak is often feminized. For instance, "Mr D Ellington heard a very high pitched screaming noise;" "Mr R J McGregor heard a 'banshee' noise[...] He stated someone in the tea room likened it to somebody strangling a woman;" "Mr J S Meanen heard a loud wailing noise, very high pitched, which lasted 5 seconds" (Cullen 1990, 49). McGregor's description is especially noteworthy; he associates the gas with the sound emitted by a female Gaelic spirit presaging death (and indeed, as subsequent events confirm, the sound *does* constitute such a forewarning). However, he also wishes to note that others perceived the sound as a woman suffering attack – rather than an ominous female spectre.[18] Either way, and as in the case of the *Ocean Ranger* rescue attempts detailed above, the nonhuman is connected with a group – women, this time, instead of Indigenous people – long othered in western discursive regimes.

Control room operator Geoff Bollands had to contend not just with disturbing auditory cues, but visual ones as well. In *Baptism of Fire*, he recalls how within the control room:

> [...]the fire and gas panel went ballistic. Numerous gas alarms for C module all came in together, low gas alarms and high gas alarms, the panel was flashing like a Christmas tree and I couldn't silence the panel, which had a loud klaxon alarm. The quantity and severity of the alarms coming in was unique. I had never seen so many gas alarms activate together before and I barely had time to think; with mounting horror, the realization that there was a massive gas leak in C module began to dawn. (Bollands 2018, 55)

Bollands provides a compelling impression of human-built systems, and his human mind, being overwhelmed. Shortly afterwards, physical bodies are overcome by nonhuman forces, too: Bollands himself is propelled across the control room. As Lord Cullen describes, in "Effects on personnel of initial explosion" (another God's eye-view sub-section, and one deploying far more impersonal diction than the control-room operator does), elsewhere on the rig, bodies are violently and abruptly thrown to the floor, knocked off their feet, lifted off chairs and beds and/or subject to abrupt rushes of cold or hot air (Cullen 1990, 50–52). The urgent struggle for all on board the *Piper Alpha* then becomes escaping a platform consumed by fire.[19] Despite experiencing completely antithetical conditions, Jon May, on the nearby *Tharos* fire-fighting ship, deploys imagery of the same personified intensity as that used by Newfoundland observers to depict dire wintery conditions offshore. May comments how the heat emanating from the *Piper Alpha* "was dry and hateful and made your skin crawl" (May 1995, 46).

Jane Bennett writes that "the image of dead or thoroughly instrumentalized matter feeds human hubris and our earth-destroying fantasies of conquest and consumption" and stops us detecting a fuller range of the nonhuman powers circulating around and within human bodies (Bennett 2010, ix). How exactly this power is evoked, however, matters considerably. In moments of offshore catastrophe, the power of nonhuman elements becomes unquestionable. It is significant, though, that the imagery and vocabulary used by offshore workers to describe even these moments of extremity implicitly brings nonhuman forces within repressive human hierarchies. The power of the nonhuman to terrify and harm is acknowledged in the anthropocentrism of personification, but also associated with oppressed human others posited as inferior, such as Indigenous people and women. This association arguably makes it easier to reassert mastery even in the wake of catastrophes as serious as the ones under study here. As Simon Estok, discussing the gendering of Hurricane Katrina notes, however, "Sexist, anthropomorphic metaphors of a malevolent nature are not going to help. This is not the kind of agency – gendered, emotional, and calculating – that material ecocriticism is talking about" (Estok 2014, 133). Thus, even while the witnesses considered here do acknowledge the vital materiality of the world that surrounds them: "Recognizing agency in nonhuman materials only through ecophobic lenses is on the same side of history as racist fears of African American agency, sexist fears of women's agency, ableist fears of the differently ableds' agency, homophobic fears of queer sexuality, and so on" (Estok 2014, 133). This ecological parable, then, provides striking evidence of just how ingrained and detrimental the closely connected processes of othering nature and othering women and Indigenous people can be. In the face of catastrophe, witnesses choose diction that simultaneously acknowledges and denies the agency of the more-than-human world.

The once-human at sea

Accounts of the *Alexander Kielland, Ocean Ranger,* and *Piper Alpha* disasters address not just collisions between the human and the nonhuman world but also the very transformation of the human itself into something other. Given the high mortality counts associated with each of these catastrophes, descriptions of dying and dead bodies recur in reports and narratives about all three of the incidents. While some men drowned in the context of the *Piper Alpha* blowout, in contrast to the *Alexander Kielland* and *Ocean Ranger* events, smoke and gas inhalation and other fire-related fatalities, predominated (Cullen 1990, 463–484). In what follows, I dwell most extensively on encounters with the once-human in the Norwegian and Canadian incidents, arguing that important keys to bridging human and more-than-human worlds can be found in accounts of bodies at sea.

Officially charged with dispassionately tracking events and consequences (notwithstanding they are also granted the de-facto role of helping acknowledge and heal the grief suffusing affected local communities), inquiry commissioners must be meticulous in their accounting of the dead. As Finn Stepputat argues, "States tend to establish a range of laws, institutions and practices to take control of the transition from life to death, including the whereabouts of dead bodies." Further, "they usually claim the ultimate authority to define and govern the dead within their jurisdiction through legislation and institutionalised procedures" (Stepputat 2014, 4). Thus, for instance, in the chapter of his report devoted to "The causes of loss of and danger to life," Lord Cullen exhaustively recounts the process of recovering bodies and assigning causes of death in the wake of the *Piper Alpha* blowout (Cullen 1990, 185–190). Rather than providing names, he allots numbers to those who died. For example, he writes: "Of the 27 deceased whose bodies were recovered from the seabed between 10 and 29 July, 4 of them (Nos 33, 43, 73 and 141) had apparently died during an attempt to escape from the platform" (Cullen 1990, 187). Appendix H of Cullen's report, a "Schedule of information relating to the deceased" does provide names, occupations, employer details, recovery details, and principal causes of death for each number (Cullen 1990, 463–484). However, the appendix's rigid and detailed structure again suggests a dispassionate, legalistic accounting. Cullen's tallies resonate quite differently to those found in Ed Punchard's and Geoff Bollands' memoirs. At the outset of his book, Punchard pays homage to his lost co-workers by simply listing their names in alphabetical order under the heading "We remember" (Punchard 1989), and Bollands offers a "Tribute to the men who died" (Bollands 2018, 128–133) at the close of his text, again providing names, although appending occupations, too. In both Punchard's and Bollands' memoirs, naming is framed as an act of memorialization and respect rather than a primarily state-authorised exercise in gathering and presenting evidence.

First-person stories of encounters with dying and dead bodies offshore attest to the trauma experienced when witnessing transitions between living and lifeless states. As Magne Kåre Sildelid, who worked on *Kielland*, comments: "The worst part is thinking of all those who sank around us. We just saw their arms disappearing" (Tungland 2020, 71). Category confusion between environmental phenomena and human bodies is also a recurring theme in these narratives. For example, in the context of the *Ocean Ranger* rescue mission, Patrick Fahey, Second Mate on the *Nordertor*, recalls how "we started to come across sparkles in the water. The realization then hit us that those were life jackets and men floating below the surface – dead" (Fahey 2009, 75). Thomas Kane, a deckhand on the *Boltentor* also describes the difficulty of determining whether people floating in the water are still alive: "In the pitch of darkness, the water-activated lights of the life jackets were like stars in the night sky[...] We went for them – that's all we had to go by" (Kane 2009, 69). Kane's simile here inverts his use of personification, mentioned above, in describing the weather on the night the *Ocean Ranger* was lost. The life jackets are metonyms for the bodies they house, and via this association, people are equated with stars in the night sky. The terms objectification or dehumanization, the typical antonyms for personification, do not seem appropriate here, though. The image is too evocative of heavenly beauty, an association freighted, too, with Kane's retrospective knowledge that there were no survivors of the *Ocean Ranger* disaster. The potential sentimentality of the simile is tempered by the reference to the very practical fact the lights were also beacons, used by the rescue crews to direct their movements.

Encountering unequivocally deceased people could be a quite different matter. Rescue divers' accounts are thus especially compelling in their evocations of the once-human at sea. Their impact is magnified by the stark loneliness of the divers and the poor visibility in which they had to operate.[20] As Per Arne Jakobsen puts it: "You exit the bell and stand there with your feet on the bottom of the sea. It's quiet, and dark. You're aware of where you are and that there's no one to talk to, or lean on" (Tungland 2020, 132). Yngve Tveit confirms: "On the bottom of the North Sea you're just as isolated as on the North Pole" (Tungland 2020, 133). In this context, encountering the dead can be particularly shocking. Tveit comments how he "Found one early in the search. I was walking on the pile of debris. Bang! Out of nowhere, there's one! I had stepped on his foot without noticing and his entire stiff body pivoted up toward me. I got a real shock" (Tungland 2020, 134). The divers devise various strategies for avoiding this kind of trauma. Per Arne Jakobsen describes his method of never shining a light directly into the hood of a survival suit: "we didn't know if there was anything inside them or not. Your pulse would be pretty high. I swam up from behind and felt the hood with my hand, squeezed it" (Tungland 2020, 155). The divers also manage the strangeness of these encounters (intensified by the fact that the bodies were eyeless, and sometimes "brutally maimed" (Tungland 2020, 138)) by talking to the deceased. Angus Kleppe, Wigulf Schjøll and Per Arne Jakobsen, for instance, ask the corpses questions about themselves and relate the mechanics of how they will be transported up to the surface (Tungland 2020, 154–156).[21] Members of the rescue crews below and above the sea were intensely respectful of the bodies they retrieved.

While sudden encounters with bodies can be deeply disturbing, the divers also comment on how water seems to become an appropriate medium for the recently human. Per Arne Jakobsen comments that: "A drowned person underwater can be a pretty calm experience. When they lie down there, their hair gently swaying, it looks peaceful. It's much worse when they come up – the weightlessness ceases, the hair sticks to the face" (Tungland 2020, 138). Harald Klinge also notes that while in the sea the bodies were "easy to deal with," but once they were brought to the surface and decompressed, "they would throw up, and there were all kinds of noises. Also, the smell was not pleasant" (Tungland 2020, 138). Men involved in the rescue efforts following the *Ocean Ranger* capsizing confirm the disturbing effects of the sea. Thomas Kane states: "For us, being seamen, the worst was seeing what the water could do to a man after just a week. It wasn't a nice sight; not a nice sight. A few of the boys quit over that" (Kane 2009, 70).

These reflections upon the most intimate of encounters with cold water seas suggest that lifeless bodies are actually potentially powerful bridges between the human and nonhuman worlds. However, as many of the accounts I have cited attest, this notion can also be very difficult to assimilate. Sorrow, grief, fear, revulsion, and even the dispassionate accounting of the kind found in the inquiry reports, often obscure the ways that dead bodies at sea can remind us of our intimate imbrication with the more-than-human world. However, in *Bodies of Water: Posthuman Feminist Phenomenology*, Astrida Neimanis makes a convincing, environmentally motivated, case for conceiving of all human embodiment as watery. While her focus is primarily on living, rather than dead, bodies, her observations hold for both. We are, Neimanis claims, "made mostly of wet matter, but also aswim in the discursive flocculations of embodiment as an idea" (Neimanis 2017, 1). Neimanis' figuration challenges ideas of "modern water" (Linton qtd Neimanis 2017, 19) that, since their development in the 18th century, have stressed water's "abstract, isomorphic, measurable quantity" (Linton qtd Neimanis 2017, 20). Concomitantly, Neimanis argues, seeing ourselves as bodies of water "stirs up considerable trouble for dominant Western and humanist understandings of embodiment, where bodies are figured as discrete and coherent individual subjects, and as fundamentally autonomous" (Neimanis 2017, 2). Such trouble is arguably even more agitated when the human bodies in question are lifeless and at sea. Men who helped recover those killed in the *Alexander Kielland, Ocean Ranger,* and *Piper Alpha* disasters were directly confronted with evidence that "our bodies enter complex relations of gift, theft, and debt with all other watery life" (Neimanis 2017, 3). In the process, there are hints of efforts, even when they are rapidly repressed, to see the human as part of multispecies entanglements, rather than as independent.

The North Atlantic and the Naufragocene

Dwelling primarily on the legacies of transatlantic slavery, and the contemporary militarization and privatization of its seas, Elizabeth DeLoughrey perceptively describes the waters of the Atlantic as "heavy" with history and material waste (DeLoughrey 2010, 1). Her categorization counters views of the ocean as a frictionless space undeserving of detailed engagement on its own terms: what Christopher Connery calls "Western capitalism's primary myth element" (Connery 2010, 686). It also anticipates and aligns with John Urry's (2014) and Hannah Appel's (2015) critiques of the contemporary oil industry's tendency to literally and metaphorically "offshore" its operations; to remove them from view in order to avoid oversight and regulation and, concomitantly, obscure ocean histories and environments.

The offshore petroleum catastrophes I address in this chapter, and the reports and accounts they sparked, are instances wherein attention *was* unequivocally drawn to the North Atlantic and capital's actions upon and beneath it. The "heaviness" of North Atlantic waters was keenly felt by survivors, rescuers and the onshore communities most closely associated with the accidents. These rig-wrecks prompted detailed official and unofficial inquiries into human activities in the zone of cold water oil. As I have argued here, accounts of the accidents reveal contemporary ecological parables of import to our current environmental predicaments. They highlight the extraordinarily powerful hold comforting routines keep on our imaginations. Even in moments of extreme crisis, men on board the doomed rigs refused, at first, to acknowledge that they were in mortal danger; even though some had longstanding misgivings about their safety, the tendency was to assume all was well. Accounts of the offshore disasters also shed light on human tendencies to personify the nonhuman, endow it with malevolent intent and characterise it in terms resembling those used to "other" groups within human society. There are disturbingly few points of sympathetic potential connection between the human and nonhuman realms when there are "enormous forces in play." However, in the aftermath of catastrophe, in the process of retrieving the lost from calmer oceans, different possibilities can emerge. In encountering the once-human at sea, humanity's intrinsic connection with processes of watery embodiment is brought to the fore. It is perhaps this realization more than any other that has the most potential to shift entrenched petro-narratives about the instrumentality of the nonhuman world.

Notes

1 The most comprehensive listing of post-1975 offshore disasters, the Worldwide Offshore Accident Database, is compiled by the classification society DNV-GL and does not make its data publicly available (https://www.dnvgl.com/services/world-offshore-accident-database-woad-1747). However, well-publicized North Atlantic events in the 1980s included the loss of a Chinook helicopter returning from the Brent oilfield to the Shetland Islands in 1986, which killed 45 (Kemp 2012, 616); the *Byford Dolphin* accident in Norway's Frigg sector in 1983 in which five divers lost their lives during a decompression explosion (Gjerde and Ryggvik 2014, 230); and the sinking of the jack-up rig the *Rowan Gorilla* while being transported from Halifax, Nova Scotia to Trinidad in heavy seas in December 1988 (U.S. National Transportation Safety Board 1989).

2 For instance, these three incidents appear among the top four in *Offshore Technology*'s 2020-updated listing of "The World's Worst Offshore Oil Rig Disasters."

3 Government probes were not confined to the countries in whose waters the accidents occurred. For instance, the *Alexander Kielland* was built in France, and the Norwegian accident was consequently also investigated by the French Government (Tribunal of Commerce, Paris 1985). And an inquiry into the *Ocean Ranger* capsizing was initiated by the United States (United States Coast Guard 1983) because the rig was owned and operated by US-based concerns.

4 First-hand autobiographical accounts of the *Piper Alpha* catastrophe include those by Ed Punchard (1989), Jon May (1995), and Geoff Bollands (2018). Significant oral testimony is also included in the archives of the *Lives in the Oil Industry: Oral History of the UK North Sea Oil and Gas Industry* project held at the University of Aberdeen. Marc Reid's (2020) recounting of his father's experiences subsequent to surviving the *Piper Alpha* is provided in the context of a broader consideration of safety and trauma-related issues. Norwegian historian Marie Smith-Solbakken and her colleagues (2019, 2020) have played a particularly significant role in recording the oral testimonies of people connected with the *Alexander Kielland* accident. In Newfoundland, Mike Heffernan's (2009) *Rig: An Oral History of the Ocean Ranger Disaster* records the testimony of rescuers, oil-industry workers and executives, journalists, family members of victims, and others.

5 For instance, fiction about, respectively, the *Ocean Ranger, Piper Alpha,* and *Alexander Kielland* accidents includes Lisa Moore's *February* (2009), Ian Maloney's *The Waves Burn Bright* (2017), and Atle Berge's *Puslingar* (2019). Plays include Joan Sullivan's *Rig: Voices from the Ocean Ranger* and Stephen Phelps' *Piper Alpha: The Human Price of Oil* (2008). Ron Hynes' song "Atlantic Blue" and Andrew Staniland's composition *The Ocean is Full of its Own Collapse* were both inspired by the *Ocean Ranger* disaster. The Norwegian Petroleum Museum in Stavanger recently opened a new exhibition dedicated to the *Kielland* catastrophe (Petroleum Safety Authority Norway 2020). Public monuments to the accidents can be found in Aberdeen, Scotland; St. John's, Canada; and at Smiodden, near Stavanger, Norway.

6 As Kian Reme, whose brother was killed in the Norwegian accident, puts it: "Kielland never ends" (Reme 2020, 12).
7 The Kielland Network, a group representing families and survivors, has long felt that the commission of inquiry into the disaster was not thorough enough. At the time of writing, their efforts had led to the Norwegian Office of the Auditor General agreeing to review the investigation of the incident. The Network is hopeful that the Office will recommend a full new inquiry when it reports back to Parliament in 2021 (Thomas 2020).
8 One of the most notorious examples of the lack of attention oil companies can give to environmental reporting was revealed during investigations into the *Deepwater Horizon* blowout. BP, ExxonMobil and a host of other companies were found to have included walruses in their emergency oil-spill response plans for the Gulf of Mexico despite the fact none, of course, exist there (James 2010).
9 Many scholars have correctly highlighted flaws with the term "Anthropocene" (see, e.g., Tsing, Haraway, and Moore). Among its most serious problems, it shifts responsibility for environmental destruction to a totalized humanity. However, I agree with Una Chaudhuri that it is still a "galvanizing concept" that has value for prompting new thinking about "the species and ecospheres that climate change is plunging into a state of emergency" (Chaudhuri 2017, 144). I also find Mentz's "Naufragocene," discussed later in my chapter, especially useful in a specifically oceanic context.
10 To complicate this picture still further, workers on these rigs also displayed environmental awareness on occasion. For instance, Greg Tiller, who died when the *Ocean Ranger* capsized, wrote a poem just before the disaster in which he likens the rig to a "man-made mechanical rapist." For more on this poem, see Dodd (2012, 110–111).
11 See, for instance, Alaimo (2017), DeLoughrey (2019), Hessler (2018), Jue (2020) and Shewry (2015).
12 This comment is most relevant to the Newfoundland and UK situations. The fact that the Norwegian inquiry was not a public one, or, in the opinion of many, exhaustive enough, has been an ongoing source of distress and agitation. Indeed, in June 2019, the Norwegian Parliament bowed to lobbying by survivors and the bereaved, academics, the media, and unions to order a new audit of the accident (Tungland 2020, 20).
13 *Kielland* quotes directly and extensively from approximately 300 interviews undertaken between 2013 and 2019 with people associated with the accident. My thanks to Marie Smith-Solbakken for sharing a pre-publication copy of the book with me. The full interviews have been published by Marie Smith-Solbakken et al. in Norwegian in the several volumes of *Minnebank Alexander L. Kielland-ulykken.*
14 See Jason Haslam's chapter in this volume for a detailed discussion of the ways in which author China Miéville conjures the monstrous characteristics of offshore rigs.
15 See, for instance, Bonita Lawrence's discussion of colonial concepts of "savagery" in her essay in *Tracing Ochre: Changing Perspectives on the Beothuk* (Polack 2018).
16 These observations are in keeping with Rachel Webb Jekanowski's assertions in her chapter in this volume that settler colonial perspectives undergird offshore oil industry activities in Newfoundland and Labrador.
17 This statement indicates that it was the disintegrating rig itself that was perceived as a greater threat, a feeling shared by Leif Jørgen Mykland who recalls thinking to himself "we'd be better off jumping into the sea" (Tungland et al. 2020, 58).
18 These feminized descriptions of the nonhuman also resonate with Greg Tiller's poem, which I refer to in a note above.
19 As men who miraculously survived jumping from the helideck of the burning rig into the ocean put it, in gallows humour-style terms, the choice was "to jump and try or fry and die" (Bollands 2018, 77).
20 The poor visibility was primarily due to the fact that Phillips had resumed mud circulation at a nearby platform and was dumping drilling mud near where the divers were working (Gjerde and Ryggvik 2014, 194).

21 Gjerde and Ryggvik note that, in a survey completed after the rescue work, around half of the divers involved stated that they experienced mental health disturbances as a result of participating in the recovery effort (Gjerde and Ryggvik 2014, 198).

References

Alaimo, Stacy. 2017. "The Anthropocene at Sea." In *The Routledge Companion to the Environmental Humanities,* edited by Ursula K. Heise, Jon Christensen, Michelle Niemann, 153–162. Abingdon: Routledge.

Appel, Hannah. 2015. "Offshore Work: Infrastructure and Hydrocarbon Capitalism in Equatorial Guinea." In *Subterranean Estates: Life Worlds of Oil and Gas*, edited by Hannah Appel, Arthur Mason and Michael Watts. 257–273. Ithaca: Cornell University Press.

Avery, Dwayne. 2020. "Death by a Thousand Spills: How Corporate Branding and Media Strategies Downplay the Risk of Offshore Oil Spills in Canada." *Journal of Environmental Media*, 1 (2): 225–244.

Beck, Ulrich. 1999. *World Risk Society*. Cambridge: Polity Press.

Bennett, Jane. 2010. *Vibrant Matter: A Political Ecology of Things*. Durham: Duke University Press.

Berge, Alte. 2019. *Puslingar*. Oslo: Samlaget. [In Norwegian].

Berlant, Lauren. 2011. *Cruel Optimism*. Durham: Duke University Press.

Bollands, Geoff, with Julie Gorton. 2018. *Baptism of Fire: Life, Death and Piper Alpha*. Leicester: Matador-Troubador Publishing.

Bolster, Jeffrey. 2012. *The Mortal Sea: Fishing the Atlantic in the Age of Sail*. Cambridge: Harvard University Press.

Brown, Andrew. 2003. "Authoritative Sensemaking in a Public Inquiry Report." *Organization Studies* 25 (1): 95–112.

Buell, Frederick. 2014. "A Short History of Oil Cultures; or, the Marriage of Catastrophe and Exuberance." In *Oil Culture,* edited by Ross Barrett and Daniel Worden, 69–88. Minneapolis: University of Minneapolis Press.

Burke, Edmund. 1792. *A Philosophical Inquiry into the Origin of our Ideas of the Sublime and Beautiful. With an Introductory Discourse Concerning Taste, and Several Other Additions*. Basil: J. J. Tourneisen. http://find.gale.com.qe2a-proxy.mun.ca/ecco/infomark.do?&source=gale&prodId=ECCO&userGroupName=sain66118&tabID=T001&docId=CW3317401787&type=multipage&contentSet=ECCOArticles&version=1.0&docLevel=FASCIMILE

Butler, Wade. 2009. "The Worst Week of My Life." In *Rig: An Oral History of the Ocean Ranger Disaster,* edited by Mike Heffernan, 84–87. St. John's: Creative.

Chaudhuri, Una. 2017. "Interspecies Diplomacy in Anthropocenic Waters: Performing an Ocean-Oriented Ontology." In *The Routledge Companion to the Environmental Humanities,* edited by Ursula K. Heise, Jon Christensen, Michelle Niemann, 144–152. Abingdon: Routledge.

Connery, Christopher. 2010. "Sea Power." *PMLA* 125 (3): 685–692.

Cullen, The Honourable Lord. 1990. *The Public Inquiry into the Piper Alpha Disaster*. 2 vols. London: HMSO.

DeLoughrey, Elizabeth. 2010. "Heavy Waters: Waste and Atlantic Modernity." *PMLA* 125 (3): 703–712.

———. 2019. "Towards a Critical Ocean Studies for the Anthropocene." *English Language Notes* 57 (1): 21–36.

Dodd, Susan. 2012. *The Ocean Ranger: Remaking the Promise of Oil*. Halifax: Fernwood.

Duckett, Adam. 2018. "Piper Alpha Monopoly Sparks Backlash." *The Chemical Engineer*. 10 April, 2018. https://www.thechemicalengineer.com/news/piper-alpha-monopoly-sparks-backlash/.

Edensor, Tim. 2005. "Waste Matter – The Debris of Industrial Ruins and the Disordering of the Material World." *Journal of Material Culture* 10 (3): 311–332.

Estok, Simon. 2014. "Painful Material Realities, Tragedy, Ecophobia." In *Material Ecocriticism*, edited by Serenella Iovino and Serpil Oppermann, 130–140. Bloomington: Indiana University Press.

Fahey, Patrick. 2009. "About Feeling Bloody Useless." In *Rig: An Oral History of the Ocean Ranger Disaster*, edited by Mike Heffernan, 62–65. St. John's: Creative.

Gjerde, Kristin Øye and Helge Ryggvik. 2014. *On the Edge, Under Water: Offshore Diving in Norway*. Translated by Rolf E. Gooderham. Stavanger: Wigestrand Forlag.

Haraway, Donna. 2016. *Staying with the Trouble*. Durham: Duke University Press.

Heffernan, Mike, editor. 2009. *Rig: An Oral History of the Ocean Ranger Disaster*. St. John's: Creative.

Hessler, Stefanie, editor. 2018. *Tidalectics: Imagining an Oceanic Worldview through Art and Science*. Cambridge: MIT Press.

Higgins, Jenny. 2018. "Response to the Ocean Ranger Disaster." *Newfoundland and Labrador Heritage Website*. https://www.heritage.nf.ca/articles/politics/ocean-ranger-disaster-response.php

Hynes, Ron. 1993. "Atlantic Blue." *Cryer's Paradise*. Universal Music Canada, compact disc.

Inquiry Commission for the "Alexander L. Kielland" Accident. 1981. *Norwegian Public Reports: The Alexander L. Kielland Accident: From a Commission Appointed by Royal Decree of 28th March 1980. Report Presented to the Ministry of Justice and Police, March 1981*. Translated into English by Translatør service A/S, Stavanger, Norway. [no publisher].

James, Frank. 2010. "Oil Execs Grilled on Copycat Emergency Plans." *NPR: The Two Way*. 15 June, 2010. https://www.npr.org/sections/thetwo-way/2010/06/15/127863551/oil-execs-grilled-for-identical-emergency-plans-walruses-and-all

Jue, Melody. 2020. *Wild Blue Media: Thinking through Seawater*. Durham: Duke University Press.

Kane, Thomas. 2009. "About Where We Were and Where We're At Now." In *Rig: An Oral History of the Ocean Ranger Disaster*, edited by Mike Heffernan, 68–71. St. John's: Creative.

Kemp, Alex. 2012. *The Official History of North Sea Oil and Gas: Volume II: Moderating the State's Role*. Abingdon: Routledge.

Lawrence, Bonita. 2018. "Unrecognized Peoples and Concepts of Extinction." In *Tracing Ochre: Changing Perspectives on the Beothuk*, edited by Fiona Polack, 297–320. Toronto: University of Toronto Press.

LeMenager, Stephanie. 2014. *Living Oil: Petroleum Culture in the American Century*. New York: Oxford University Press.

Maloney, Ian. 2017. *The Waves Burn Bright*. Glasgow: Freight Books.

May, Jon. 1995. *Divers: From Piper Alpha to the Gulf War: A Deep Sea Story*. London: Hodder Headline PLC.

Mentz, Steve. 2015. *Shipwreck Modernity: Ecologies of Globalization 1550–1719*. Minneapolis: University of Minnesota Press.

Moore, Jason. 2017. "The Capitalocene, Part 1: On the Nature and Origins of Our Ecological Crisis." *The Journal of Peasant Studies* 44 (3): 594–630.

Moore, Lisa. 2009. *February*. Toronto: Anansi.
Morton, Timothy. 2013. *Hyperobjects: Philosophy and Ecology after the End of the World*. Minneapolis: University of Minnesota Press.
Neimanis, Astrida. 2017. *Bodies of Water: Posthuman Feminist Phenomenology*. London: Bloomsbury.
Nixon, Rob. 2011. *Slow Violence and the Environmentalism of the Poor*. Cambridge: Harvard University Press.
Offshore Technology. 2020. "The World's Worst Offshore Oil Rig Disasters." https://www.offshore-technology.com/features/feature-the-worlds-deadliest-offshore-oil-rig-disasters-4149812/
Petroleum Safety Authority Norway. 2020. "Kielland at 40: New Exhibition on the Disaster." *Petroleum Safety Authority Norway: Technical Subjects*. https://www.ptil.no/en/technical-competence/explore-technical-subjects/features/2020/kielland-at-40-new-exhibition-on-the-disaster/.
Phelps, Stephen. 2008. *Piper Alpha: The Human Price of Oil*. BBC Radio 3-commissioned drama. https://www.bbc.co.uk/programmes/b00cdbpc
Punchard, Ed. 1989. *Piper Alpha: A Survivor's Story*. London: W.H. Allen.
Reid, Marc. 2020. "The Piper Alpha Disaster: A Personal Perspective with Transferrable Lessons on the Long-Term Moral Impact of Safety Failures." *ACS: Chemical Health and Safety* 27: 88–95.
Reme, Kian. 2020. "Foreword." In *Kielland: Based on the Stories of Those Who Were There*, edited by Else Tungland, Marie Smith-Solbakken and Ellen Kongsnes. Translated by Elisabeth Tungland. Sola: Tungland Media.
Ritson, Katie. 2019. *The Shifting Sands of the North Sea Lowlands: Literary and Historical Imaginaries*. Abingdon, Routledge.
Royal Commission on the *Ocean Ranger* Marine Disaster. 1984. *Report One: The Loss of the Semisubmersible Drill Rig Ocean Ranger and its Crew*. Ottawa: Canadian Government Publishing Centre.
———. *Public Hearing*. 1983. St. John's, Newfoundland, 23 March, 1983. Volume 36. Centre for Newfoundland Studies Archives, Memorial University.
Shewry, Teresa. 2015. *Hope at Sea: Possible Ecologies in Oceanic Literature*. Minneapolis: University of Minnesota Press.
Simpson, Mark, and Imre Szeman. 2018. "On Energy Impasse: Cultural and Social Barriers to Energy Transition." Paper presented at Petrocultures 2018, Glasgow, Scotland, August 29, 2018.
Smith-Solbakken, Marie et al., editors. 2019. *Bind 1: Minnebank Alexander L. Kielland-ulykken: Vi Som Overlevde*. Stavanger: UiS Scholarly Publishing Services. [In Norwegian]
———. 2019. *Bind 2: Minnebank Alexander L. Kielland-ulykken: Vi Som Reddet, Berget Og Etterforsket*. Stavanger: UiS Scholarly Publishing Services. [In Norwegian].
———. 2019. *Bind 3: Minnebank Alexander L. Kielland-ulykken. Vi Som Mistet*. Stavanger: UiS Scholarly Publishing Services. [In Norwegian].
———. 2019. *Bind 4: Minnebank Alexander L. Kielland-ulykken. Vi Som Arbeidet Og Vi Som Var Arbeidsgivere*. Stavanger: UiS Scholarly Publishing Services. [In Norwegian].
———. 2019. *Bind 5: Minnebank Alexander L. Kielland-ulykken. Vi Som Bestemte, Støttet Og Var Tilstede*. Stavanger: UiS Scholarly Publishing Services. [In Norwegian].
Staubitzer, Carl. 2009. "About Looking and Looking." In *Rig: An Oral History of the Ocean Ranger Disaster*, edited by Mike Heffernan, 90–93. St. John's: Creative.
Stepputat, Finn. 2014. *Governing the Dead: Sovereignty and the Politics of Dead Bodies*. Manchester: Manchester University Press.

Stoddart, Mark, and B. Quinn Burt. 2020. "Energy Justice and Offshore Oil: Weighing Environmental Risk and Privilege in the North Atlantic." *Environmental Sociology* 6 (4): 390–402.

Sullivan, Joan. 2017. "Voices from the Rig: NQ Editor Recounts Creation of Play from Ocean Ranger Tragedy." *MUN Gazette*, 13 February. https://gazette.mun.ca/public-engagement/voices-from-the-rig/.

Sutton, Ian. 2013. *Offshore Safety Management: Implementing a SEMS Program*, 2nd ed. Amsterdam: Elsevier.

The Ocean Is Full of Its Own Collapse, composer Andrew Staniland, performers Duo Concentrate, Suncor Energy Hall, Memorial University, St. John's, 15 March 2019.

Thomas, Alistair. 2020. "Alexander Kielland: Norway Still Fighting For Answers 40 Years after Its Worst Industrial Disaster." *Energy Voice*, 27 March, 2020. https://www.energyvoice.com/oilandgas/north-sea/230923/alexander-kielland-norway-still-fighting-for-answers-40-years-after-its-worst-industrial-disaster/

Tribunal of Commerce, Paris. 1985. *Platform Alexander L. Kielland Accident of 27th March 1980: Expert's Report*. Paris: Tribunal of Commerce.

Tsing, Anna Lowenhaupt. 2015. *The Mushroom at the End of the World*. Princeton: Princeton University Press.

Tungland, Else, Marie Smith-Solbakken and Ellen Kongsnes, editors. 2020. *Kielland: Based on the Stories of Those Who Were There*. Translated by Elisabeth Tungland. Sola: Tungland Media.

United States Coast Guard. 1983. *Marine Casualty Report: Mobile Offshore Drilling Unit (MODU) Ocean Ranger, O.N. 615641. Capsizing and Sinking in the Atlantic Ocean on 15 February 1982 with Multiple Loss of Life*. Washington, DC: United States Coast Guard.

United States National Transportation Safety Board. 1989. *Capsizing and Sinking of the Mobile Offshore Drilling Unit Rowan Gorilla in the North Atlantic Ocean, December 15, 1988*. Washington, DC: National Information Service.

Urry, John. 2014. *Offshoring*. Cambridge: Polity.

Wilson, Sheena. 2014. "Gendering Oil: Tracing Western Petrosexual Relations." In *Oil Culture,* edited by Ross Barrett and Daniel Worden, 244–263. Minneapolis: University of Minneapolis Press.

4

DYNAMIC POSITIONING

North Sea petroculture's backwash

Graeme Macdonald

On 8 August 2016, at around 4 a.m., the semi-submersible drilling rig *Transocean Winner*, lost its tow in the Inner Atlantic, during a seasonally untypical storm and drifted, crewless, towards the Outer Hebrides in the North West of Scotland. Approximately four hours later, the 17,000 tonne rig ran aground at Dalmore Bay, a remote surfer's beach on the east coast of the Outer Hebridean Island of Lewis, and became international news.[1] For a number of days, it languished on the rocks unclaimed, generating mounting calls for transparency on the decommissioned rig's ownership and undisclosed breaker destination. It was eventually revealed to belong to Transocean, the US owners of the infamous *Deepwater Horizon*, with their company brass-plates in Zug, Switzerland.[2] Last leased to Marathon Oil, the 33-year-old behemoth had begun its journey to scrap off Stavanger in a southern Norwegian Fjord and was en-route to a staging point in Malta. The possible destination beyond that lay unrevealed and became an issue in the days following the platform's beaching.

One week after the accident, the journalist Richard Cresswell (2016) published an op-ed in the online daily trade journal *Energy Voice* accusing Transocean of being "unnecessarily secretive" about the rig's final destination. While inadvertently acknowledging some form of open secrecy as an operational strategy of the offshore petroleum firm and its financing infrastructures, Cresswell's accusation echoed a more general concern in and beyond the offshore energy industry about the growth of a trans-national and neo-liberalized market in marine scrappage. The practice of outsourcing contracts to breaker yards in Southern Europe and the Global South had become increasingly common, motivated by cheaper costs and locally looser regulatory regimes, extending from environmental standards and labour conditions to safety criteria and insurance policy.[3] These corporate practices also facilitate the offshoring of waste disposal protocols

DOI: 10.4324/9781003024125-4

and the offsetting of pollution indexes, helping massage the constricting demands of Northern carbon budgets. All of these, in turn, help economize the recycle value of the steel and other components. In turn, scrappage arrangements like Transocean's support the credit commissioning of new rigs and generally assist the entire global oil infrastructure – from platform to pipeline to forecourt to trader terminal and tax haven – to remain onstream and offshore in a time of diminishing cheap inputs, low barrel prices, and unprecedentedly voluble environmental pressure to transition away from oil altogether.

Amidst the confusion surrounding the *Winner*'s provenance, lay questions over its remains and its potential afterlives; existential questions that projected onto larger, already existing debates about the future of oil itself as an outdated fuel source and petrochemical stock. Such concerns, as this chapter contends, have been intrinsic to North Sea oil and the discourse surrounding it since its inception, but they essentially define its contemporary condition as the era of climate consciousness has dovetailed with that of decommissioning. The fateful story, following in the watery wake of the stricken *Winner*'s journey inshore, was its inevitable transmutation into another definition of wake: that of the ritual enacted after a death. This was realized in the form of a protracted battle not only over the means of the rig's disposal but also in the symbolic domain over the very meaning of its carcass. The chapter explores how, in the wake of the violent onshoring of an out of control oil behemoth, appeared larger questions of the terminal beaching of North Sea oil altogether: was this *un*dynamic position on the rock of Lewis a broader signifier of the deathbed of the industry? Was it at the very least a horizonal signal of its inevitable end? As I will argue, these questions are central to petroculture because they are inherent to oil as an evaporative material form, but they are particularly resonant in the hostile environments of cold water oil. The examples below highlight how questions of the end are present throughout a North Sea (particularly Scottish) petroculture consistently sensitive to depletion, becoming irresistible as climate change intensifies.

A notable divergence emanates from and veers constitutively between those mourning the end of oil who seek to prolong its life and those actively campaigning to hasten its demise. And, as the chapter demonstrates, the tools and domains of the literary: narrative, symbol, performance, genre, metaphor, image, poetics, speculative-work, etc. are pressed into service by vested parties from multiple sides engaged in a dynamic war of position over the past, present, and future form and meaning of oil – not only in the North Sea but also around the fossil-world-system in general.[4]

The strange manifestation of this vessel in the Hebridean littoral zone immediately begged a degree of explicability that, in turn, generated wider affordances with a range of literary, artistic, and cultural work, registering in various ways the precarious condition of late cold water oil and its environments. A range of examples from across the North Sea oil imaginary appear below, drawn from literary fiction and creative non-fiction, film, theatre, visual art, and documentary.

They all, in their different ways, register this 'sense of an ending.' They also retain a consonance of sorts in their shared revelatory focus on the murky world of oil, driven by a critical tendency identifiable at the heart of most petrofiction,[5] bent as it is on exposing the deleterious effects of petroleum and petromodernity on a range of social and environmental fronts.

Events revealed the fate of the *Winner* as determined by the familiar abstract realms of the "offshore:" the energo-political entanglement of the material infrastructures of oil with the geo-political economies of state-territorial regimes and the trans-national realms, corporate extractive operations and derivative products of a dizzying range of financial services and systems. The discovery and development of the North Sea as a relatively late oil play in the petroworld-system did so much to create and grow these tentacular relations as it came onstream in the early 1970s and developed through the neoliberal policy consolidations and alterations of the futures trading infrastructures and financial markets of the 1980s (Harvie 1994; Kemp 2014). It was, however, the primal, turbulent geophysical power of the offshore – the marine environment and its drag, amidst unseasonably atmospheric storm conditions – that imposed itself as the principal undercurrent of this disaster.[6]

The running aground of the *Transocean Winner* became a public saga of the age of late petroculture in the North Sea and beyond. The irruptive manner of the rig's very appearance onshore was critical to its subsequent aesthetic and interpretive iteration: colossus-like, unexpected, unignorable, it burst into media and public consciousness, instantaneously becoming part of a wider and longer field of critical and cultural representations of energy production and circulation

FIGURE 4.1 Grounded *Transocean Winner*
Source: *Murdanie Macleod*

from the North Sea and beyond. Here also was a rude reminder of the literal and metaphorical definitions of "offshore," referencing a world obscured from view, either stubbornly insensible and disarticulated in public life or belonging to a rather restricted mesh of mediations. In many respects, this was apt demonstration of a critical shibboleth in the first wave of Petrocultures/Energy Humanities scholarship emphasizing the relative absence of oil and energy in the spectrum of concerns animating cultural history; viewed as 'unconscious' or hidden, even inconsequential (Boyer and Szeman 2017; Yaeger 2011). A predominant rationale for this cites the limited visibility of petroleum sites, part of the general abstraction from common experience of those systems and infrastructures that sustain high-carbon and energy-intensive lifeworlds across the world. This discretion is historically consistent with various lacunae created and structurally reproduced by the politico-economic and corporate interests of oil; those representative agencies, institutions, and discrete assemblages that have worked, since its inception, to enhance its estimable position in part by concealing, hedging and controlling knowledge about its operations and effects (Hitchcock 2010).

In literary contexts, the most notable articulation of this condition of absence is Amitav Ghosh's now seminal and much-cited 1992 essay "Petrofiction," which pondered oil's relative non-appearance in modern world literature, belying its heavy saturation in modern global life. This, Ghosh argued, was not simply due to the peripheral spatiality of oil's extraction zones ("a space that is no place at all, a world that is intrinsically displaced, heterogeneous and international") but also a feature of shame (Ghosh 1992, 30). This he saw as emanating in oil's historical energopolitics: "a matter of embarrassment verging on the unspeakable, the pornographic" (Ghosh 1992, 30). The general tendency, in this view, is determinative: that culture, in a society that tends to rely on oil and its attendant array of products – consciously or otherwise – unsees or is prevented from seeing its perennial entanglements with imperialism, violence, corruption, disaster and especially its destructive social and environmental effects. This critical position is increasingly challenged by raised public awareness of the necessity of energy transition, but it remains critical to comprehend the historical cultural registrations of oil in order to understand its future determination. By this, I mean realizing the objective will of the energy system of fossil capital to continue to reproduce, in full knowledge of climate's demands. This means a requirement not only to articulate transition but also to navigate it in full view of oil's dynamic positioning.

Ghosh's (2014) later reassessment of some of his claims acknowledges the findings and assertions in the emergent field of petroculture (Barrett and Worden 2014; LeMenager 2014a; Szeman 2012, 2019; Wilson et al 2017), which has identified a broadly comparable trans-national body of petro-literature and theatre from the mid-19th century to the present. As this archive has expanded, it has become clear that the issue is not necessarily the absence of petrofiction, but rather how to appropriately conceive, define, and house it: comparatively, methodologically, and interpretively. There is an imperative, as Jennifer Wenzel argues, to identify "protocols of reading and modes of inquiry that can perceive the pressure that energy exerts on culture, even and especially when energy is not-said: invisible, erased, elided" (Wenzel 2017, 11). These changing reading contexts and methods, it should be emphasized with an emphasis on disclosure and revelation, operate in accordance with – and oftentimes in opposition to – oil's general and discrete manifestations and representations in time and space.

Spatial and political geography remains critical to this consideration of energy's reading contexts and protocols. The argument is well established (Balkan and Nandi 2021; Barrios 2021; Deckard 2021; Macdonald 2012; Niblett 2020) that oil is always to some degree an extra-national resource produced and impacting within and beyond the original extractive region or nation and that 'worlding' critical methodologies should follow accordingly. A legitimate question arises, however, on the specificity and difference between different sites, methods, and territories of oil's extraction, production and circulation and how this might manifest in *particular* cultural representations and connections. There is at a general scale a level of comparability and aesthetic affinity between the literary genres and artistic renderings of North Sea oil, for example, and that, say, of the Permian basin or the Niger Delta. But there are also clear, distinguishing and localized features of a specific mode of oil in discrete territories and regions, as Stephanie LeMenager has pointed out (LeMenager 2014a, 11). Deep watery sites have a specific cultural codification, as arguments in the developing fields of the Blue humanities and critical ocean studies point out (Campbell 2019; Campbell and Paye 2020; DeLoughrey 2019).[7] Relative distance from common view and perception impacts, in turn, on cold water oil's articulation in cultural responses and becomes a characteristic feature of its content. Whether repressed, unconscious or deliberately hidden in securitized and/or peripheral or extreme locations, (in)visibility remains a powerful expression in the extrapolation of oil as both ontology and infrastructure. This is particularly resonant in a North Sea petroculture shaped by a turbulent onshore political context of competing nationalisms and economic philosophies. It also resounds in an oil region consistently haunted, for reasons historical, geological, technological, financial, and political, by its terminal horizons.

It is not unreasonable to note that cold water oil has been somewhat marginal in the general critical literature on petrofiction mentioned above. There are qualified reasons for this, not least that a core critical constituency in the disciplinary field of petroculture has been geographically focused on North American, largely (though not always) land-based oil cultures. A separate but probably more significant rationale for this deficit is that writing (especially forms of reportage) about and from distant offshore rigs and platforms is subject to extremely challenging access issues. Notable offshore accounts from writers unattached to the industry or within the realms of its political geography remain relatively rare.[8]

Around the same time as Ghosh's original intervention, the Scottish historian Christopher Harvie, in his 1994 book on North Sea oil, also linked oil's "discreet, geographically withdrawn" (Harvie 1994, 259) sites with his stress on the strange "non-critique" of oil in metropolitan Anglo-British culture, where it fails to significantly register in proportionately explicit fashion in metropolitan and middlebrow cultural production.[9] "Where was the novel or the poem of North Sea oil? Or the play?" asks Harvie (Harvie 1994, 267). Reasons for this cultural absence extend beyond physical geography. The retreat from industrialism in the services headed economy of post-1970s Anglo-Britain, and the increasing compartmentalization of onshore life were intensified as the Thatcher regime of the 1980s systematically deregulated the industry and left it increasingly a rentier matter for the companies to operate. In many respects, keeping oil a 'remote' issue, 'out there' also worked in the favour of the Thatcher regime. The early equipment sourcing and engineering skill was heavily reliant on US capital and expertise, meaning that the initial promise of an immediate national panacea (for Scottish nationalist, devolutionist and unionist alike) floated in the 1970s was neither palpable nor forthcoming in anything like a uniform or prolific fashion. Writing in 1986, Danny Hahn identified different approaches to extraction and development that had characterized the industry in accordance with wavering Government policy, uncertainty over proven reserves, and opposed forms of political ideology, veering between rapid exploitation and steady control over depletion and disposal policy. Harvie also notes an "inchoate" and "unpredictable" government strategy, on extraction rate, taxation and depletion, regulatory control, unionization and nationalization, and emphasizes the obstructive and "notoriously secretive" nature of the oil company (Harvie 1994, 124). Offshore meant that oil did not blend into the general landscape, like the railways, and thus:

> One can't avoid the conclusion that because the oil business did not occur in, and did not create, a community, and was largely American in ownership, it seemed almost a colonial imposition: something inherently alien. (Harvie 1994, 142)

A strong sense, of course, that the industry itself was volatile, mobile, and exploitative by nature and necessity has always been resonant. Nothing like a mainstream oil culture appeared in the UK. There remains no national museum of oil and gas, for example, and only now, in the industry death throes, amid the advent of petroculture as a knowledge-base and epistemological orientation, and the unignorable issue of climate change, has there been an impetus for curation and exhibition.[10]

It should also be underlined that, as Timothy Mitchell (2011) has argued, the political right in Britain, the US, and elsewhere ultimately utilized the boon from the supply of domestic petro-energy to destroy the kinds of carbon democracy created by collectivist modes of coal and steel. Oil's significant post-war role in the architecture and implementation of neoliberal, market-oriented forms of political economy and culture is now confirmed (Huber 2013), though as Harvie notes, it was curiously underplayed, even "disremembered" (Harvie 1994, 289) in Margaret Thatcher's memoirs of her time in Downing St. The revival of Scottish nationalism through the 1970s was important to manage in Westminster, where petro-nationalist claims surrounding the potential sovereign economic benefits of 'Scotland's' oil would implement a difficult political balancing act for the unitary union state that has persisted to the present. Questions about the specific ideological orientation of modern Scottish nationalism were considerably energized by oil (Jackson 2020, 72, 171). The now evident fundamental significance of the onstream revenues to the onshore de-industrialization policies driven by the nascent Thatcher government – never as popular in Scotland as in England – ensured a different impression and magnified import for offshore oil in Scottish constitutional politics and public consciousness that has not receded (Brotherstone 2012; Christophers 2020). Oil's characteristic figuration as a futural entity offering a prosperous economic platform for an autonomous future state has retained a significant role in Scotland since the 1970s and featured heavily in the literature surrounding the campaign for Scottish independence in the 2014 Referendum (Common Weal 2014; Foley and Ramand 2014; Ramsay 2014). Credible or otherwise, any 'pot of black gold' sensibility is now irrevocably diminished by the global environmental movement's arguments to radically and rapidly curtail oil and gas extraction, one now largely accepted – albeit reluctantly and with qualifications in some quarters – by a majority in Scottish politics and civil society (Maxwell 2020).

Even in a predominantly Anglified British literary firmament, Harvie's thesis perhaps rests on an overly localized reading of petro-literary production. Writers living close to the infrastructures of oil should not necessarily be expected to write more about it, and this is not always the case, even in those facets of Scottish literary culture that underwent a late 20th-century renaissance. Nonetheless, his point arguing that the explicit registration of petroculture in the UK was somewhat limited to Scotland has lasting salience. Also prescient is Harvie's early realization that the "complex transformations" wrought by cold water oil has registered across various genres and especially in experimental and speculative literary and cultural forms. The technological superimpositions of a dazzlingly futuristic architecture of industrial terraforming in often rural and peripheral sites of eastern Scotland and the northern archipelagos, and in deeply inhospitable zones of the marine environment, together with the intensifying realization of the dark ecology of oil as the century progressed has formed a repertoire of writing and filmic work dominated by irrealist, unorthodox and/or combinatory forms, producing gothic, epic theatre, concrete poetry, science-fiction, experimental film, musical theatre and other speculative, fantastic, and experimental petro-imaginaries. In the age of decommissioning, post-oil and climate fictions continue in this fashion.[11]

"Bullshitters on the beach"

Entangled themes of visibility and secrecy return again and again in the cultural work of North Sea oil. These are often bound up with a documentary impulse discernible regardless of form or genre. Conventional media has tended to emphasize the techno-heroic and swashbuckling economic entrepreneurialism side, with the windfall narrative a fulcrum for a spectrum of political positions on North Sea oil's historical import. The Labour Prime Minister Jim Callaghan famously made the point, in a speech at the Forties Field in 1977, that "God has given Britain her best opportunity for one hundred years in the shape of North Sea oil" (Harvie 1994, 2). It is sometimes understated that a chief factor in the late-imperial spasm of the 1956 Suez crisis was a rooted anxiety about Britain's imported oil supply. This had been present since the beginning of the century but had become severely pronounced after World War II and the era of decolonization. The rise of OPEC also boosted Britain's sense of fortune on the discovery of a large domestic supply. In his seminal history of oil, *The Prize*, Daniel Yergin relates the words of the drilling superintendent on Phillips Petroleum's *Ocean Viking*, the exploratory rig that made the first significant find in 1969, in what became the Norwegian Ekofisk Field: "What the astronauts have done is great, but how about this?" (Yergin 2008 [1991], 650). Yergin, like many other classic oil historians, emphasizes the risk and danger but also the host of new engineering technologies that allowed penetration to depths deeper than hitherto drilled in a "nasty and vicious sea and some of the worst weather in the world." The permanent platforms

> had to stand on mud, quicksand, clay and sand wave bottoms, but also had to withstand the fury of a '100-year wave', ninety feet high, as well as winds of 130 miles per hour. Altogether, the development of the North Sea was one of the greatest investment projects in the world [...]. It was also a technological marvel of the first order. (Yergin 2008 [1991], 651)

This tone of confidence, accompanied by an awe-inspired techno-utopianism reappears across documentaries, industry pronouncements and artistic responses. For example, the first episode of James Naughtie's three-part 2015 BBC Radio 4 documentary, *Oil: A Crude History of Britain* is typical, headlining it as "a black gold rush," and a "story of a technical and engineering miracle, that took place against the odds." The by-line of the 2021 BBC Scotland television documentary, *Rigs of Nigg*, about the 1970s effort to construct North Sea oil infrastructure on Nigg Bay on the Cromarty Firth, declares it a "story of how over 5,000 men, sought locally and from across the globe, came together to create structures previously thought to be unbuildable and to withstand elements considered unsurvivable" (Coutts 2021). Similar rhetoric is ironically redeployed in the narrative of the (fictional) Scots Tory MP David Eddlestane, one of the characters in James Robertson's *And the Land Lay Still* (2010), a historical novel spanning 60 years of post-war Scotland. Sometime in the mid-1980s, Eddlestane, who harbours his own personal secrets, ponders the Thatcherite revolution, ruing its ruptures on "the quiet continuity" of the country:

> He thought of the oil pumping ashore north of Aberdeen. Astonishing feats of engineering had been performed to extract oil and gas from the depths of the sea. Amazing ingenuity combined with hard sweat and great braveness had achieved a daily production of two and a half million barrels of North Sea oil. Without that the public finances would be a disaster. Nobody mentioned this fact. The SNP whined about how prosperous an independent, oil rich Scotland could be, and the other parties rubbished their claims, partly because they didn't dare admit that without the oil revenues the whole of the UK would be bankrupt. *Don't mention the oil!* It was the great unspeakable of the age. (Robertson 2010, 552)

The temporal concertina-ing here (a 2010 novel covering 60 years reimagining a repressed character in the mid-1980s considering how to best repress the effects of oil struck in the 1970s) manages a retrospective critique of oil's historically significant and ongoing role in the UK's political economy and constitutional debate, but frames it as a phenomenon producing unevenness (for one thing, Eddlestane's points are subtly delivered in the wake of the 1984 miners' strike). The passage channels a number of ideological positions tinged with promise, subterfuge, and speculation, even prising open a counterfactual where nationalists assume full control of oil revenues and – "unspeakably" for unionists like this – independence becomes a palpable threat. Placing aside the Anthropocene perspective that cannot but read such exuberance for oil as deeply ironic from a climate anxious present, the historicity of Robertson's contemporary novel allows the contemporary reader to reconsider the reality of Britain's cold water oil and its contested political iterations from various staging points. We consider Eddlestane's sentiments from a 'not-quite-yet-after-oil' position where its 'unspeakability' has an altogether different hue. But also in the knowledge that as he thinks these thoughts the developmental boom years were already moving over the horizon, leading to him fretting: "What if Great Britain plc collapsed?" (Robertson 2010, 552).

The historian Richard Finlay describes UK oil as ultimately an "economic failure" (Finlay 2004, 333). Contending that "the discovery of North Sea oil gave a greater boost to the economy of Texas than to that of Scotland," Finlay underlines elements like geography, foreign capital with *de facto* contracts, rapid extraction, and erratic fiscal and depletion policy as offering little chance for a domestic industrial revival to bloom. Others (e.g. Harvie 1994, O'Neill 2004) and nationalist opinion, in general, agree that oil's heady economic promises were not realized uniformly, with squander and mismanagement reiterated. In this view, the opening lines of Edwin Morgan's sixth Glasgow Sonnet, composed at the inception of the oil era, in 1972, proved somewhat prophetic on the relatively uneven share of regional benefits: "The North Sea oil-strike tilts east Scotland up/and the great sick Clyde shivers in its bed" (Morgan 1996, 290).

It is not coincidental that, in Robertson's narrative published at the other end of the oil era, the page previous to Eddlestane's views on oil relates his personal inflated wealth from property speculation in London and Edinburgh, amid the rise of "unfettered" (Robertson 2010, 550) financial capital. This means of acquisition, where "people seemed to be able to make vast amounts of money out of nothing" was, of course, fostered by and during the initial era of North Sea oil and was a major feature of the neoliberal turn under Thatcherism. Robertson, rather unsubtly, connects oil to new modes of finance and wealth, gained not by the "hard sweat" of energy workers noted above but by personal privilege, networked connection, and capital advantage that the financial industries and service economy oil helped bring into life. There are echoes here of Alasdair Gray's controversial 1984 novel *1982 Janine*, which presents the fragmented memories of its dissolute, troubled, and perverse central character, Jock McLeish, in an experimental narrative form, blending subjective fantasy and memory with the larger historical contexts of McLeish's Scotland. The squandering of oil registers alongside the squandering of McLeish's life and youthful idealism. Again, this is conveyed in ironic retrospective, tinged with counterfactual:

> The euphoria of the North Sea oil boom, the depression of the British coal industry, the depression of the British car industry, the depression of the British steel industry, the discovery that North Sea oil benefitted hardly anyone in Britain but the shareholders – all this, though partly conceivable, had not been conceived (Gray 1985, 230).

There has long been an obscuring of the ("unspeakable") downsides of oil-rentier neoliberalism and the forms of uneven modernization and credit-finance cultures it has wrought onshore, not to mention the environmentally unjust ethics of sustaining an oil-based "risk" economy in the era of climate breakdown.[12] This chapter concludes with an analysis of contemporary texts that register this concern. Yet, it is important to note that cultural responses to North Sea extractivism and its petro-capitalism have offered pointed critiques from its beginnings and throughout. An entire section of John McGrath's hugely successful 1973 "ceilidh play," *The Cheviot, The Stag and the Black, Black Oil*, for example, is a savage satire on the manner in which the British Government becomes a naïve and willing plaything for US-style carbon democracy. The appointed minister for oil negotiations, Lord Polwarth, is literally turned into a puppet by a frenetic American oilman, "Texas Jim," in league with a character named "Whitehall," who declares amidst all of the string-pulling that "we don't believe in fettering private enterprise: after all this is a free country" (McGrath 1973, 149). The arch response to this from Jim – "Never known a freer one"– exudes all of the regulatory concessions demanded by extractivist capitalism and underpins McGrath's play's angry local politics.[13] Polwarth, in the prototypical language of neoliberal lubricity (Simpson 2017, 155) so often deployed in oil company rhetoric declares his appointed role as "a knot-cutter, a trouble-shooter, a clearer of blockages, and a broad forum to cover the whole spectrum." The broadened meaning of "offshoring" (Shaxson 2011; Urry 2014) as a term that has come to be associated with a spectrum of deregulated financial instruments and practices, in addition to political cover-up, aided and assisted by the developments around offshore oil were already being excoriated in this 1973 play.

The best early instance of official British culture's obfuscatory response to the coming of oil, in fact, is a story of censorship around *The Cheviot*.[14] Archival scholarship by Robin Nelson in 2005 uncovered the deep levels of anxiety in the BBC over the television adaptation of McGrath's play – on the popular weekly *Play for Today* Sunday evening show. An experienced film and television editor as well as an acclaimed Marxist dramaturg, McGrath had to fight hard to get his controversial anti-imperialist treatment of oil's injurious social effects (on and offshore) shown on the state-broadcaster, guaranteed to be watched by millions of viewers in June 1974. This occurred at the same time that British state and capital interests were repeatedly subject to onshore forms of collective action from that older and more powerful nexus of fossil fuel, the coal mining unions.[15]

Arguments for a greater share of oil's resources also saw the resurgence of resource sovereignty and sub-state nationalism amid the decline of the industrial economy of a postimperial UK beset by the oil shock embargos of OPEC. The desire in many quarters to become an oil-producing state as quickly as possible required that disempowered labour conditions, unevenly spread property surges, and the generous rents and indebtedness to US oil capital had to be kept very much offshore in the public consciousness. McGrath's inspired piece of travelling folk theatre conjoined modern and historical eras of clearance, private land consolidation, and capital accumulation in the Highlands and Islands to the 'live' issue of oil as a new round of resource grabbing by 'absentee' landlords. Its tremendous success was a major contribution to the public apprehension of this new cold water industry 'out there,' in their territorial waters (Macdonald 2015).

The BBC, ever mindful of what such a damning critique of oil might do to their much vaunted neutrality, but equally cognisant of their duty to support artistic and cultural freedom, in the end somewhat neutralized *The Cheviot*'s politics by broadcasting it sandwiched between a boosterish documentary film and a discussion show on the decent prospects for Britain's new oil find.[16] McGrath was convinced that the documentary sought to discredit his 7:84 theatre company, framing them as partisan radical leftists (which, to be fair, many of them were). The then controller of BBC 1, Alasdair Milne, provided passages for excision (which nonetheless remained in the broadcast) and is quoted by Nelson as saying to McGrath that "this piece sounds as if you don't want the oil to be dug out of the sea" (Nelson 2005, 125). Nelson's view notes that rather than maintaining this position as a result of the Broadcasting Corporation's charter mandate for liberal consensus, in fact, the BBC's position was that the play "inconveniently undermined the position which the government of the day was trying to construct": i.e., that under no circumstances was North Sea oil to be viewed as anything other than a gift from Heaven (Nelson 2005, 129).

Later petro-theatre followed McGrath's example in making ever more explicit attempts to bring the privations of the cold water world directly into the onshore senses. A considerable section of Duncan MacLean's 1993 play *Julie Allardyce*, for example, focuses on the tribulations of an Aberdonian woman working on rigs as an ROV operator. Her position in control of an ocean exploring viewfinder mechanism emphasizes her privileged viewpoint as focalizer of the murky world below and on the life of the platform, especially in relation to gender. Repeated performative acts, gestures and themes of control entangle class, constitutional, and feminist politics on and off rig. Julie's description of her workplace skills makes the question of both her perceptive visibility as a (rare) female specialist rig worker connect to her uncovering of questions of safety and indispensable virtuosity:

> Headlights on in the dark black depths, and I see things that no one's seen, the floor of the sea, the rocks and the sand, the fish and the weed and the reefs and the wrecks. I've eyes as sharp as a shark's, see landshift under pipelines or cracks in concrete platforms through water dark as squid ink. In control, in front of the monitors, the joystick and the instrument panels. I'm the one in the driving seat. (McLean 1995, Scene 3, 186)

Operational opacity and the ability to navigate and understand potential dangers are inflated into the wider narrative of the play which, seen from the framing of the fate of the *Transocean Winner*, is ironic to say the least.[17]

Fiona Polack and Danine Farquharson make the resonant point that "in the absence of direct contact, the offshore rig lends itself to myriad reimaginings beyond the oceanic frontier" (Polack and Farquharson 2017, 253). Placing aside the very real question, touched on above, of the inaccessibility of rig experience and its relation to the ethnographic authority of the artists who take cold water oil as their subject, the form and theme of (im)palpability between offshore-onshore is a central feature in *Julie Allardyce*. A non-naturalistic, multi-platform view is insisted upon and conveyed in the staging and production. Offshore focalizes onshore life in the play, and many of the performative and technical effects of its stylistically heterogeneous form present deliberate assaults on an assumed insensible audience onshore. *Look at this world, pay it attention* is the none-too-subtle subtext, in form, scenario, and dialogue. The playwright himself argues that the drama of offshore oil is best rendered in its accompanying visual and affective abstractions, in all of the new shapes and hues and sounds and angles of vision the industry and its attendant infrastructure brings to place and setting. McLean views it as like "a dramatic equivalent of a Suprematist painting" (McLean 1995, 237) where different tones operate in fragmented harmony. In between, a deliberate didacticism pushes such abstraction as a theme. In one exchange, for example, Finn, a frustrated rig worker, claims to Julie that:

> Here's what I think: I think there's a lot of bullshitters on the beach that don't give a toss about us working out there. They drive about in their

> cars, they switch on their central heating, they fry up their bacon and beans – and the thought of folk in the middle of the North Sea drilling up the oil and gas they're using, it never enters their head. And then one day, a chopper ditches, or there's a gas leak, or a platform comes adrift in a gale, and oh dear me, suddenly they're up in fucking arms, getting themselves in a fuck of a state, shooting their mouths off! And why? Cause they feel fucking guilty! Cause ninety-nine per cent of the time they don't even think of us. Cause a hundred per cent of the time they're gobbling up the stuff we produce – they're keeping us out here! Fuck them! (McLean 1995, Scene 10, 213)

This is a deliberate attempt to restlessly assault the senses of the energy unconscious audience to the notion that oil is *here*, among us, in the theatre, on the city streets, because of the oil play out there, in the sea. The protocols that would come to underpin the energy humanities 20 years later are very much in play. The staging and dialogue loudly berate onshore publics for failing to adopt some form of moral and political awareness about both oil's pervasive presence in their lives and the difficult conditions of those working to extract it. A deliberately estranging yet immersive quality is sought to complement this in parts of the production: the opening scene, for example, turns the theatre into a helicopter ride to a rig, turning off lights, with roaring engine sounds, the shaking of audience seats, projecting northern light illuminations. Drummers move round the stalls, providing a noisy chorus, battering on petrol cans. You cannot avoid the offshore's pervasive presence on various orders and faculties of your experience is the signal order of McLean's play, and it transitions through multiple registers and styles to make it.[18] The question it generates time and again concerns the perceptual gaps and material distance between oil work, artistic and cultural work, consumer affects, and the abstract world of deep water oil with all its refined products conjoining them.

This material stress on the palpable and visible reminders of offshore oil's onshore presence reappears in work in other artforms. The visual artist Ernst Logar's multi-format project *Invisible Oil* (2011), for example, includes sculptures of oil rigs constructed from plastic bottles and other flotsam washed onto Aberdeen's beaches. Each model rig is set on a beach and named after the poorer districts of "oil-rich" Aberdeen. This signature gesture publicly diverts attention to the offshore industry "out there" and sets up a social critique of the unevenness of the world *with* more oil in it. This is described in Karin Kneissl's introduction to the exhibition book as oil's material "widespread and often invisible effects" (Kneissl 2011, 15). The effects directly referred to here are social: creating greater inequality and instability in the very societies oil is purported to enrich. But the invisibility is perhaps even more (non)apparent in the atmospheric sense from GHGs to engine particulates. The rig sculptures additionally enact provocative points about accretive plasticity and waste created by that petro-world – actually "beaching" it for those "bullshitters on the beach" to realize their (lack of) social and ecological attachments to its material backwash.[19] The scale and emphasis on climate and environmental degradation has deepened from McLean's play's context, with the necessity of understanding oil by realizing it differently, ecologically; 'reimagining' in order to transition from it: to make it an artefact. Ernst Ulrich von Weizsäcker puts it beautifully in his foreword to Logar's book: "How, after all, is one to photograph the barrel of oil that did not have to be pumped?" (Weizsäcker 2009, 9).

It is one thing, however, for a small-scale artwork to make a forceful impression of the precarious feedbacks of cold water oil in the 21st century, and entirely another when a crewless, 100 metre high, 17,000 tonne behemoth, carrying 300,000 litres of diesel crashes into the shoreline, leaking 52 tonnes into the sea.[20] A detailed documenting of *Transocean Winner*'s history and biography by Tom Lamont (2017) for *The Guardian* newspaper records the public spectacle the rig proved at its various continental staging points during the drama.[21] The salvage operation took time to organize and carry out. The longer the high-profile, photogenic wreck squatted incongruous and uninvited on the scenic shoreline of Scotland's periphery, caught in the flash, the more it became exposed

FIGURE 4.2 *Tillydrone*
Source: Logar, Ernst. *Invisible Oil*. 2011.

to all manner of global media and alternative mediation. Lamont notes the locals becoming attached to it. The site attracted sightseers, giving a sudden boost to tourism.[22]

This offers an opportunity to open out some questions around what are now familiar claims made in petrocultural criticism and in energy humanities more broadly, with regards to the ambition of making energy and its infrastructures more explicit and visible in cultural work and in public awareness; namely, what happens *after* we see that which has been invisible, offshore? When energy unconscious surfaces, when the wires, pipes, and ducts are exposed for all to see and contemplate, what then? What angle of vision or perspective determines the best way of seeing and why? These reflexive questions will no doubt be grappled with in the coming waves of critical scholarship, but it goes without saying that any identifying act of representation is always incumbent with the subsequent interpretive challenge of making meaning and, especially in the consequential environmental politics of petroculture, constituting action and activism.

"Another entity altogether"

The *Winner*'s unexpected onshoring afforded creative *re*interpretations of the representational infrastructure of this offshore behemoth. Connections to other oceanic petro-narratives, especially those of a fantastic, speculative, and irreal nature were apposite, as we shall see below.[23] The immediate and long-term future of this monster from the deep became a question that exceeded the singular event, spilling across multiple contexts of representation and constituencies of concern. There was a tempered case for the argument that this was another disaster amid a litany of historical examples: oil's irregular surfacing into public consciousness tends to occur in traumatic moments of spectacular spill or explosive accident (see Polack in this volume).[24] These tragic events, with their deep local resonance and memorialized afterlives, usually impact on alterations to regulations around safety but have nonetheless had little discernible overall effect on system disruption. The fact, however, that *Transocean Winner* was bookended by two world-resonating oil-critical events punctuated the heightened response to it. The first was the 2010 Mexican Gulf disaster, but the second was the ratification of the UNFCCC Paris Agreement a few months after the crewless rig became unmoored. As the definitional orbit of 'safety' around the functional capacities and limits of the oil world system has shifted and expanded onto existential planetary terrain, the generic understanding of 'oil-related disaster' has come under severe scrutiny. Here, it deployed in tandem with the unavoidable public envisioning of an unmanageable rig and the recycling and rebooting of the system it encapsulated. The *Winner* affair thus materially and symbolically demonstrated this contemporary contradiction between offshore oil and capital's attempts to salvage and regenerate itself and voluminous calls for accelerated energy transition. It presented in miniature John Urry's assertion that "offshoring and powering down are in strict contradiction with each other" (Urry 2013, 234).

Appel, Mason and Watts (2015) emphasize the role of oil as a metonym: seamlessly representing money, power, corruption, excess, etc. But here, it became effortlessly metaphoric. Cresswell worried that the "dead rig" was becoming "another *entity*" (my emphasis) altogether, inadvertently revealing how petroculture as a transformative critical method and aesthetic pursuit might enable a means to see, write, and think differently and dialectically about the meaning, import and even the style of this particular rig's marooning. It also provided opportunity to consider how any symbolic economy measures up against a predominant energopolitics generated by an established media-petroleum nexus. This latter entity has been long lashed to the rocks of the larger, seemingly immovable entity the post-war carbon complex helped create: *the* economy (Huber 2013; Mitchell 2011).

To reiterate a point made earlier, the *Winner*'s skewing into littorality provided an iconic gateway for thinking ecocritically and materially about the wider cultural form and meaning of the offshore in late petroculture; a means not only for thinking through and apprehending the ends and afterlives of oil in the 21st century but also its unwelcome appearances. The angle of vision changed the meaning as much as the material context. In a time of intensifying decommissioning of oil infrastructure and voluble debate about the viability of extending the life of the North Sea fields, the coroner description of the *Winner* resonated. One of the geographically biggest and actively drilled cold water territories in the oil world system is under unprecedentedly extreme pressure to transition and drawdown, though any proclamations of the timing of its eventual outright demise remain unwise. The *Winner* revealed decommissioning as an increasingly lucrative business with a globalized set of entanglements: rigs remain assets to the end. But what happens if the event of 'death' involves a process of rebirth and reincarnation? In a post-peak age, despite the Paris Agreement, and despite historically low barrel prices and storage glut, fossil demand remains globally projected and offshore drilling remains a vanguard element of oil's constantly renewed determination, either in new territorial waters such as the Arctic or in sucking out the dregs of extant fields and exploring for new finds. Commentators are entitled to ask if bold pronouncements concerning the passing of the age of oil are greatly exaggerated or wishful thinking.[25]

Since its inception, the oil system has been characterized by recursion: the tidal and evacuative nature one might expect from a non-renewable liquid source of energy, presenting ebbs and flows, evaporations and returns (Watts 2012). This is often a result of technological advances, from geological prospecting to drilling, but the late 20th-century assemblages in petro-financing and capital infrastructure maintain and reproduce "offshore" activity in "paper oil" as much as (and often way more than) "wet oil" (Diamanti 2021). The recent shale gas revolution in the US is a case in point here, with the revival of various sites from the Bakken to Pennsylvania signifying the physical, financial, and cultural ability of the oil and gas industry to revive its fortunes in once drying (or "mature") sites and begin anew in others.

This is accentuated in the zones of cold water oil and in fact epitomizes and explains the very existence of such a complex and expensive industry operating in harsh and forbidding oceanscapes across the world. It would not exist or last without the scientific, technological, and financial developments enabling and sustaining the age of tough oil. The irony here is that such extreme operational environments are testament to oil's long waves of "success," signifying the extent to which the world of the 20th century became (and remains in the 21st) petro-reliant. Any junkie will suffer duress and construct elaborate and difficult forms of procurement for another fix. As Gerry Canavan (2017) has argued, addiction and its bywords, such as tolerance, withdrawal, relapse, and substance abuse, are legitimate conditions and categories to apply to the oil business. The incontrovertible fact – and one that wends itself into most petrofiction if we are prepared to look for it – is that the oil system is and has always been a business of extending and delaying the inevitable: exhausted wells, dry fields. Throughout all of the frenetic prospecting activity, discovery wells, spectacular gushers, and feverish enthusiasm for oil finds in Upton Sinclair's 1926 early classic of petrofiction, *Oil!*, for example, are notable signs that it runs out in the short and long term. The paradox of over-reliance on a non-renewable substance is apparent in Ross the oilman's repeated stress on the search for more. This is a staple feature in many other petro-narratives, which can be read to reveal what Appel, Mason and Watts almost a hundred years later will describe as the world of oil's saturation with "declensionist language and dark futurity. Can we find enough? How long will it last? How quickly is it running out?" (Appel, Mason and Watts 2015, 6).

The last paragraph of *Oil!* focalizes an oilfield surrounding a small cemetery and forebodingly notes that "someday all those unlovely derricks will be gone, and so will the picket fence and the graves" (Sinclair 1926, 548). Such a reflection, call it either the longue durée or deep-time view, of petro-geology and extractivist capitalist development, reminds the reader not only of the inevitability of depletion but also of fiction's speculative prowess and temporal elasticity. The same 'death-in-life' mood is apparent in one of the very first works of North Sea petrofiction, George Mackay Brown's 1972 novel *Greenvoe*, where the accelerated development of petro-modernization on a small island in the Orkney archipelago (in many ways reflecting the rapid extraction strategy of the early years of UK oil policy through narrative compression and exaggeration) is suddenly and almost inexplicably ceased and the island abandoned. The somewhat rarefied final scene is set a decade later where descendants of the empty island's former populace reappear on the land and perform a pre-oil ritual that suggests a form of post-oil sustainable relations.[26]

In fact, the pervasive 'after-oil' sensibility demonstrated in *Greenvoe* remains marked throughout North Sea petroculture through various phases of the life of the industry. We might expect a recent novel concerned with North Sea oil, like Ian Maloney's 2016 *The Waves Burn Bright*, for example, to make explicit comments regarding the certainty of the end of oil. This it does by building its narrative around a generationally fraught father-daughter relationship and their historical investment in the history and future of energy extraction. Both are expert geologists and the latter's lecture in Aberdeen, advocating a future of geothermal energy production in the context of climate change makes a stark point to a gathered audience of students, oil industry representatives, and academics: "The oil industry is in its endgame [...] the age of petroleum is over" (Maloney 2016, 269). Earlier in the novel, the same point is posed in a lecture on "Oil Field Redevelopment" by her father, a petroleum geologist and survivor of *Piper Alpha*. This time, though, it is in the form of a question: "What do you do when all the oil and gas is gone? Do you pack up and fuck off or are there options? These questions would become real in their working lifetimes" (Maloney 2016, 67). On these stark terms, they conjoin and eventually reconcile, though the reaction of others and their future relations are left somewhat hanging.

This sense of the likely end of oil in the North Sea appears again and again, sometimes attached to a marked consciousness of a boom and slump system and endemic industry volatility in price fluctuations and political crises. In Gray's *1982 Janine*, for example, McLeish expresses the view, in a conversation about why Britain failed to nationalize its oil assets, that "most of our oil goes to Americans who pay less for it than we do, and they'll have exhausted the supply within twenty-five years" (Gray 1985, 145). In another example, Jamieson's *Thin Wealth*, a key protagonist is an old crofter, Lowrie, whose way of living is never fully eclipsed by the amazing speed of petro-development in his Shetlands. He views the oil era as exactly that – an era – from its inception, one "utterly manufactured and unnatural [...]. He knew that this pattern was only a temporary stop in the greater flux, and that it would be sharply destroyed once the money stopped flowing" (Jamieson 1986, 130). Such declensionism is partly economic, but it meets with an ever-intensifying environmental argument in later texts, such as Ian Rankin's 1997 crime novel *Black and Blue*, where a murder victim is a worker on a platform due to be decommissioned. This spurs an investigation in oil territory, including a visit to an oil rig itself, where time and again the reader is reminded that "very little gas and oil were left to exploit" (Rankin 1997, 432). The question of undetected agents of exploitation and the "useful life" of oil is wended into a murder plot. The subtext is that the deathly business of the substance, despite its decreptitude, remains criminally profitable and exploitable in more than one sense of the word. The murky tussle between an industry anxious about its decline, torn between what one character describes as "the economics of retrieval" (Rankin 1997, 385) and the possibility of opening another frontier in the West ("Atlantic oil"), proves an effective setting for the whodunnit genre in an era where the economic boom of the mid-1980s had definitively peaked.

As Rankin's detective learns the challenges of decommissioning, he notes that despite the huge impressive effort to extract and pipeline it, "still the oil proved stubborn" (Rankin 1997, 425). That final adjective denotes a certain doubleness between a dogged refusal to give itself up and a refusal to let go; a contradiction of sorts that encapsulates where the North Sea oil industry is in the 21st century.

As the threats posed by the process of global heating have not only promoted the necessity of the demise of fossil fuels everywhere but overtly demanded its acceleration, there should be little surprise at the fact that, in the age of late-late oil, a zombified petro-system has sought to retain some kind of residual and transformative capacity. This comes in and against the compelling scenarios of low-carbon futures made by multiple agencies vested in energy transition and climate stabilization. The *Winner* affair highlighted this contemporary condition, but other significant literary and cultural registrations of cold water oil bear all its attendant elements if we are prepared to read for oil petroculturally.[27] In moving towards my conclusion, I want to examine the specific uses of the speculative genre in revealing some of the conditions late North Sea oil faces.

North Sea revenants

In exposing the long entangled "offshore" industries of financial services and fossil fuels, the tale of the *Winner* also helped demonstrate how the oil industry's expansionist inclinations *and* its intrinsic sense of its finitude is mediated via a marshalling of competing valences of 'realism' and fantasy in both cultural production and political economy; valences that stretch any consideration of how we think and define how the international field of petroculture uses, remixes, refurbishes, and sometimes rejects the realms, genres, and protocols of the fictive, speculative, and the fantastic. Any cultural and environmentally sensitive expression of a low-carbon future, for example, must necessarily extrapolate oil's dangerous entanglement with the reality of warming. It must also elicit the means through which the industry attempts to freight and entrench its own mode of "climate realism" as a bulwark against any radical (some might say apocalyptic, others rational) alternatives. As I have noted above, this it does in part by offshoring, in geographical space and financial instrument, as well as in fragmentation of responsibility and ethical accountability, via split ownership models, regulatory avoidance, data obfuscation, scenario reporting, etc. (Appel, Mason, and Watts, 2015).

But, more significantly for my purposes here, the vested interests of oil attempt representative ownership of a futuristic narrative of a "sustainable" future *with* oil; a future that in climate science and so many post-oil or climate fictions is catastro-realistically projected to be volatile and unpredictable. Here, industry and creative narratives can appear oddly co-constitutive: alarming and alarm*ist*, reliant on fantasy, underpinned by pleas for realism. They compel us to ask what the world of, say, *Mad Max*, notionally viewed as a kind of cautionary tale of the Anthropocene rather than an attractive rendition of a future low-carbon imaginary does for advancing climate activism. But they also ask if post-apocalyptic imaginaries (no doubt unintentionally) might also work for a corporate petro-realism presently advocating oil's second and third life at the same time as advocating a mode of transition from oil to "energy" company. (Would you, they might ask, want to live in a world like *Mad Max*, where a low oil supply creates violent neo-feudalist societies?). The stricken *Winner*, then, is an archetypal tale of salvage, clearance, abandonment, cover-up and return (one so common to Scotland's modern history with resource capitalism); but is also a story which implicates the very future of the oil system itself as is confronts its event horizon. Can we, therefore, view the disaster and its aftermath as a positive spur to imagining a low carbon future?

Contemporary responses to oil in the age of climate change now reflect restless and anxious (often tonally mournful or hectoring) concerns on how to effectively manage decline and best react to the pressures of energy transition (Winchester 2018; Baker and Wood 2019). The era of scaled-up decommissioning, however, remains beset by companies seeking to extract from increasingly diminishing returns. In this context, the very existence of cold water oil proves exemplary terrain for perennial concerns with petroleum's twilight, with its extreme geography betraying both the thirst for the commodity and the drying plays of easy oil. The North Sea, as I have argued, has fretted over its ends since its beginnings. "An asshole of an oil field. Worse than the Canadian East Coast, worse even than fuckin' Alaska" was the considered view of one American roustabout after oil was discovered in 1969 (Harvie 1994, 67). And yet, it happened with great speed, for a number of overlapping factors: from EROI finally coming good to OPEC's rise, the willingness of US oil capital to finance deals, to the designs of Thatcherite neoliberalism and the situation of the British economy in the 1970s (Devine 2000, 585–586; Edgerton 2018; Marriott and Macalister 2021).

I have noted above the lacunae created by oil and government, additionally nervous about constitutional pressures on the British State from nationalist forces. Perpetual shifts between the exposure and concealment of oil and its limits have repeatedly registered in North Sea petroculture. The possibility of new "secret" fields and new explorative and extractive technology has been an accompanying feature of oil's own representational apparatus, but in the last decade, this has become a fixation in aspects of national and local media, particularly around all sides of the constitutional debate in the United Kingdom, tied to the particularities of a future independent Scotland's economic prospects with and without extractable oil fields. The interest in expanded exploration and life extension is a feature in late oil's local and worldly prospects and is, as we might expect, particularly acute in the North Sea.[28] Striking, for example, about the *Winner* accident's reporting was the palpable anxiety around its refloating and removal from the scene. Fears of fuel leaks from the diesel in her tanks were one thing, but as Lamont (2017) notes, the company had a policy of discretion in such night scrappage operations, and here, "the sun was up on a fiasco." A 300 metre land, sea and air exclusion zone was enacted and a drone watch set-up. The rig's beached position was clearly a focus of public embarrassment for the company and offered a languishing symbol of the contemporary debilitations of the North Sea oil industry, in general, now facing existentially difficult questions every day in a heated debate about the projected life of many fields and how to deal with the already defunct platforms and infrastructure that sits corroding in the sea.

This lost and unclaimed rig was just too damn apparent; a highly visible symbolic sore. The stress on getting the vessel back out to sea was the perfect evocation of a desire to return to the structurally limited visibility of petroleum, to avoid further questions about ownership, future direction of travel and ultimately, oil's diminishment.

If a strain of offshore North Sea oil culture sought to deal with making visible the transformative experiential, political, and capital practices of offshore oil, the second wave, while not at all dispensing with that critique, has registered the violent determinations of oil's persistence despite the magnified acknowledgements of its toxic and climatic consequences.[29] Harry Pitt-Scott (2020) has persuasively shown how the tendency towards mystification in offshore practices and infrastructures, from shipping to banking to drilling, has a correlative aesthetic registration in world-literary forms of the mystery and thriller. I am interested here, however, in the recurrence of the speculative and fantasy genres as a means to realize the debate about the end of oil in the North Sea. These of course involve encoded forms of mystery amid elements of the irreal. They combine and entangle orthodox forms of realism with elements of SF, horror, and fantasy. In the final part of this chapter, I want to focus in some detail on a couple of examples.

Michel Faber's novel *Under the Skin* (2000), set on the North East oil coast, is not ostensibly about oil at all, but it must be read in part as a work of petroculture; a road narrative of secretive resource extraction. In this case, the resource is human males drugged and fatted as foie gras for distant aliens living on an ecologically stressed dying planet with shrinking resources and liveable habitats. The principal character, Isserley, is another courageous, skilful, and compromised alien/female crushed by exploitative class and gendered conditions on her home planet, where a toxic atmosphere has spatially split classes into two worlds, above and below surface; pointedly called the subterranean estates. Her job on Earth is to fit herself into a man's world of refining and processing precious resources: working for a powerful alien force secreting away a valuable domestic commodity with their superior technology and expertise. It can't *not* be read as an allegory of North Sea oil.

In Jonathan Glazer's remarkable 2013 film of Faber's novel, there is a resonant scene of littoral distress, set on a beach that looks out on the horizon of the North Sea. Out there lurk drilling platforms and supply ships, going about their business unseen. On the beach, the alienated alien Isserley (played by Scarlett Johannsen) stands watching possible prey: an itinerant diver. The only other people are a family: mother, father, and young baby. Their dog – like Goya's dog – becomes stuck in the swirling surf. The mother chases it, the father follows her. Both drown. The diver, failing to save them, reaches the surf, but is knocked insensible by the alien, who drags him across the beach and off camera, to some undisclosed location. The camera sees it aslant, the action initially stuck on the alien's focalization, surveying and waiting, offering no assistance through the drama of drowning purely to see what can be gained from it. The violent rhythm of the turbulent waves is drawn out, forceful and recursive. Simultaneously energetic and entropic.

North Sea culture, like most maritime culture is full of drownings, abandonments, salvage and trauma in a hostile environment, but I am interested here in how the determination (in both senses of the word) of oil is difficult to exact in this film yet encoded in an extractivist storyline. The alien industry is pressured by diminishing returns and increasing dangers.

The beach scene bears out the indifferent morality, the insensibility to any local attachments amid the necessity and will of accumulation and extraction; the indifferent neglect of the vulnerable is registered in the screaming despair of the abandoned baby, the coming generation. All that matters in this hugely difficult to watch scene is what remains to be dragged from the sea onshore in order to remove it to an altogether more secretive plant for refinement in an exclusive offshore. That which is lost or has no "mature" value is expendable. The act of dragging the commodifiable human onto the beach, to land the unexpected bounty, is shown as a long and arduous effort, in a dangerous environment, for absurdly exceptional (and, provocatively, relatively scarce and unnecessary) surplus value. The film, following the novel, shows the alien worker's own repeated alienation and exploitation, having to go out again and again, placing herself at risk for relatively minimal returns (she is eventually replaced by a copy). The baby's distress makes us worry for its immediate welfare, its likely short-term future, but also its exploitation in and out of the fiction: how can anyone use it that way?

Here, in all its ambivalence, is a rendition of abstracted resource extraction and with it the forceful resolve of energy imperialism: a hidden but awesome force using astonishing, incomprehensible technology, backed up by a mysterious and faceless onshore and off-planet conglomerate, preying on human futures: their very innards harvested, processed and refined into a viscous liquid on an industrial conveyor belt to some vague but starry destination elsewhere, as another notable scene makes apparent. Unsuspecting male victims are lured and suspended in a dark tank of oily liquid until they eventually implode into what we presume is the precious abstract matter.[30]

What Glazer's subjunctive plotline withholds, by non-sequiturs, and its defamiliarized Scottish setting is key to the offshore reading we can make. The viewer has to work hard to see these victims as petro-subjects, furtively enclosed within a larger, darker liquidity, aware only at the last to their inevitable, horrific fate. Such scenes command a form of interpretive dragging, a dynamic positioning, if you will, alert to a world system under pressure from depletion. What both film and novel of *Under the Skin* show is that no matter how arduous or minimal the struggle of dragging the load onshore, of processing and refining it; no matter the risk, it remains worth it. It remains worth it even when that energy expenditure is damaging to those doing the labour of extraction *and* perpetuating a toxic environment elsewhere. There is some confluence here, between classic theories of cognitive estrangement (Suvin 1979) developed in science fiction criticism and the manner in which an estranged offshore world is realized in apparently non-petrolic cultural forms. In this view, a variety of onshore settings require a fluid form of energo-critical cognition.

Returning monsters

The *Winner* haunted the shoreline of Lewis for around a fortnight. Ghosh's sense of oil's embarrassment was writ large. On 23 August, news reports emerged of the installation's successful refloat. An audience of islanders and salvors cheered and applauded, scenes of relief reminiscent of the saving of beached whales or porpoises. Amid these buoyant scenes of relaunching, a detail from Lamont's (2017) report noted the retention of the exclusion zone placed around the rig for any locals interested in viewing it, and a 1-km maritime exclusion limit imposed as it set back out on its ignoble journey to eventual scrap. The shoulders of the rig lurched inwards and seemed impacted by a kind of abject humiliation.

To reiterate: does the petrocultural reading of this stricken rig ultimately offer an alternative salvor's tale? One parsing the official story of successful refloat and rescue? The alternative to that would be a tale that demonstrates contemporary oil's objective containment strategy, requiring minimum fuss in the name of system preservation. It would repoint the semantics of environmental stewardship of human, plant, and animal endangerment and safety, where the securing of the onward journey rights is exclusively parsed by the long reach of petro-PR. For the petrocultural eye though, the *Winner* signified the end(s) of oil. A ghost rig from a drying field broke free its bonds and temporarily hugged the land, as if announcing its vestigial presence; a near corpse leaking out its insides one last time before its final journey. An additional element here was that, in the days the *Winner* sat immobile on the rocks, it was widely reported in the world's press (Press Assoc. 2016; Witte and Aldag 2016) that tidal and wind-powered renewables had managed to utilize those same powerful winds that broke the *Winner*'s bonds to power the equivalent of all of Scotland's electricity needs for the day. This growing renewables generation, much of it offshore, in marine and wind power is slowly changing the valence of that term in summations of the energy culture of the North Sea (Watts 2019). Was this all a symbol, in microcosm, of non-renewability; of the projected and inevitable end of petroculture in lieu of a transition to 'zero carbon' life?

Well, yes and no.

As I have noted, such post-oil fantasies are continually reality-checked by ongoing petro-development. Another less-publicized launch celebration five days prior to the relaunch of the grounded *Winner* proved an apt example. Lady Gretchen H. Watkins, CEO of Maersk Oil, was naming its newest "state-of-the-art" platform, the 400 ft *Maersk Highlander*, acquired from the subsidiary Hercules Offshore, at a ceremony on the North East rig yard at Cromarty Firth, attended by Maersk Oil, BP, JX Nippon, and reps from Jurong Shipyard, Singapore, where the jack-up rig was assembled. The Maersk drilling CEO, Claus V. Hemmingsen, gave a speech emphasizing safety as a key priority and highlighted recent years' significant fleet growth within Maersk Drilling: the acquisition of four deep-water drillships and four ultra-harsh environment jack-ups in response to the increasing demand for new rigs in an ever-evolving industry. "Now the tide has turned," Van Hemmingsen noted "and a new oil reality has arrived. Opportunities can emerge even when facing adverse market conditions" (Maersk Drilling 2016). Here was confirmation, borne out in global data since, that oil bounties remained in some of the most difficult places yet drilled and that they were irrevocably bound to the stock market. Evidently, still, very much worth it.

Such statements reveal the whack-a-mole reality of the contemporary global oil assemblage and its galvanizing "new oil reality." Such petro-reincarnation is, to recall Michael Watts' (2012) point in relation to Nigerian oil, the historical character of the oil frontier in microcosm: recursive and ever-shifting, enclaving, returning like a tidal system. It is, however, increasingly designed *because* of the geophysical and eco-political pressures of oil's eco-social and fiscal volatility, and despite the original irony of its non-renewability, to operate in "harsh environments." There, deep water is the new black, and flags of convenience make regulation easier to avoid, safety conditions negotiable and international labour loopholes apparent.

If there was ever a tale of "the new oil reality" imitating art (or perhaps the other way round, given oil's repetitive and imitative geo-historical form and the world-intersections of its cultural expressions), it is in the *Winner*'s story of scrappage, salvage, and proud rebirth. This is the "new oil reality" in the terms of industry, but we can also see its inverse as the old oil absurdity: the continuation of the unsustainable, the anxious reincarnation of the oil frontier. This is what Jeff Diamanti calls a "terminal landscape" (Diamanti 2021), where the terminal is both material and conceptual, a site of processing and process, between energy and raw commodity, climate and capital. It does and does not signify an end in the shape of a finality, rather a switching point that illuminates the oppositional yet entangled "modal logics of climate and capital" (Diamanti 2021, 6):

> The terminal is both a state of things and a condition in which things relate [...] the concept and logistical infrastructures of the terminal focalize these oppositions between globe and planet, form and formlessness, *in medias res*–that, in other words, worlds are made and unmade at the terminal. (Diamanti 2021, 4)

In this light, the terminal landscape of the *Winner* seems intensely yet logically absurd, to the point of being weird, offering surreal forms of incongruity alongside the general consensus of the need to transition in line with the Paris Accords.[31]

It is with an example of this "weird oil reality" that this chapter concludes. As Haslam demonstrates in detail elsewhere in this volume, the 2011 short story "Covehithe" by the contemporary fiction writer China Miéville offers warped confirmation that the genre of the weird and fantastic, with its unexpected, off-angular figurations of the monstrous and irruptive is, in fact, fantastically refractive registration of the carbon-web's offshore reality.[32] Revenant rigs thought to be long sunk and scuttled, resurface in the world's oceans. They smash into the infrastructure of mainland cities and spawn baby rigs, forming a new world-oil-ecology. Oil's metastasizing nature, the story appears to suggest, will not be sunk without a major and unified – and unprecedented – effort, one that is in many ways intrinsic to the work of petroculture as an academic field.

The Cthulthu-cene ecohorror politics of "Covehithe" is clear: we should be horrified at what Oloff (2017) identifies as "petro-zombies," but also expect the unexpected extrusions of oil's infrastructural tentacles. We should, in fact, be prepared for all forms of "natural" returns from the sunken monsters we have constructed from the ocean's bounty. "Perspective is peculiar," the narrator comments, and we can here see the dovetail with the weird image of the *Winner*.

"These were clumsy steps with which it came," we are told of the fictional *Petrobras* returner, "a few waves-width from the beach it hesitated … [with] strange ungainliness, it moved like a quadruped skiing" (Miéville 2015). This is the *Winner*, in a weird post-oil imaginary: a precarious, oddly incongruous automaton that has stumbled, unexpectedly out of the depths and been dragged by the current onto the beach, as if the sea wants to throw it back onshore, to project its remaining and returning power, proving the obsolescence of this directionless object without steerage, without evident claim or agency, shackled to the rocks and exposing its corporate sponsors, its would-be disassemblers.

FIGURE 4.3 *Transocean Winner*
Source: *Alex Boyd (https://imgur.com/r/pics/miXOBdN)*

It squatted and endangered, but it also became petro-littoral potential. Perhaps, it should have been kept on the site of its grounding as a museum installation, the perfect pedagogical object reflecting our once-we-were carbon-history? Children could have come to witness this impressive ancient beast of the ocean, responsible in so many ways for the social and energetic lifeworlds of their fossil-y parents.

Deckard and Oloff (2020) have identified "the oceanic weird" as a world-literary genre moving across the chronological boundaries of the old to new expressions of the genre, which some commentators (Canavan and Hageman 2016; Shapiro 2016) have described as registering monumental moments of crisis and insecurity in capitalism and society. The Lovecraftian old weird Deckard and Oloff associate with the violence of imperial oil and the contemporary new weird produces novel work, such as the Dominican Rita Indiana's *Tentacle* (2015). We can add important texts like Miéville's story or Nnedi Okorafor's Nigerian novel *Lagoon* (2014). This work is ecologically anticipatory, pointing towards a world of radical difference in the midst of climate's unknown future effects. In Miéville's story, rigs lost in accidents resurface as "revenants." The inference is that petro-infrastructure has taken on some autonomous mega-identity; figured in zoomorphic organic form, as oil infrastructure's socio-natural feedback loops. At one point, they are described as "hydrocarbon Ragnaröks," the Norse mythological prediction of a momentous coming future battle and violent events, covering the world in water. This is now an identifiable scenario in the newly flooded worlds of climate fiction, where watery landscapes are identified as the inherited spaces of a humanity that refused to turn off the oil spigots, to leave undrilled fields alone, refused to leave the carbonizing structures of the oil era as scrap. Worlds such as the future British dystopia of John Lanchester's *The Wall* (2019), where the central protagonists are exiled to a life adrift on the newly unmappable and expanded cold water of the North Sea, which has risen to cover most of the European continent and conjoins with the Atlantic and Mediterranean. Their fate seems certain death until, as the novel concludes, they discover a structure in the middle of the seemingly endless ocean that offers them shelter and supplies. It is an oil rig, from which a ladder extends.

Notes

1 Severin Carrell, "Oil rig runs aground in storm off Western Isles of Scotland." *The Guardian* August 8, 2021 https://www.theguardian.com/uk-news/2016/aug/08/uk-coastguard-and-salvage-experts-draft-plan-to-refloat-oil-rig; Angela Fritz, "This giant oil rig broke free from a tugboat during a storm and ran aground." *The Washington Post*, August 9, 2016. https://www.washingtonpost.com/news/capital-weather-gang/wp/2016/08/09/this-giant-oil-rig-broke-free-from-a-tugboat-during-a-storm-and-ran-aground/; "Una tormenta arrastara una plataforma petrolera de 17.000 tonnas." *El Correo*, Martes, Augusto 9, 2016; https://www.elcorreo.com/bizkaia/internacional/201608/09/tormenta-arrastra-plataforma-

petrolera-20160809144900-rc.html; Catherine Thorbeck, "Massive oil rig washes ashore in remote Scotland." *ABC News,* August 9, 2016. https://abcnews.go.com/International/massive-oil-rig-washes-ashore-remote-scotland/story?id=41246971

2 A year later, on 7 September 2017, the UK Government Marine Accident Investigation Branch published an official investigation report revealing that the rig was registered in the Marshall Islands, well-known as an offshore tax haven. Details can be found at: https://www.gov.uk/maib-reports/grounding-of-semi-submersible-rig-transocean-winner-after-the-loss-of-tow-from-tug-alp-forward

3 The NGO Shipbreaking Platform campaigns for a clean, just and safe recycling system to be initiated globally. It provides informative overviews, tracker systems for individual ships and annual reports on the problem. See: https://shipbreakingplatform.org/

4 I am here shamelessly punning on the type of vessel – a dynamic positioner – used to place rigs and platforms over new wells and onto salvage ships.

5 In this chapter, I am stretching the provenance of this term to incorporate other modes, notably creative non-fiction, film and theatre.

6 My thanks to Ernst Logar, Alex Boyd, and Murdanie Macleod for their generosity in granting permission to reprint their images.

7 This multi-disciplinary field of study centres the oceans and seas of the earth as a primary location for the various work committed under the rubric of environmental humanities in the context of the Anthropocene. Its practitioners take issue in particular with a predominant representation and constitution of oceanic space as an empty or 'free' site; a mode subject to centuries of militarization, colonial conquest, national/corporate territorialization and, increasingly, industrialization. The field seeks to expose and critique the toxic social and environmental effects of so much of these practices. It also seeks and fosters alternative approaches and views on oceanic space and its multifarious organisms and materials, with an added emphasis on non-anthropogenic perspectives.

8 There is a subgenre of worker written accounts, many of which tend to emphasize the difficulty of working conditions. See Sutherland (2004); Shepherd (2015); Mackie (2004). The related themes of heroic adventurism and technological marvel are never that far away from many narratives. Al Alvarez's (1986) account of his time on a platform remains the most notable literary nonfiction rendering. Ian Rankin's famous detective John Rebus takes a chopper ride to a platform, where he investigates a murder case in the novel *Black and Blue* (1997), though Rankin (2008) emphasizes his rendition of conditions on a "paraffin budgie" was second-hand and reliant on valuable accounts by workers. The artist Sue Jane Taylor (2005) has also produced notable work from her time on various platforms. See Taylor's chapter in this volume, and https://suejanetaylor.co.uk/ An exceptional resource giving voice to actual offshore oil workers was the newspaper *Blowout* edited (and with written contributions) by Neil Rothnie, an experienced rig worker and lifelong socialist. The paper began in reaction to the 1988 *Piper Alpha* disaster and the activist formation of the Oil Industry Liaison Committee to co-ordinate union activities. An interview with Rothnie and PDF archive of the issues are available on the People and Nature blog site: https://peopleandnature.wordpress.com/site-contents/the-north-sea-1-the-reaction-to-piper-alpha/

9 Harvie humorously describes the moment when a forlorn search for a middlebrow book about oil is temporarily buoyed by the discovery of Penelope Fitzgerald's 1979 Booker Prize winning novel *Offshore,* only to find a narrative of middle-class life on houseboats on the Thames. Oil does register in a range of British popular music, from Dr Feelgood to Gang of Four to Slovo, as Marriott and Macalister (2021) note.

10 The National Museum of Scotland curated an "Age of Oil" show, featuring Sue Jane Taylor's work in 2017. https://www.nms.ac.uk/national-museum-of-scotland/things-to-see-and-do/past-exhibitions/age-of-oil/. The 2017 acquisition of the Flare

Tip from the decommissioned *Murchison* platform, for a permanent National Museum exhibit (donated by Canadian Natural Resources International (UK) Ltd), exemplifies the contemporary sense of North Sea oil as historical artefact. See Robertson (2017). The art exhibition "Crude", curated by Rachel Grant for Fertile Ground took place at Look Again Project Space, Aberdeen, from 28 August – 12 September, 2021. It took a more artistically adventurous and critical social and ecological approach to Scotland and Aberdeen's relationship with oil and its transformative capacities, weaving together conceptual art works with scholarly material from academic petroculture, public talks on after oil imaginaries and Just Transition, and displays of archival pamphlets from oil workers and activists. Details and some images from the exhibition can be found at: https://www.fertileground.info/crude.html

11 George Mackay Brown's *Greenvoe* (1972) and Robert Alan Jamieson's *Thin Wealth* (1986) relate oil's dramatic transformation of the island archipelagos of Orkney and Shetland, respectively. Both novels blur genres, threading a conventional realism with elements of what might be called modern saga and myth. They are tinged with SF figurations, used to render the sudden appearance of a massive, technologically advanced oil infrastructure swiftly grafted onto a largely rural setting and economy based on agriculture and fishing. A protagonist of *Thin Wealth*, for example, notes the oil terminal of Sullom Voe on Shetland's landscape as "a striking spectacle, vivid and alive in the mirk of the night like a downtown area of some lost city – a myriad lights shine out of the gloom, surrounded on all sides by blankness and above all shines its beacon, the gas flare. [...] it is a gigantic alien, constructed out of nothing natural, by hands which never belonged to faces" (11). Other notable novels located within the genre of SF and fantasy such as Alasdair Gray's totemic *Lanark* (1981) and Michel Faber's *Under the Skin* (2000) register the imperial power and political struggles for energy control and the 'alien' qualities of oil development. A depleted or post-oil world and/or climate changed Scotland form the scenarios of Ronald Wright's *A Scientific Romance* (1999), Ken MacLeod's *The Night Sessions* (2008), Michael Russell's *Lie of the Land* (2015), Jenny Fagan's, *The Sunlight Pilgrims* (2016) and Mary Talbot et al's graphic novel *IDP: 2043* (2014). North Sea oil features in a range of genre fiction from the crime writer Ian Rankin's *Black and Blue* (1997) and Christopher Brookmyre's *One Fine Day in the Middle of the Night* (1999) to popular thrillers such as Hammond Innes's *North Star* (1974) or Richard Douglas's *The Rig* (1975). Petro-financial thrillers also exist from Jeffrey Archer's *Not A Penny More, Not A Penny Less* (1976) to Jane Eddie's *Black Blood* (2020). There has been prominent and much lauded experimental theatre, from John McGrath's *The Cheviot, the Stag and the Black, Black Oil* (1973) to Duncan McLean's *Julie Allardyce* (1999) to Clare Duffy's *Arctic Oil* (2018). Cinema has ranged from Blockbuster trash such as Andrew V. McLagen's *North Sea Hijack* (1980) to Bill Forsyth's acclaimed comedy fantasy *Local Hero* (1983), adapted into musical theatre by Forsyth and David Greig for the Lyceum Theatre in 2019, to the arthouse experimentalism of Lars Von Trier's *Breaking the Waves* (1996) or Erik Skjoldbjaerg's Norwegian conspiracy thriller *Pioneer* (Pionér) (2013). Television series also range from early British dramas such as John Eliot's *The Troubleshooters* (BBC, 7 seasons, 1965–1972), Gerard Glaister's *Oil Strike North* (BBC, 1975), Kieran Prendiville's *Roughnecks* (BBC, 2 seasons, 1994–1995) to Mette M. Bølstad's Norwegian period drama *State of Happiness* (*Lykkeland*) (2 seasons, 2018, 2021). Prominent recent drama with a speculative premise includes the Norwegian series *Occupied (Okkupert)* (three seasons, 2015–ongoing), created by Karianne Lund and Erik Skjoldbjaerg's, based on an idea by the popular crime writer Joe Nesbø. *Rig* 45, by Ola Norén and Roland Ulvselius is another science fiction thriller series set on a North Sea rig (2018–2020). In production in 2021 is *The Rig* a supernatural thriller series commissioned by Amazon directed by John Strickland. In 2018, History Radio ran a four-part podcast drama series, "At the Bottom of the Sea", written by Michael Henrik Wynn, relating an experimental sustainable-living community submersed in the

North Sea, attached to an oil rig, who experience a terrifying alien invasion. Last, are notable work by artists directly concerned with energy transition and the end of oil in the North Sea. Experimental films such as Emily Richardson's *Petrolia* (2005) and Stephen Hurrel's *Dead Reckoning* (2012) and *The Sea, The Sails and the White White Blades* (2016); the photography of Tanja Engleberts http://www.tanjaengelberts.eu/cities-of-desire-1 and the lino-cut prints and etchings of Ade Adesina http://www.adeadesina.com/prints.html.

12 The deep contradiction between a policy of radical cessation and extended exploration and extraction remains a feature in British government actions. This was sharply illuminated in the year when the United Kingdom prepared to host the UNHFCC Cop 26 in Glasgow. In March 2021, *The BBC*'s Roger Harribin reported the Conservative Government's decision to grant new exploratory licenses. "UK seeks to drill more oil and gas from the North Sea." BBC Online, 24 March 2021. https://www.bbc.co.uk/news/science-environment-56503588. During the composition of the final draft of this chapter, a story broke in the London *Times* that Ministers were set to approve the 150MB Cambo field in Shetland: Oliver Wright, "UK prepares to approve oilfield despite COP26 climate conference." *The Times* 23 June 2021. https://www.thetimes.co.uk/article/uk-prepares-to-approve-oilfield-despite-cop26-climate-conference-0wlxjm7tv. Such media stories are a veritable speculative genre in themselves and are strewn over the 21st-century age of depletion. Particularly resonant are those that make headline promises of new and as yet unproven fields ready to be drilled both in the North Sea and in the Atlantic West of Shetland.

13 A decade later, Nigel Lawson, the Thatcher Government's Chancellor of the Exchequer (and arch neoliberal), who was first appointed to her Department of Energy in 1981, would mirror such language in various speeches, declaring, for example, that "Our Task is rather to set a framework which will ensure that the market operates in the energy sector with a minimum of distortion". And "the whole outstanding success of the North Sea has been based on the fact that it is the freest oil province in the world, in which decisions on levels of output are a matter for the companies and not for the government" (Harvie, 300–301). Lawson's government privatized (and disbanded) the British National Oil Corporation (BNOC), finally ending any slim remaining hopes of a potential parallel with Norwegian state oil policy.

14 The most well-known and controversial example of this is the so-called "McCrone Report", a UK Government paper/memo on the tax receipts/balance of payments potential of an independent Scotland gaining a boost from the new oil play in the North Sea. Written by the then civil servant Prof. Gavin McCrone, the report was made available in 2005. Supporters of Scottish independence have claimed that it was suppressed because of its politically explosive potential in British politics. The most quoted excerpt remarks that a petro-sovereign Scottish economy might have had a "chronic surplus to a quite embarrassing degree." See Hassan (2019), Kemp (2012, 569).

15 Two major strikes in 1972 and again in 1974 were seismic events in British politics and society, causing power shortages and shortening the working week. A greater post-war energy mix (especially in oil and nuclear power) had proved a threat to the declining wages and conditions of coal mining, with multiple pit closures across the country, but the insecurity of oil-dependency amid volatile price shocks from events in the Middle East provided the National Union of Mineworkers with an opportunity for (successful) action on their demands. See Gibbs (2021, 44–48).

16 The 'British' in BBC was also crucial here, in accentuating the corporation's nervousness to be seen to be broadcasting any separatist claims for oil, in the context of the rise of the SNP and their iconographic slogan "It's Scotland's Oil!"

17 McLean's play was written in the wake of the *Piper Alpha* disaster, when on 6 July 1988 an Occidental platform in the Piper oilfield exploded 120 miles NE of Aberdeen, killing 167 men. To date, it remains the worst offshore oil catastrophe. *Julie Allardyce* was first performed in the city, where a memorial stands. The disaster and

the generational trauma enacted on survivors and their families is the subject of Iain Maloney's 2016 novel *The Waves Burn Bright*.

18 Echoing Bertolt Brecht's noted aphoristic assertion that "Petroleum resists the five-act form".

19 The refusal of oil companies to make their operations more transparent is encapsulated in Logar's decision to exhibit – with multiple redactions – letters from various offshore oil production and services companies refusing him access to their premises for his "non-public spaces" work.

20 In a wholly unironic and thus revealing report about the short-term thinking in petroculture and the resilient status of oil as a precious if volatile commodity, the BBC Online News report (2016) noted that although two tanks on the *Winner* had been "compromised", happily two remained intact, adding that the stormy conditions ensured that the spilled oil was quickly dispersed and evaporated into the wider ocean, reducing the threat to local wildlife and ecosystems. What petroculture has provided is a means to read such statements from an entirely different approach, especially one wrought by an 'After-Oil' perspective. Never has the satirical Newspaper *The Onion*'s 2010 headline: "Millions of Barrels of Oil Safely Reach Port in Major Environmental Catastrophe" been so salient.

21 Transocean, probably spooked by the increased attention and notoriety garnered by the 2010 Deepwater Horizon disaster and a leak off the Brazilian coast by its *Sedco 706* drill rig in November 2011, opted to use Turkish breakers at Aliaga, rather than opt for the more lucrative prices on offer at less regulated sites in China or South East Asian sites at Chittagong or Gadani. Lamont (2017) notes that the *Winner* passed 18 EU-approved demolition sites on its sea journey from Norway to Turkey, in search of greater profit margins.

22 See LeMenager (2014a, 2014b), who details the Oil Museums of East Texas as sites for memorializing oil. One includes The Ocean Star Offshore Drilling Rig and Museum in Galveston. The notion of the *Winner* being left on the rocks as a pilgrimage point for future sightseers in the post-carbon age was attractive. A museum of the fossil world's passing echoed with a wistful line in the closing section of McGrath's *Cheviot* where tourists come to the Lochs of West Coast Scotland to view the oil rigs in the sound.

23 The historical image-bank of water-borne oil spills resound with gothic horror and the petro-weird. See, for example, the photographic work of Daniel Beltra on Deepwater Horizon: https://www.theguardian.com/environment/gallery/2013/dec/09/spill-daniel-beltra-in-pictures.

24 To date, two North Sea rig catastrophes hold the unfortunate record of having the highest mortality rate for world oil disasters. The *Piper Alpha* platform is detailed in Note 16. The Norwegian platform *Alexander L. Kielland* capsized on 27 March 1980, killing 123 people. The *Winner* struck close memories of the tanker MV Braer, which ran aground on the Southern Shetland Islands in 1993, after losing control during a cyclonic storm. It broke up a week later, dispersing 85,000 tonnes of Gulfaks crude into the sea. The supertanker SS *Torrey Canyon*, registered and chartered between Liberia, the US and Japan ran aground off the Cornish Coast in 1967. It spilled over 100 million litres of Kuwaiti crude, creating an environmental disaster over an area that reached the French coast. The very first British platform to find gas deposits, the *Sea Gem*, a barge converted to a jack-up rig, capsized in 1965. Thirteen people lost their lives as a result.

25 Oil companies through the 2010s repeatedly vaunted exploration despite acknowledging the need for transition. At the time of writing, the latest IAE report acknowledges the exceptional drop in demand in 2020, but presents two dramatically varying future scenarios. The Stated Policies Scenario notes that "there is no definitive peak in oil use overall, as there are continued increases in petrochemicals, trucks and the shipping and aviation sectors" (IEA "Oil" 2021). This is in stark opposition to the Sustainable Development Scenario. The UK Government Oil and Gas Authority,

meanwhile, hedges its bets. Its Decommissioning Strategy begins: "The OGA is committed to achieving the maximum economic extension of field life" (OGA 2021, 3). See also Hanieh (2021), who argues that the imbalance in scholarly focus on oil as fuel/energy tends to occlude the crucial historical and contemporary importance of petrochemicals as the declared future arm of an industry of producers and refiners seeking to maintain and extend the life of oil as the central stock for synthetic materials. "This paradox" Hanieh asserts, "must be fully confronted if we are to move beyond oil" (Hanieh 2021, 51).

26 John Burnside's fabulistic 2017 novel *Havergey*, set on a utopian, ecologically aware island community after an earth scale 'fall' event (and a "Great Epidemic" of 2024) works well as an imagined narrative of the community of *Greenvoe*'s young reclaimers. The inhabitants consciously reject the world of the "The Machine People" (Burnside 2017, 17) who preceded them and precipitated over collapse.

27 The journalist Tabitha Lasley's 2021 auto-fictional memoir *Sea State* relates her destructive relationship with a rig worker while researching a book on offshore oil. The narrative deliberately blurs emotional and infrastructural lines between an oil industry in "increasingly hostile territory", where "the age of easy acquisition is over. Every new prospect has its problems, every windfall its own set of caveats" (39) and the affective narrative of a doomed yet compulsive relationship resonates with LeMenager's (2014a, 11) description of the story of this "unsustainable" and yet "charismatic" energy system as "essentially destructive attachment, bad love". Lasley registers the intersections between the private and public realms of "loving oil" while at the same time recognizing her shipwrecked state from the present. Dealing with the sharp pain of inevitable abandonment in a context where there is a reluctance to relinquish the relationship is the underlying metaphor: "Once, we had loved each other, with the force to wreck a home. As the weeks and months slipped by since, that love had taken on the sliding, treacherous texture of a dream. At the time it was all too real" (Lasley 2021, 193).

28 The literature on this is copious and often sharply contrasting, with reports from Environmental NGOs urging a faster drawdown. The 2019 RSE Report on Scotland's Energy Future emphasized the "difficult trade-offs" that had to be made to meet the Scottish Government's purportedly ambitious Climate Change Act (2008). See, for example, Carr (2017) and Keane (2017). News reports on major new finds, such as the one off Shetland in 2017, are typically given considerable column inches as are stories about so-called 'secret fields' in the West and North. The UK OGA nonetheless continues to grant new licenses to oil and gas and, in 2020, offered 113 license areas over 260 blocks to 65 companies (OGA 2020).

29 For an authoritative series of reports on the historical and contemporary condition of energy, including oil and gas in Scotland in the context of the campaign for Just Transition, see Christopher Silver (2018–2019).

30 The film, like the novel has a clear intertext in the 1954 B Movie, *Devil Girl From Mars* (dir. David MacDonald, UK).

31 Such awkward contradictions are encapsulated in exploration contexts. An OGA podcast from 2019, for example, considered "Revitalising Exploration" in the UK sectors of the North Sea. Involving representatives from major companies and petro-geologists, it is notable for the tone of excitement and anticipation of new finds in both mature basins and unexploited territories. "Value" is the principle driver and only late in the discussion does the host introduce the topic of transition and sustainability. This does not sound like a field ready to wrap-up any time soon. See: https://www.ogauthority.co.uk/news-publications/media-library/podcast-streaming/.

32 The story was initially exclusively published in *The Guardian* in 2011, as part of a series on Oil Fictions. See: https://www.theguardian.com/books/2011/apr/22/china-mieville-covehithe-short-story. See Katie Ritson (2019, Ch.6) for a wider, environmental contextualization of the story in North Sea literature and culture.

"Covehithe" riffs from a long world history of monstrous representations of subsurface carbon; irruptive beasts from the resource-laden depths range from Greek myth to the coal pitheads of Zola's *Germinal* (1885) to Melville's white whale, H.P. Lovecraft's Cthulthu, even Godzilla. The petroleum complex has its archetype in Udo Keppler's famous 1904 octopod cartoon satirizing Standard Oil's tentacular reach across the country's political, financial and industrial institutions. 'The Great Ocean' canto of Pablo Neruda's 1950 *Canto Generale* offers an ecosocialist figuration of that company as a bloodsucking creature operating the "implacable intestines" of their drilling platforms, forming a "scaly system" of oil-imperialism in Latin America; producing a world of "sinister depth; subterranean estates". It is not unironic that the most recent new field opened up in the North Sea in 2017 is named the Kraken oilfield. The subsurface weirdness of extractivist infrastructures is a common trope in science fiction from the abundant energy of the mysterious gas giants of space opera or the unscrupulous energy ocean of Stanislaw Lem's 1961 novel *Solaris*, a novel, like many others in the alien landscape genre that offers a correlative re-reading in terms of the fortunes and processes of cold water oil. In this, the strange ocean, as forbidding as it is alluring to human knowledge and profit, remains a transformative and unknowable phenomenon with seemingly autonomous agency, unwilling to cede its potential riches despite the astonishing efforts of humans to master it.

References

Alvarez, Al. 1986. *Offshore: A North Sea Journey*. London: Hodder & Stoughton.

Appel, Hannah, Arthur Mason and Michael Watts. 2015. "Introduction: Oil Talk." In *Subterranean Estates: Life Worlds of Oil and Gas*, edited by Hannah Appel, Arthur Mason and Michael Watts, 1–26. Ithaca: Cornell University Press.

Baker, Keith and Geoffrey Wood. 2019. "Our Time Is Up". In *The Palgrave Handbook of Managing Fossil Fuels and Energy Transitions*, edited by Geoffrey Wood and Keith Baker, 617–620. Cham, Switzerland: Palgrave Macmillan.

Balkan, Stacey and Swaralipi Nandi, editors. 2021. *Oil Fictions: World Literature and Our Contemporary Petrosphere*. Pennsylvania: Penn State Press.

Barrett, Ross and Daniel Worden, editors. 2014. *Oil Culture*. Minneapolis: Minnesota UP.

Barrios, Elizabeth. 2021. "This Is Not an Oil Novel: Obstacles to Reading Petronarratives in High-energy Cultures." *Textual Practice* 35 (3): 363–78.

Boyer, Dominic and Imre Szeman, editors. 2017. *Energy Humanities: An Anthology*. Baltimore: Johns Hopkins University Press.

Brecht, Bertolt. 1929. "Über Stoffe und Formen." In *Brecht on Theatre: The Development of an Aesthetic*, 30. 1964, edited and translated by John Willett. New York: Hill and Wang.

Brotherstone, Terry. 2012. "A Contribution to the Critique of Post-Imperial British History: North Sea Oil, Scottish Nationalism and Thatcherite Neoliberalism." In *Flammable Societies: Studies on the Socio-economics of Oil and Gas*, edited by John-Andrew McNeish and Owen Logan, 70–97. London: Pluto.

Burnside, John. 2017. *Havergey*. Dorset: Little Toller.

Campbell, Alexandra. 2019. "Extractive Poetics: Marine Energies in Scottish Literature." *Humanities Special Issue: Environment, Ecology, Climate and 'Nature' in 21st Century Scottish Literature* 8 (1), 16. https://doi.org/10.3390/h8010016.

Campbell, Alexandra and Michael Paye. 2020. "Water Enclosure and World-Literature: New Perspectives on Hydro-Power and World-Ecology." *Humanities Special Issue: Blue Humanities* 9 (3), 106. https://doi.org/10.3390/h9030106.

Canavan, Gerry. 2017. "Addiction." In *Fueling Culture: 101 Words for Energy and Environment*, edited by Jennifer Wenzel, Imre Szeman and Patricia Yaeger, 25–27. New York: Fordham University Press.

Canavan, Gerry and Andrew Hageman. 2016. "Special Issue: Global Weirding." *Paradoxa* 28.

Carr, David. 2017. "Beyond the McCrone Report – Changing Scotland's Cultural Dependence on Oil." *Source: Common Weal News*. April 7. https://sourcenews.scot/david-carr-beyond-the-mccrone-report-changing-scotlands-cultural-dependence-on-oil/.

Christophers, Brett. 2020. "How North Sea Oil Shaped Britain's Economy." *Tribune*. December 13. https://tribunemag.co.uk/2020/12/how-north-sea-oil-shaped-britains-economy.

Common Weal. 2014. *All of Us First*. Glasgow: Scottish Left Review Press.

Coutts, Donald, dir. 2021. *Rigs of Nigg*. BBC Scotland Television. https://www.bbc.co.uk/programmes/m000yybp

Cresswell, Jeremy. 2016. "Transocean Winner: Who Really Owns this Wreck?" *Energy Voice*. August 16. https://www.energyvoice.com/opinion/116975/opinion-transocean-winner-really-owns-wreck/.

Deckard, Sharae. 2021. "'This Oil thing Touches Everything': World-Literary Crime Fiction and Fossil Capital." *Études Anglaises* 74: 34–52.

Deckard, Sharae and Kerstin Oloff. 2020. "'The One Who Comes from the Sea': Marine Crisis and the New Oceanic Weird in Rita Indiana's La Mucama de Omicunlé." *Humanities* 9 (3): 86. https://doi.org/10.3390/h9030086.

DeLoughrey, Elizabeth. 2019. "Toward a Critical Ocean Studies for the Anthropocene." *English Language Notes* 57: 21–36.

Devine, T. M. 2000. *The Scottish Nation: 1700–2000*. Harmondsworth: Penguin.

Diamanti, Jeff. 2021. *Climate and Capital in the Age of Petroleum*. London: Bloomsbury.

Edgerton, David. 2018. *The Rise and Fall of the British Nation: A Twentieth-Century History*. Harmondsworth: Penguin.

Faber, Michel. 2000. *Under the Skin*. Edinburgh: Canongate.

Finlay, Richard. 2004. *Modern Scotland: 1914–2000*. Profile: London.

Foley, James and Pete Ramand. 2014. *Yes: The Radical Case for Scottish Independence*. London: Pluto.

Ghosh, Amitav. 1992. "Petrofiction." *The New Republic*. 2 March: 29–33.

Ghosh, Amitav. 2014. "Petrofiction and Petroculture." August 27. https://amitavghosh.com/blog/?p=6441.

Gibbs, Ewan. 2021. *Coal Country: The Meaning and Memory of Deindustrialization in Postwar Scotland*. London: University of London Press.

Glazer, Jonathan, dir. *Under the Skin*. London: Film 4/BFI.

Gray, Alasdair. 1985 [1984]. *1982 Janine*. Harmondsworth: Penguin.

Hahn, Danny. 1986. *Government and North Sea Oil*. NYC: Palgrave MacMillan.

Hanieh, Adam. 2021. "Petrochemical Empire." *New Left Review* 130: 25–51.

Harvie, Christopher. 1994. *Fool's Gold: The Story of North Sea Oil*. Harmondsworth: Penguin.

Hassan, Gerry. 2019. "The Importance of the McCrone Report and Scotland's Future." *Scottish Review*. March 6. https://www.gerryhassan.com/blog/the-importance-of-the-mccrone-report-and-scotlands-future/.

Hitchcock, Peter. 2010. "Oil in an American Imaginary." *New Formations* 69: 81–97.

Huber, Matt. 2013. *Lifeblood: Oil, Freedom, and the Forces of Capital*. Minneapolis: Minnesota UP.

IEA. 2021. *Oil.* https://www.iea.org/fuels-and-technologies/oil. Paris

Jackson, Ben. 2020. *The Case for Scottish Independence: A History of Nationalist Political Thought in Modern Scotland.* Cambridge: Cambridge University Press.

Jamieson, Robert Alan. 1986. *Thin Wealth: A Novel from an Oil Decade.* Edinburgh: Polygon.

Keane, Kevin. 2017. "Oil and Gas Industry Backs New Exploration Licences." *BBC News* Online. March 23. https://www.bbc.co.uk/news/uk-scotland-scotland-business-39357122.

Kemp, Alex. 2012. *The Official History of North Sea Oil and Gas: Vol. I. The Growing Dominance of the State.* London: Routledge.

Kemp, Alex. 2014. *The Official History of North Sea Oil and Gas: Vol. II: Moderating the State's Role.* London: Routledge.

Kneissl, Karin. 2011. "In the Name of Oil." In *Invisible Oil*, edited by Ernst Logar, 11–15. Vienna: Springer-Verlag.

Lamont, Tom. 2017. "Where Oil Rigs Go to Die." *The Guardian.* Tuesday May 2. https://www.theguardian.com/business/2017/may/02/where-oil-rigs-go-to-die.

Lasley, Tabitha. 2021. *Sea State.* London: Fourth Estate.

LeMenager, Stephanie. 2014a. *Living Oil: Petroleum Culture in the American Century.* Oxford: OUP.

LeMenager, Stefanie. 2014b. "Fossil, Fuel: Manifesto for the Post-Oil Museum." In *Oil Culture*, edited by Ross Barret and Daniel Worden, 309–327. Minneapolis: Minnesota UP.

Logar, Ernst. 2011. *Invisible Oil.* Vienna: Springer-Verlag.

Macdonald, Graeme. 2012. "Oil and World Literature." *American Book Review* 33 (3): 7–31.

Macdonald, Graeme. 2015. "Commentary." In John McGrath's *The Cheviot, The Stag and the Black, Black Oil*, edited by Graeme Macdonald, 17–67. London: Methuen.

Mackay Brown. George. 1972. *Greenvoe.* London: Hogarth Press.

Mackie, Bill. 2004. *The Oilmen: The North Sea Tigers.* Edinburgh: Birlinn.

Maersk Drilling. 2016. "Newest Rig Got Its Name." Online Press Release. August 18. https://www.maerskdrilling.com/news-and-media/press-releases/newest-rig-got-its-name.

Maloney, Ian. 2016. *The Waves Burn Bright.* Glasgow: Freight.

Marriott, James and Terry Macalister. 2021. *Crude Britannia: How Oil Shaped a Nation.* London: Pluto.

Maxwell, Jamie. 2020. "Scottish Nationalists' Love of North Sea Oil Sours." *Politico.* https://www.politico.eu/article/scotland-nationalists-rethink-their-love-of-north-sea-oil/.

McGrath, John. 2015 [1973]. *The Cheviot, the Stag and the Black, Black Oil.* London: Bloomsbury Methuen.

McLean, Duncan. 1995. *Julie Allardyce.* In *Made in Scotland: An Anthology of New Scottish Plays*, edited by Ian Brown and Mark Fisher. London: Methuen.

Miéville, China. 2015. "Covehithe." *Three Moments of an Explosion: Stories.* NYC: Del Rey: 299–311.

Mitchell, Timothy. 2011. *Carbon Democracy: Political Power in the Age of Oil.* London: Verso.

Morgan, Edwin. 1996 [1990]. Glasgow Sonnets vi. In *Collected Poems 1949–1987*, 290–291. Manchester: Carcanet.

Naughtie, James and Dave Howard (producer). 2015. *Oil: A Crude History of Britain.* BBC Radio Four.

Nelson, Robin. 2005. "The Television Adaptation of *The Cheviot, the Stag and the Black, Black Oil*." In *Freedom's Pioneer: John McGrath's Work in Theatre, Film and Television*, edited by David Bradby and Susanna Capon, 115–130. Exeter: U of Exeter Press.

Niblett, Michael. 2020. *World Literature and Ecology: The Aesthetics of Commodity Frontiers Frontiers, 1890–1950*. London: Palgrave.

Oil and Gas Authority UK. 2020. "Offer of Awards for the UK's 32nd Offshore Licensing Round." September 3. https://www.ogauthority.co.uk/news-publications/news/2020/offer-of-awards-for-the-uk-s-32nd-offshore-licensing-round/.

———. *Decommissioning Strategy*. https://www.ogauthority.co.uk/media/7538/decommissioning-strategy-may-2021.pdf

Oloff, Kerstin. 2017. "From Sugar to Oil: The Ecology of George A. Romero's *Night of the Living Dead*." *Journal of Postcolonial Writing* 53 (3): 316–328.

O'Neill, Michael. 2004. "Politicising Territorial Identity: Theorems of Change." In *Devolution and British Politics*, edited by Michael O'Neill, 67–89. London: Pearson.

The Onion. 2010. "Millions of Barrels of Oil Safely Reach Port in Major Environmental Catastrophe." August 11.

Pitt-Scott, Harry. 2020. "Offshore Mysteries, Narrative Infrastructure: Oil, Noir, and the World-Ocean." *Humanities Special Issue: Blue Humanities* 9 (3), 71. https://doi.org/10.3390/h9030071.

Polack, Fiona and Danine Farquharson. 2017. "Offshore Rig." In *Fueling Culture: 101 Words for Energy and Environment*, edited by Jennifer Wenzel, Imre Szeman and Patricia Yaeger, 252-254. New York: Fordham University Press.

Press Association. 2016. "Scotland's Wind Turbines Cover all Its Electricity Needs for a Day." *The Guardian*, August 16.

Ramsay, Adam. 2014. *Forty Two Reasons to Support Scottish Independence*. London: Commonwealth Publishing.

Rankin, Ian. 1997. *Black and Blue*. London. Orion.

Rankin, Ian. 2008 [1997]. "Introduction." *Black and Blue*. EBook. London: Orion. https://www.orionbooks.co.uk/titles/ian-rankin/black-and-blue/9781409107644/

Robertson, Hayleigh. 2017. "Blog: Collecting Contemporary Energy." National Museum of Scotland. https://blog.nms.ac.uk/2017/12/14/collecting-contemporary-energy-murchison-flare-tip/.

Robertson, James. 2010. *And the Land Lay Still*. London: Hamish Hamilton.

Royal Society of Edinburgh. June 2019. *RSE Inquiry: Scotland's Energy Future*. https://www.rse.org.uk/inquiries/scotlands-energy-future/.

Ritson, Katie. 2019. *The Shifting Sands of the North Sea Lowlands: Literary and Historical Imaginaries*. London: Routledge.

Shapiro, Stephen. 2016. "The Weird's World-System: The Long Spiral and Literary-Cultural Studies." *Paradoxa* 28: 256–277.

Shaxson, Nicholas. 2011. *Treasure Islands: Tax Havens and the Men Who Stole the World*. London: MacMillan.

Shepherd, Mike. 2015. *Oil Strike North Sea: A First-Hand History of North Sea Oil*. Edinburgh: Luath.

Silver, Christopher. 2018–19. "A Just Transition: From Fossil Fuels to Environmental Justice." *DesmogUK*. 5 parts, Oct 2018–May 2019. https://www.desmog.co.uk/2018/11/07/part-two-city-oil.

Simpson, Mark. 2017. "Lubricity: Smooth Oil's Political Frictions." In *Petrocultures: Oil, Politics, Culture*, edited by Imre Szeman, Adam Carlson and Sheena Wilson, 287–318. Montreal: McGill-Queens University Press.

Sinclair, Upton. 2008 [1926]. *Oil!* Harmondsworth: Penguin.
Sutherland, John. 2004. *The Fields of Dreams*. Victoria, BC: Trafford.
Suvin, Darko. 1979. *Metamorphoses of Science Fiction: On the Poetics and History of a Literary Genre*. Yale: New Haven.
Szeman, Imre. 2012. "Introduction to Focus: Petrofictions." *American Book Review*. March–April: 3.
Szeman, Imre. 2019. *On Petrocultures: Globalization, Culture and Energy*. Morgantown: WVU Press.
Taylor, Sue Jane. 2005. *Oilwork: North Sea Diaries*. Edinburgh: Birlinn.
Urry, John. 2013. *Societies beyond Oil: Oil Dregs and Social Futures*. London: Zed Books.
Urry, John. 2014. *Offshoring*. Cambridge: Polity.
von Weizsäcker, Ernst Ulrich. 2019. "Foreword" In *Invisible Oil*, Ernst Logar, 9. Vienna: Springer-Verlag.
Watts, Laura. 2019. *Energy at the End of the World. An Orkney Islands Saga*. Cambridge, MA: MIT Press.
Watts, Michael. 2012. "A Tale of Two Gulfs: Life, Death, and Dispossession along Two Oil Frontiers." *American Quarterly* 64 (3): 437–467.
Wenzel, Jennifer, Imre Szeman and Patricia Yaeger, editors. 2017. *Fueling Culture: 101 Words for Energy and Environment*. New York: Fordham University Press.
Wilson, Sheena, Adam Carlson and Imre Szeman, editors. 2017. *Petrocultures: Oil, Politics, Culture*. Montreal: McGill-Queens University Press.
Winchester, Dick. 2018. "We Need to Decommission Absolutely Everything as Fast as We Can." *Energy Voice*, November 6, 2018. https://www.energyvoice.com/opinion/185340/we-need-to-decommission-absolutely-everything-as-fast-as-we-can/.
Witte, Griff and Jason Aldag. 2016. "In Scotland, Gusts of Wind Usher in a Quiet Energy Revolution." *The Washington Post*, October 15.
Yaeger, Patricia, et al. 2011. "Editor's Column: Literature in the Ages of Wood, Tallow, Coal, Whale Oil, Gasoline, Atomic Power, and Other Energy Sources." *PMLA* 126, no. 2, March: 305–326.
Yergin, Daniel. 2008 [1991]. *The Prize: The Epic Quest for Oil, Money and Power.* NYC: Free Press.

5

THE CALL OF CTHOILU

The weird subject of cold water oil

Jason Haslam

In his recent work, "New Weird" English author China Miéville both engages and analyses the relations between energy, the Gothic, and exploitation in the North Atlantic, updating earlier Gothic and Weird fiction tropes for our own period of petroculture and climate crisis. Miéville's short story "Covehithe" was first published in a section of *The Guardian* dedicated to "oil stories" (2011)[1] and then in his collection of short stories, *Three Moments of an Explosion* (2015). Not every piece in this large collection deals with environmental concerns, but a significant number do, always refracting contemporary anxieties through a Weird lens, rendering the material and cultural effects of climate change as defamiliarizing images of a world where the "natural order" has become singularly disordered. Melting ice sheets become large icebergs floating, and breaking up, over London, while the build-up of garbage, pollution, and concomitant effects on the environment become a course assignment on a syllabus that asks students to examine the fact that "history is littered with the trash left by time-travelers (Miéville 2015, 153)."[2] "Covehithe" similarly reframes offshore oil extraction in the deadly waters of the North Sea and, primarily, the North Atlantic.

One reading of these tales could follow a relatively traditional path of SF criticism: by defamiliarizing issues of extraction, pollution, and climate change, Miéville offers a "shock of misrecognition," so to speak, that forces one out of complacency. Andrew Hageman, in his own piece on "Covehithe" (published while this collection was in process), offers a strong version of such an analysis, arguing that "a story like Miéville's provides a speculative encounter with petroleum energy infrastructure that through its weird estrangement [...] forces us to confront a vision of the future we might prefer to ignore" (Hageman 2019). But, such a reading can come up against some problems. Specifically, the SF subgenre of "climate fiction" or "cli-fi" (as coined by Dan Bloom) is fraught.[3] It may

DOI: 10.4324/9781003024125-5

be a means of rethinking climate change in generative ways, but still the fantastic element of cli-fi threatens to render its defamiliarization into disconnection, its ethics into entertainment that glosses real problems. As Andrew Kalaidjian deftly argues, "[w]hile environmental efforts can reach a larger audience than ever before, they are not immune to the paralyzing effects of spectacle"; "environmentalism," he continues, can be "complicit with global capital" (Kalaidjian 2017, 20), and the often spectacular fantasies of SF and cli-fi are perhaps singularly vulnerable to this problem.[4] Emanuel Herold, Sina Farzin, and Susan M. Gaines frame this danger differently, writing of the conclusions of their 2016 "Fiction Meets Science" workshop that "dystopian and apocalyptic narratives could generate political paralysis" and that "despite some activists' claims, climate-change novels have no instrumental or prescriptive value"; however, they still defend the form, concluding that such fictions do "create what Johns-Putra called an 'affective space' in which the abstractions of complex scientific discourses are opened to critical thinking without the oversimplification and polarization typical of other forms of mass media" (Herold et al. 2016, 611). Their concluding suspicion of "other forms of mass media," however, leaves hanging a spectre of a political failure due to the negative influence of spectacle upon the audience: the *spectacles* of climate fiction risk interrupting the *action* needed to mitigate the effects of climate change itself.

This contentious issue, one that haunts especially non-realist fiction, is one of Miéville's sites of analysis in "Covehithe." In addition to portraying offshore oil extraction as a destructive monster, the story also discusses the efficacy and ethics of rendering such events as entertainment, be it on the news or in Weird fiction. What does it mean when offshore oil disasters become spectacle, rendered as objects to be observed for pleasure ("scopophoilia," if you will)[5] rather than terrors that demand physical and immediate reaction? More philosophically, the story poses ontological questions, querying the relation between fictional objects and material reality, asking if it is humanity itself that needs to be displaced, rendered as pure object rather than active subject. Taken to its extreme, the story asks if humanity is capable of understanding our species' own inevitable extinction as a reality, or if we can only imagine it as fantastic entertainment? And if the latter, can those fantasies still be put to use within a social framework in a manner that allows us to debate how best to ethically live in this world? Miéville's story argues yes, that to reframe our lived relation to the objective world – a world seemingly constantly mediated by our subjective imaginations – we must first fantasize about humanity's extinction, or vulnerability thereto. Gothic fantasy and science fiction, wedded together in the Weird, allow Miéville to dislocate humanity from its too easy assumption of its own primacy in the world, presenting a world in which humans are meaningless, while other objects are presented as exceeding their relationship to us.

As such, this chapter glances towards, but ultimately moves away from, object-oriented ontology and new materialism, instead engaging some of the debates over the ethics of these methods. Such approaches, including work by, for example, Ray Brassier and Timothy Morton (both of whom are discussed below), tend to share a vision of a reality that exceeds or stands apart from human understandings thereof. While Miéville's story does ascribe an apparent agency to the inorganic world and plays with the idea of humanity as object rather than agent, it also suggests – albeit, perhaps, against the grain – that it shouldn't end there: a recognition of our own extinction can be just as narcissistic as its denial. Instead, this reading would posit that the true goal should be to reform ethical relations to what is so easily referred to as "our surroundings" and should include a radical rearticulation of who "we" are. The privileging of certain subjects and their perspective, as Mulvey articulates so well, risks the category of "humanity" being reserved only for a few privileged subjects, and risks that privilege being used to bait-and-switch the "cause" of global warming to the transcendent subject of "humanity," which, in turn, makes necessary local actions within particular contexts seem inadequate or impossible. Miéville's story allows for the disassembling of both anthropocentrism and of the centrality of particular forms of the "anthro-," including those within ecocritical discussions of the "Anthropocene."[6] This is a debate centred in the history of the North Atlantic, where dominant epistemologies are framed and maintained through colonialism, slavery, capitalism, and the exploitations embedded therein.[7]

Generic energies

The historical and formal relationships between the Gothic, science fiction, fantasy, the Weird, and other subgeneric terms are as convoluted as the categories themselves. The study of science fiction's role, specifically, in analysing and offering alternatives to contemporary energy cultures (if not sources of energy themselves) is long standing in the albeit short history of energy humanities studies. As Graeme Macdonald argues in his foundational work, science fiction, in particular, "allows us not only to realize the nature of our long and ongoing addiction to petroleum but also the manner in which we have continually sought to occlude or sublimate the monstrous nature of our petromodern fantasy in order to drive an increasingly unsustainable petrolic life ever onwards" (Macdonald 2014). Citing Imre Szeman, Macdonald reads "SF as the genre that might be best placed to 'shake us out of our faith in surplus,'" even as he notes that Szeman "admits that this is not guaranteed" (n.p). Hageman expands on this argument, noting that one can "extend [an] infrastructural reading approach to speculative fiction – science fiction, fantasy, the weird, and other genres that feature futuristic and/or non-realist narrative elements – a literary realm that foregrounds infrastructural objects and systems among the techno-scientific developments" (Hageman 2019). Indeed, the history of SF is entangled with that of the Gothic, at least from Mary Shelley forward, and just as contemporary SF deals with our current petrocultural moment, so too can energy be said to be a primary site of contestation for the Gothic *per se*. From *Frankenstein* to *Dracula* and their various monstrous offsprings, energy, and its relation to social power, is a constant site of discursive struggle in the Gothic. Further, at moments – like now – when energy regimes seem in flux, the Gothic can effectively show how that struggle becomes one of signification itself.[8] The processes involved in a society moving – however partially – from wind or gas to electricity, from coal to petroleum to green energies, spark such vicious debates not solely because of capital, but because entire systems of meaning, at the level of the symbolic, have allied themselves to those energy regimes.[9]

Weird fiction becomes a way of both joining and troubling the relationships between science fiction and the gothic, on the one hand, and the relations between these non-realist forms and energy cultures on the other. Discussing the development of "science fiction" and "weird fiction" in the early 20th-century pulp market, Roger Luckhurst notes how they both claim similar Victorian roots: Edgar Allan Poe, Jules Verne, and H. G. Wells (Luckhurst 2017, 1042). However, he continues, the Weird "was not science fiction [...] and never became so. It was also not just a continuation of the late Victorian Gothic revival" (Luckhurst 2017, 1043). Luckhurst argues that while the Weird relies heavily on gothic tropes and lurches toward the science-fictional fixation on the alien other, it arguably goes beyond both genres in addressing a form of radical alterity. The Weird, he writes,

> is not reducible to the Gothic's economy of the uncanny or its compulsion to repeat. For Freud, the *unheimlich* was coiled inside the *Heimlich*, making the uncanny "that class of the frightening which leads back to what is known of old and long familiar." The uncanny is a series of displacements that always leads back to the ultimate familiar home: the womb. The interpretive machinery of the uncanny thus inherently domesticates. In contrast, the monstrous breaches of the weird do not return us to something familiar but repressed, but instead veers away to invoke a dread that is irreducible, that cannot be reductively interpreted, translated or returned. (Luckhurst 2017, 1052)

This figure of a radical other that is untameable by reality, even as it reshapes what we consider to be "real," makes the Weird an ideal form for the analysis of cold-water energy culture and its epistemologies. For Luckhurst, this is an explicitly queer. Building on Timothy Morton's etymological vision of the weird as a "twist" on so-called normalcy that allows for new perspectives, Luckhurst argues that the "ability to sketch out a wholly other trajectory reconfirms that the weird is a labyrinthine mode rather than an easily definable genre. The weird can veer in the sense of exemplifying a *queer errancy*," which Luckhurst defines, by quoting Anne E. Duggan, as "a 'literal and figurative errancy, which allows [characters] to move in and out of closed spaces and ways of thinking'" (Luckhurst 2017, 1051–1052; Duggan 2013, 93). Dominant forms of energy, after all, weird our world in both material and psychic ways, rendering themselves invisible as ideology even as they power our very material and intellectual beings.

Energy is weird, but the Weird can pull apart the too-mundane, straight acceptance of energy systems.[10]

Rigging the spectacular

"Covehithe" imagines a world – and specifically a North Atlantic world – made weird by oil. More specifically, it imagines that world ruined by abandoned and wrecked offshore oil rigs – rigs that after their sinking gain a form of undead life and wander the oceans, occasionally coming ashore to trample all in their path. With traces of *Godzilla* and other *kaiju*,[11] *The X-Files*, *Pacific Rim*, and *Transformers*, Miéville's short story is, however, more than a petrocultural updating of H. P. Lovecraft's works, which (to varying degrees) lie underneath those more recent Weird tales. In imagining our future relation to a world destroyed by petroculture, Miéville is able to rearticulate human relations with the surrounding world and with each other.

The story follows a man named Dughan and his unnamed daughter; Dughan was one of the only survivors of early military attempts to board the animated wrecks of petroculture, and he is clearly traumatized by the event. His daughter, meanwhile, is both protector of the traumatized father and a figure of a largely disaffected future, unclear on the importance of the oil-rig apocalypse, misrecognizing its being and significance as spectacle.

But lurking behind these characters and the fantastic figures of the rigs is a more realistic representation of the environment, one that goes almost unnoticed in the face of the spectacular wrecks. Most of these descriptions focus on the small seaside hamlet of Covehithe, in Suffolk, on the North Sea: "The air smelt"; "the smell should have been sappy and muddy of the sea," indicating it smelt of something else; and "the soil felt greasy, heavy, as if someone had poured cream onto loam" (Miéville 2015, 300–301). Elsewhere, Miéville relies on the actual landscape of the region: there is "A clutch of dead trees [that] jutted from the scrub like bleached markers of where Covehithe ended," with "barkless branches" (Miéville 2015, 307, 308); he describes an extant road that "stopped abruptly, rag-edged, fell into nothing [...] 'right off the cliff'" because "The sea's taking it all back." When the daughter asks her father if it is "still going [...] Being eaten?" he shrugs (Miéville 2015, 301).

All of these descriptions are secondary to the plot and, in some ways, are secondary to the narration itself. Written in the third-person narrative voice, the text is heavily focalized through the central characters and especially Dughan. As such, these descriptions are only ever background information, secondary to the characters' central quest to find the rigs. In this way, the ruined environmental world exists as an unquestioned ground of human experience, even as that ground is literally falling out from under them. A pre-petro history, and a pre-petro environment, is non-existent, even in memory. No one questions the existence of a dead landscape.

But *is* it dead? While the prevalence of a "greasy," oil-ridden, smelly landscape would seemingly signify a polluted lifelessness, instead what is being described here is a human inability to render the landscape and the landscape's primary humanlessness (Miéville 2015, 301). The land may be greasy, but it is "like milk poured onto loam," images of fertility piled on fertility, while the sea is presented as an agent, eating the human-made section of the landscape and taking it back for nature (Miéville 2015, 301). Likewise, while the small "clutch of trees is dead," they mark the end of the human habitation of "Covehithe" and sit on a "cold marsh, a roosting place for birds" (Miéville 2015, 307).[12]

This landscape, then, "serves figuratively [...] as a liminal space to frame the story" (Ritson 2019, 133); Covehithe seems poised between a petroleum-fuelled end and new life, a juxtaposition that finds its culmination in the vivified oil rigs. Into the marsh marches the remains of *Petrobras 36*, which sank in 2001[13]:

> It closed on her and she stopped breathing. It came near enough that she felt the envelope of cold air it brought, smelled abyssal rot and chemical cracking. Spray hit her. The weary factory's spray. It giant-walked by her hiding place sending all those Suffolk birds away, hauled into the fens to squat like a monument that had always been there. (Miéville 2015, 307)

Again, one can read this image as one of the human exploitation of offshore petroleum destroying the more naturally formed environment around it, its "spray" "sending all those [...] birds away" in an image of extinction that counters the fertility of the natural landscape offered before. In this reading, it is the human perspective, the mediation of human limitation, that misreads the "greasy" landscape earlier, while the oil rig becomes not so much a metaphor as a blunt synecdoche for the return of the repressed effects of human petroculture on the planet. As Katie Ritson nicely articulates this reading, the story "plays on a latent fear that oil extractivism has created monsters beyond human control" (Ritson 2019, 132), a connection that, as Ritson notes, Ursula K. Le Guin also noted in her review of Miéville's collection (Ritson 2019, 134). It is, in the terms of the story, one of the "petrospectral presences" that haunt our present. In this reading, the image of father and "daughter hug[ing] the trunks" of trees as they "watch[] the oil rig sway in their direction" (Miéville 2015, 304) is an apocalyptic image of the inevitable end as petroculture turns against those using it in ways even literal tree huggers will be unable to stop: think an updating of the nuclear blast scenes in *The Day After* (1983) or *Terminator 2* (1991) or maybe more in keeping with the tone, *Dr. Strangelove* (1964).

But it's not so simple. The oil rig itself shares the figures of fertility with the landscape. After the weirdly undead rigs come to shore, their drills shoot into the ground, the rigs remain motionless for some time, and then return to the ocean. Researchers, we learn,

> expressed as much shock as the rest of the world when, over a year after the *Ocean Ranger*'s visit, up from the still-recovering Newfoundland ground

> into which it had pushed its drill, the first clutch of newly-hatched oil rigs had unburied themselves. They had emerged into the night, shaking off earth. Stood quivering on stiffening metal or cement legs. Tilted tiny helipads. Tottered finally for the sea. (Miéville 2015, 309)

This newly formed uncanny life is inexplicable to humanity: people can't raise any baby rigs in captivity, while the wild births are nearly impossible to stop: "Only the most violent post-return decommissioning," we are told, "could stop all this, only second deaths, from which the rigs did not come back again" (Miéville 2015, 310). Again, one could read simple allegory here: without a transformation of social infrastructure, petroleum dependency will reassert itself. It's not enough to decommission just individual rigs, or pipelines, or what-have-you: petroculture is a rhizomatic, fractal entity that will continually resurrect.

But I want to return to Kalaidjian's suspicion of spectacle (one that echoes the worries of Teresa Goddu and Fredric Jameson – worries, indeed, that go back to Frederick Engels), concerning the efficacy of the gothic, science fiction, utopian fiction, and other non-realist modes in relation to political action.[14] Extrapolating from these concerns, it is possible that the disconnect from material action and political thought renders this Weird cli-fi into entertaining escape from, rather than critical analysis or intervention into, petroculture. The metaphor here is so fantastic as to venture into the absurd, and the metaphorical target – petroculture "itself" – threatens to remain hidden behind the spectacle.

But, the story analyses this conclusion: people arrive from nearby hotels, in tour buses and with a guide, to be entertained by the spectacle of the arrival of *Petrobras 36*. A guard who finds the main characters and escorts them back from the danger zone around the rig's nest also tells the daughter about a kid's club dedicated to the rig. When she asks what the club is like, he responds: "Stuff like, you can see the eggs on a live feed [...]. Sometimes you can even see movement through the shells. And there's colouring books and games and that" (Miéville 2015, 310). Likewise, upon arriving at the guards' checkpoint, they find "the few tourists lucky to have been nearby, at accredited hotels, when Petrobras's heavy steps had registered on the scanners"; a guide accompanying the tourists "was in the middle of a spiel. 'We'll come back in the morning, when it's finished laying,' she said. 'You can bring your cameras then – no danger then if you forgot to turn your flash off.' People laughed" (Miéville 2015, 311). And, finally, the daughter, close enough to the real trauma her father faced on the wrecked rigs at ground zero of the petro-apocalypse, finds herself upset, but only at a story of a dog stomped by the oil rig. As if echoing the concerns that popular culture can strive to awaken us to ecological crisis but only ever fail as entertainment, she asks, "'Do you think it's true what [the guard] said? [...] About the dog? That's horrible.' She made a face. [...] 'Maybe he was lying to scare us'" (Miéville 2015, 311). In the face of the ecological destruction caused by a reliance on petroleum that continually reproduces itself, stories of its danger and images of its reproductive destruction can only be slotted into pre-existing genres that serve the sole purpose to delight and entertain, and so cli-fi serves to perpetuate, rather than challenge.

Entertaining extinction

But that is not what I set out to argue. After all, not all images are necessarily spectacle, and – according to Jacques Rancière, at least – not all spectacle is necessarily mind-numbing. Fiona Polack makes this very point in discussing images of the actual *Ocean Ranger* and other offshore oil rig disasters. She argues that the general dearth of images of working rigs reinforces "corporations' proclivity for concealment," which can serve to perpetuate the dominant "energy unconscious" (Polack 2019, 154). Drawing a distinction between corporate and worker images, however, as well as between corporate and disaster images, Polack argues while workers' photographs "do not inevitably undermine the visual grammar of corporate photography [...] they can have the potential to be powerfully subversive" (Polack 2019, 153–154). The corporations strive, she notes, to "control the visual field surrounding offshore oil," especially at moments when spectacle threatens to undermine their carefully controlled invisibility:

> When spectacular catastrophes – blowouts, fires, and large-scale spills – occur, different kinds of images come into circulation. These typically include but also extend beyond now iconic photographs of oil-soaked wildlife. However, such expansions of the visual realm can still meet considerable corporate resistance. (Polack 2019, 155)

Likewise, following oil-rig disasters, even the corporate images undergo a semantic transformation: "After the Ocean Ranger's demise," she writes, "especially as the extent of the oil industry's negligence emerged, the affective impact of ODECO's key promotional image shifted radically. Instead of conjuring optimism, it became associated with grief. In the immediate aftermath of the catastrophe the shot acquired a kind of obscenity" (Polack 2019, 160). Image and spectacle are not monologic units, always overpowering viewers' ability for critical thought; instead, they exist in a dialogic tension with their material and affective contexts. As Rancière more poetically puts it, "spectators see, feel and understand something in as much as they compose their own poem, as, in their way, do actors or playwrights, directors, dancers or performers" (Rancière 2009, 13).

SF and the Weird are also inherently spectacular genres, often focusing on the visual recognition of a radical alterity, either the "sense of wonder" created by the technological or other novum of science fiction or the beyond-uncanny disturbing visions of the Weird. Because of this generic kernel, these forms have often been read as escapist. Following the argument of the previous section, then, Miéville may be critiquing this particular escapist strain of the Weird, parodying the reinforcement of the status quo offered by such authors as Lovecraft.

However, while a self-reflexive denunciation of his own chosen genre might well be possible, Miéville has too often written about the political nature of SF, and how new forms can challenge sedimented ideologies to end the story there. In a lecture at the University of Kansas, Miéville argues that the "shared estrangement" of science fiction and fantasy is defined by a "fundamental logic of otherness, of alterity" (Miéville 2009). To build on Polack's notion of the "obscenity" petro-images can develop in varying contexts, it's important to recall that, for Luckhurst, the Weird is "a mode that offers a formal rendition of perversity" in a queer sense, "twisting" the heteronormative order of things (Luckhurst, 2017, 1051) in order to pry open previously closed and closeted knowledges. The sense-dulling aspect of the spectacular can be twisted by both context and genre, offering an "obscene" transformation away from the normative.

While "Covehithe" clearly does demonstrate the ways in which fantasy can serve as mere escapism, still a radical alterity is present in the central descriptions of the oil rigs. The melding of oil rig and gothic oceanic monstrosity allows for a new understanding of the agential elements of environmental destruction. Specifically, the oil rigs as sentient creatures embody the central terrifying otherness of death, in a rewriting of Lovecraft's Cthulhu mythos.[15] In being presented as an other beyond our control and comprehension, an agency that "is indifferent to our existence" to quote Brassier (2007, xi), these sentient oil rigs represent the knowledge of human extinction, of humanity's "objectness" in a land of other objects beyond its supposedly agential control.

The beginnings of this recognition of our extinction take the form of an allusion to another fantasy text on the subject. The rigs themselves are presented in terms that many science fiction fans would recognize:

> On an autumn evening in the earlyish years of the 21st Century, a fishing boat south-east of Halifax radioed an SOS, under attack – the transmission was unclear. Rescuers found two traumatised survivors in a wash of scattered debris. As they did, the cause of the catastrophe reached the coast. Authorities could not suppress civilian footage of what had come back. It was the Rowan Gorilla I. That was the first. No Piper Alpha, no Deepwater Horizon; an undistinguished disaster. A tripod jack-up rig lost to storms and hull-fracture in 1988, on its way to the North Sea. Scattered surely by its capsizing and by thirty years below but there, back. Cramped-looking for all of its enormity, latticed legs braced halfway through its platform, jutting above it and below into the sea. In the videos the three skyward leg-halves switch and lean creaking towards each other, sway away again like cranes triple-knitting, as it walks the muck on spudcan feet. It staggered like a crippled Martian out of the water and onto Canada. (Miéville 2015, 304)

As the final line makes clear, the oil rigs are less Godzilla or Cthulhu than they are H. G. Wells' Martians from *War of the Worlds*, a title that invokes thoughts of human eradication as well as narratives of people taking fantasy very seriously.

But this is also a tale of uncontrollable spectacle. Just as corporations can lose control of the image, in Polack's argument, so too are authorities here unable to "supress civilian footage." This is an alterity that terrifies, now in the frame of petroculture run rampant.

The allusion to extinction in the rewriting of classic SF is redoubled in a rewriting of Romanticism, where Miéville's radical decentring of the "human" takes on a particular epistemological flavour. Like William Gibson's oft-cited "new-romantic" *Neuromancer*, Miéville's story highlights, in places, that humans have always been merely "meat" in this universe, or at most one kind of object among others, despite pretensions otherwise. While the living rigs obviously fall charge to simple personification, Miéville explicitly attempts to disallow this reading:

> They laid eggs, so, many people said, they must have sex. There was no logic there. They were oil rigs. Dughan thought the belief exoneration of the strange prurience that endlessly turned on monoliths rutting miles down. An inhuman pornography of great slams and grinding, horrified whales veering from where one rig mounted another, warmed by hydro-thermal vents. (Miéville 2015, 310–311)

Personification here – and fantasy – are both denied in a weird literary realism: these monstrous beings cannot have sex, they are just oil rigs, and imagining anything else is just giving into perversity: an "inhuman pornography" generated by the "strange prurience" of human fantasy.

But what is it that characterizes this *weird* perversity? For the main character, it is not simply the prurience of it; it is instead the positioning of human experience and perspective as the centre of the universe. The "exonerating" fantasy here is not Miéville's; it is the literary realist's implicit demand that the world itself mimic the vicissitudes of humanity's limited representation of it. Just before they see P-36, in the middle of the observations about the milky loam and odd smell, we are told that "Perspective was peculiar" (Miéville 2015, 301). Coming just before the focus of the story shifts to the living rigs, we have here Luckhurst's perverse twist, a skew in perspective, away from articulated expectations.

This skew continues in the representation of the constant failing of human creations within the landscape. In addition to the road cut off by erosion – a clear image of the ending of a human-focused petro-future – the story also includes several descriptions of St. Andrew's Church, a well-known historical site in Covehithe, consisting of a medieval ruin that surrounds a smaller, 17th-century church that is still in use, on the edges of which, of course, sits a graveyard. St. Andrew's is a quintessential ruin. If, for Romantic writers, ruins can generally both invoke the history of human relations and power struggles, they do so only obscurely, offering a sublime understanding of human endeavour within the physical and spiritual world that transcends the trappings of a mere moment, either past or present. As Nick Groom writes, certain ruins and graveyards become sites for "meditations – the meditation being a form of restoration through hallucination" (Groom 2008, 43); that is, ruins, and graveyards, as symbols of both human will and destruction, time's march and its end in death, are sites of tense opposition that can give way to transcendent revelation and so spiritual renewal. Such is, it seems, the St. Andrew's ruin when first introduced: "He stopped her by the roofless ruin of the church, pointed, heard her gasp. She stared while moonlight got past the clouds to the holed and broken walls, onto a low newer church inside the nave of the old. He smiled" (Miéville 2015, 300–301). The contradiction of old and new churches folded together, death and rebirth, leads to this gasp of the recognition of the sublime. This transcendent moment translates into a misrecognition of the site's importance: "When eventually she was done looking they continued through the graveyard," the text reads, continuing, "There was nothing at all frightening about the graves" (Miéville 2015, 301). Ruins no longer signal death, but instead allow the observer to feel as if they transcend the merely physical. Ruins, like a naïve Romantic vision of nature, lead to a sense of personal growth and transcendence that is nonetheless still centred in the individual's spirit or mind.

However, it is after the ruins, and as they get closer to "the waves," that, in the daughter's words, "the land" begins to feel "misbehavicious" (Miéville 2015, 301). Misbehaving land – that is, land that does not subscribe to the control of human expectations – becomes vicious in this portmanteau. And it is as objects outside of human understanding that the rigs take on their contradictory significance. After the rigs first started moving, Dughan remembers seeing the first images of this lack of control: "the Ocean Ranger off the Labrador coast, the tip of its tower a dorsal fin. The stomach-dropping video was proof for which no one had known they were waiting: that below the waves the rigs also walked" (Miéville 2015, 306). The vastness of space under the cold ocean waves – as in Hemingway's formulation of literature – becomes a repository for meanings and actions we cannot control or comprehend. The world is no longer human; even human creations deny "our" centrality.

Conclusion: the end of the world (as "we" know it)

But this decentring of humanity can also be read as a decentring of particular definitions of humanity. While I am theoretically inclined towards a nihilistic approach, the generally object-oriented ontological argument that I have been invoking here has been accused of misrecognizing the centrality of, for example, colonialism, Eurocentrism, and other historical forms of oppression in framing who is, and is not, figured as "human" within dominant epistemologies. Andrew Strombeck phrases this succinctly, writing that "by positing a vast, unknowable material realm, in which humans participate minimally, the ontological turn obscures historical hierarchies relating to race, poverty, and gender, even as climate change intensifies the pain caused by such hierarchies" (Strombeck 2019, 347–348). In this regard, theoretically toying with extinction as an essentially freeing, anti-petrocultural concept risks ignoring the ways in which petroleum-fuelled climate change threatens underprivileged and oppressed groups with the realities of physical and cultural annihilation that are anything but freeing.[16] As Kyla Wazana Tompkins has argued:

> A non-human centered ontology and ethics; a sense of the biological and non-biological world as vital and alive; an idea of the body as having a life and conversation of its own, with itself; and, most centrally and crucially, the idea that planetary life should, must be, and will be at the determinative center of political world-making: these are epistemologies and ontologies that can hardly be said to have recently been invented but rather are familiar to, among others, First Nations and Indigenous peoples; to those humans who have never been quite human enough as explored, for instance, in post-colonial and revolutionary black thought; to some strands of feminist thinking, for instance, de Beauvoir's thinking about the objecthood of women; and to other non-Western medical and spiritual modalities. (Tompkins 2016)

Responding to Tompkins directly, Chad Shomura also notes that "In this way, new materialism suppresses different lived experiences of power to ontology, neglects the insights of feminist and queer theory as well as indigenous cosmologies, and stumbles when it comes to race" (Shomura 2017). Such lacunae within the "new materialism," echoing, as they do, lacunae in other Eurocentric philosophies, risk reproducing definitions of "human" that exclude some, reforming a white privilege that ultimately undoes the new materialists' own logic and their ecocritical goals.

While I have been framing Miéville's object-oriented critique of the Anthropocene in this ideologically suspect "universal" way, the story does offer entry points into a more nuanced decentring of the human as instead a decentring of the history of colonialism, capitalist exploitation, and Eurocentrism that lies behind dominant epistemological definitions of "the human."[17] As all of the examples above make clear, Miéville's critique is centred in a European and explicitly British, literary and cultural tradition. It is English Romantic poetry, and the transcendental epistemology lying behind it, that is failing the world, not "humanity" as such. The story further makes this clear by demonstrating the global political inequities of petroculture. While the initial attacks occur in Canada, the global oceans are peppered with revivified rigs. As part of the "UN Platform Event Repulsion Unit," or "UNPERU," the main character attempts to "decommission" rigs around the planet:

> After the semi-sub Sea Quest retreated under heavy attack, descending back into the Gulf of Guinea, UNPERU turned its attention and resources to the Ocean Ranger, stalking the Atlantic seabed. So when, shortly after its brief first appearance, the Sea Quest reemerged, and continued its interrupted journey into the oil-fouled Delta, they were not there to intercept.

> Word reached Dughan and his crew en route to Canada. They came back fast, turning their plane in mid-air.
>
> They were escorted inland by ex-M.E.N.D. guerrillas with peerless local savvy, hastily pardoned by the Nigerian government. They followed the oil rig's mashed-up trail, the rainbow-filmed liquid spoor, the tripod crater prints. In retrospect, certain qualities of the disturbed interior foreshadowed more dramatic instabilities that later petrospectral presences would bring. (Miéville 2015, 305–306)

M.E.N.D. is the acronym of the real-world "Movement for the Emancipation of the Niger Delta," whose primary mission is the cessation of oil drilling in the Niger Delta and the exploitation of local people and decimation of the environment that both accompany that extraction. Where they are currently criminalized, in Miéville's world where extractive violence is given its own agency, separate from that of neo-colonial and capitalist power, they can be recognized as necessary to the survival of the state and planet. Thus, Miéville's choice to make Nigeria a site of a rhizomatic birth of a baby oil-rig and to triangulate that event in a military-industrial framework between the UK, Western Africa, and Canada is on point. The story reminds us that the North Atlantic is not just the site of the so-called transcendent Enlightenment but also the site of the transatlantic slave trade, a mobile system of the extractive exploitation of humans and material that continues to effect current forms of extraction and exploitation. An object-oriented ontological reading could miss the singular importance of this text being set in the North Atlantic. Extractive petroculture taking on a life of its own cannot happen outside of those historical abuses; likewise effective resistance to petroculture needs to attend to the details of history and the particulars of the local.

Dughan's trauma, left over from trying to stop the rigs, is the trauma of dominant global powers, who have heretofore identified themselves with the only "humanity" that matters, suddenly realizing they are not in control, realizing that they are finite not transcendent, realizing that they will be extinguished. The story ends with a non-depiction of the ruins:

> Dughan turned and took in the length of Covehithe beach. They were out of sight, but he looked in the direction of the graveyard, and of St. Andrew's stubby hall where services continued within the medieval carapace, remains of a grander church fallen apart to time and the civil war and to economics, fallen ultimately with permission. (Miéville 2015, 311)

Ritson reads in this scene a comparison between the ruins and the rigs as hybrid, cyborg figures (noting Miéville's use of "carapace"), arguing that the ruins indicate "that the remains of global petroculture will leave a mark on the landscape that will outlast its significance for human culture" (Ritson 2019, 135). Instead, I see the ruins of the church and the unfrightening graveyard here not as representing the uncontrollable world: they are the reminders of the particulars of history lost in the distance to a gaze too centred on its own power; they are reminders of endings that are allowed to happen because the material conditions that lead to their end are ignored. The oil rig – petroculture – is different, however. It acts outside of permission because it is a cause and signifier of a universal extinction that nonetheless has disparate local effects: in the story, M.E.N.D recognizes the need to combat it even as tourists on the English countryside imagine it as entertainment. This is why, as the main characters first wait for the rig, "Somewhere some insane bird, not a nightingale, was singing" (Miéville 2015, 302). The allusion expands Keats' lament for art in the face of death. Nature becomes not a sign of human transcendence, nor even of human struggle in the face of individual finitude, but instead becomes a sign under erasure, a recognition of the failure of a Eurocentric symbolic order. It is "insane" because it is not renderable as meaning within a dominant epistemology that treats environmental destruction as entertainment.

"Covehithe" would seem at first to bring us back to the object-oriented ontologists and new materialists with which I opened this argument. Specifically, it would seem to agree with Morton that "appearance is always strange," and that a weird spectacle can transform our vision of our relations to each other and the planet (Morton 2016, 6). But what the story emphasizes is that this weird transformation necessarily takes (disparate) effect on the local level, and that it is those local actions that matter. In insisting on a moral imperative to "think at Earth magnitude," Morton's weird vision does risk flattening that which the Weird recognizes as depthful: that human thought, with all of its vagaries and horrors, may be just another object, still it is the object that has physically transformed the world, and it is, in turn, that specific object which must be changed (Morton 2016, 32). For Morton, "Ecological awareness is uncanny like that. Freud thinks of being 'lost in a forest' and returning again and again to the same spot" (Morton 2016, 118). I would suggest that the Weird reminds us that there can be new places, or places to which "we" have never been. Defining who "we" are in epistemological terms is thus absolutely necessary. It is both more and less than "humans" who have destroyed the planet; to ignore the hegemonic interlocking of, and internecine battles between, capitalism, colonialism, and other forms of extractive activity – and thus to ignore the specific origins of the "Anthropocene" in them – is to misdiagnose how the human species functions as a species.

Brassier does make a compelling argument for the necessity of imagining human extinction through a reading of Jean François Lyotard. Lyotard, he notes, focuses on the ultimate end of the universe:

> The extinction of the sun is a catastrophe [...] because it blots out the terrestrial horizon of future possibility relative to which human existence, and hence philosophical questioning, have hitherto oriented themselves. Or as Lyotard himself puts it: "[E]verything's dead already if this infinite reserve from which you now draw energy to defer answers, if in short thought as quest, dies out with the sun." *Everything is dead already.* (Brassier 2007, 223)

If "everything is dead already" what matter if oil rigs, sentient or not, destroy the habitable planet? Does this just give license to continue on in a-ethical behaviour? Brassier, via Lyotard and Freud, argues not: "Extinction turns thinking inside out, objectifying it as a perishable thing in the world like any other" (Brassier 2007, 229). But still, this argument tends to flatten a human landscape that is instead composed of difference and hence interacts disparately with "its" surroundings.

Turning this argument back onto the specific philosophical traditions Brassier, Lyotard, and Morton invariably invoke, however, it could be read not as calling for an extinction of humanity, but as an extinction of a particular mode of thinking; the dominant, capitalist and colonial extractive epistemology becomes "a perishable thing," giving way to the new. We need to imagine the end of a Eurocentric world in order to imagine the end of extractive capitalism and vice versa. As Shomura writes:

> the effects of matter will become increasingly difficult to ignore as the seas rise and swallow land masses, as weather patterns and storms become ever more erratic and destructive, as droughts intensify, and as ecosystems destabilize and even collapse with the mass extinction of species. Those who are already deeply precarious will be even more harshly affected, others will experience newfound hardship and loss, and new opportunities for connection, creativity, and care will arise. (Shomura 2017)

The cultural imagination of the Weird offers one way to reflect on the viciousness of received worldviews. But this is precisely a call to attend to the historical and ideological specifics of certain epistemologies and a call for "new opportunities for connection, creativity, and care." To misrecognize or misvalue the "Anthropocene" as a flat event, rather than an ongoing series of events sharing in specific places and histories – such as the mobile frame of the North Atlantic and cold water extraction – is to misrecognize its causes, "goals," and cures. Conversely, imagining a new weird world born out of the end of petroculture can serve as a reminder that there are other (and othered) traditions of care in the regions of the North Atlantic that can move us beyond extractivism. New combinations and articulations of traditions can make getting "lost in a forest" a hopeful reality rather than a return of the repressed; it is time to invoke a Weird imagination that breaks through the uncanny return to the ruins of the same old same old.

Notes

1 See https://www.theguardian.com/books/2011/apr/22/china-mieville-covehithe-short-story
2 Topics for discussion for that assignment include "Benches: what will they be originally?" and "Can trees be recuperated?" (Miéville 2015, 154).
3 Dan Bloom is generally credited with coining the term "cli-fi" and has since curated a website dedicated the genre, *The Cli-Fi Report*, http://cli-fi.net/.

4 One can think here of the ecological allegory of *Mad Max: Fury Road* which reportedly caused significant damage to the ecosystems of the area of Namibia where it was shot. On *Fury Road* and its environmental impact, see Margonelli (2013).
5 My horrible pun aside, the use of Freud's notion of scopophilia, as read through Laura Mulvey's original analysis of its gendered role in cinema, is a purposeful one. As Mulvey reads it, the scopophilia, or "pleasure in looking" at another human as an erotic object, reinforces a patriarchal symbolic order, and renders any threat to that order silent (Mulvey 2009, 15). Given the link, made by Val Plumwood and others, between patriarchal and ecologically exploitative epistemologies, one can translate this analysis to popular renditions of petroculture and the threat of global warming. Identifying with the (primarily) male heroes as they bring the world back to the "order" of the status quo in such films as *The Day After Tomorrow*, or, more allegorically, in the *Avengers: Infinity War* and *Endgame* diptych, both climate change and environmentalists become represented as chaotic threats that are effectively – and easily – silenced. For a brief overview of these films and the argument, see Mattes (2019).
6 The term Anthropocene was popularized by P. J. Crutzen to define the "human-dominated, geological epoch, supplementing the Holocene – the warm period of the past 10–12 millennia," generally to describe the period from the late 18th century to now (Crutzen 2002, 23). This term, as will become clear later in this chapter, has come under more critical scrutiny, for reinstating the very anthropocentric logic it ostensibly critiques to flattening differences between human populations and ignoring the roles of, among other institutions, colonialism, racism, and other forms of oppression that are deeply imbricated with environmental exploitation and climate change.
7 On the exploitations, identities, and resistances formed upon, through, and around the Atlantic Ocean as well as "critical ocean studies," more generally, see, for example, Blum (2008), DeLoughrey (2019), Gilje (2004), and, of course, Gilroy (1993). For a series of articles that frame these discussions in relation to the gothic, see the articles in *Nautical Gothic*, a special issue of *Gothic Studies* edited by Emily Alder (2017).
8 Eric Savoy frames this issue in useful psychoanalytic terms, writing that "the gothic 'turn' toward compelling but unthematisable narrative might be conceptualized as the emergence of the Lacanian Real, which, according to Judith Butler, 'is that which resists and compels symbolization'" (Savoy 1998, 8). Rather than an irruption of the Real, however, I see the Gothic as struggling with its realization that the Real *is* always impossible to symbolize; the Gothic thus becomes an analysis of the ideological failures of the symbolic order.
9 I'm also building on Michael Niblett's useful connection between the rise of the Gothic as a cultural mode and the rise of coal to suggest that the Gothic converges with energy culture not only because of the latter's ability to express various anxieties, as Niblett gestures to, but also because the Gothic is a genre predicated on energy as a concept. The Gothic relies on forces that both enable and disrupt human interaction with the material world. In this sense, the Gothic can serve as a cultural map of energy debates.
10 In turning to Luckhurst for my literary definition of the Weird as genre, I am, to some extent, turning away from Timothy Morton's "dark ecology" definition of the "weird" functions of the Anthropocene (and I capitalize "Weird" when speaking of the literary genre to emphasize this). My analysis differs from Morton's in its approach to the social and cultural functions of the Weird and the weird. Morton is concerned with reframing human existence as a primarily species-relationship to the surrounding world, denouncing those would pay attention to human-to-human relationships; I try to find a third ground, where to understand those species effects, we need to pay attention to how our species functions on an intra-species level. To put it in Morton's terms, where he sees humanity as a "hyperobject," I see it as functioning at both a hypo- and hyper-levels; the quantum levels of human interaction can have massive effects, even if they only seem to weakly interact. I return to this discussion at the end of the chapter.

11 "Kaiju" is a Japanese word that is now generally used in English-speaking film analysis to refer to the genre of Japanese films featuring giant monsters, monsters which are, in turn, also referred to as "kaiju."

12 The area surrounding the actual Covehithe is also known as a roosting area; most of Miéville's descriptions of the setting (including the ruins and the road that has been eroded and now "ends" on the cliff face) are accurate, lending to the "weirdness" of the later action.

13 P-36, built in Italy, was reportedly "the biggest floating oil platform in the world" when it "slipped beneath the waves of the Atlantic" off the coast of Brazil in March of 2001 (Gibb 2001). As Brian C. Black details, "explosions on board killed 11 crew members and compromised the structure," leading to a "total loss" (Black 2012, 154).

14 See Goddu 1997, 132; Jameson 2005, xv; and Engels 1968, 399–410.

15 H. P. Lovecraft (1890–1937) is often cited as an originator of the Weird tale (though he is heavily indebted to earlier authors, including Edgar Allan Poe). He is best known for a series of tales now collectively referred to as the "Cthulhu Mythos," which generally focus on a series of ancient, pre-historical gods (chief among them Cthulhu) who are later awakened and, to varying degrees, destroy humanity, as part of Lovecraft's "vision of his own species as something ephemeral and irrelevant when seen against the sublime immensity of the cosmos" (Alcalá González 2017, 86). This mythos was then built upon by other writers, including Robert Howard (the creator of Conan the Barbarian). Cthulhu and the other "old ones" are often associated with the ocean; for a discussion of Lovecraft's works and the "nautical Gothic," see Antonio Alcalá González (2017). Lovecraft is equally known for his vitriolic racism and xenophobia, which is inextricable from these works, an imbrication that resonates with the current argument.

16 This debate is much wider and more complex than I can frame it here, with some theorists arguing that object-oriented ontology can undo such dominant epistemologies, and others arguing that it only reproduces the violence inherent in dominant systems. For a range of positions within this debate, see the conversation between Tompkins, Huang, and Shomura in their articles in *Lateral.*

17 Of course, while "Covehithe" is my focus here, Miéville is not the only SF or Weird author whose works push back against this. N. K. Jemisin's work leaps to mind, for example.

References

Alcalá González, Antonio. 2017. "From the Sea and Beyond: Lovecraft's Sea Monsters." *Gothic Studies* 19 (2): 85–97.

Alder, Emily, ed. 2017. *Nautical Gothic.* Special issue, *Gothic Studies* 19 (2).

Black, Brian C. 2012. *Crude Reality: Petroleum in World History.* Updated Edition. Lanham, MD: Rowman and Littlefield.

Bloom, Dan. *The Cli-Fi Report.* http://cli-fi.net/.

Blum, Hester. 2008. *The View from the Masthead: Maritime Imagination and Antebellum American Sea Narratives.* Chapel Hill: University of North Carolina Press.

Brassier, Ray. 2007. *Nihil Unbound: Enlightenment and Extinction.* New York: Palgrave.

Carter, Chris, creator, 1993–2002. *The X-Files.* Ten Thirteen Productions and 20th Century Fox Television.

Crutzen, P. J. 2002. "Geology of Mankind." *Nature* 415 (23). https://doi.org/10.1038/415023a.

DeLoughrey, Elizabeth. 2019. "Toward a Critical Ocean Studies for the Anthropocene." *English Language Notes* 57 (1): 21–36.

Dr. Strangelove or: How I Learned to Stop Worrying and Love the Bomb. Directed and written by Stanley Kubrick, Hawk Films, 1964.

Duggan, Anne E. 2013. *Queer Enchantments: Gender, Sexuality, and Class in the Fairy-Tale Cinema of Jacques Demy*. Detroit: Wayne State University Press, 2013.
Engels, Friedrich. 1968. "Socialism: Utopian and Scientific." In *Selected Works*, by Karl Marx and Frederick Engels, 399–410. New York: International Publishers, 1968.
Gibb, Tom. 2001. "Post Mortem into Rig Disaster." *BBC News*. March 21. http://news.bbc.co.uk/2/hi/americas/1233083.stm.
Gibson, William. 1984. *Neuromancer*. Ace.
Gilje, Paul A. 2004. *Liberty on the Waterfront: American Maritime Culture in the Age of Revolution*. Philadelphia: University of Pennsylvania Press.
Gilroy, Paul. 1993. *The Black Atlantic: Modernity and Double Consciousness*. London: Verso.
Goddu, Teresa A. 1997. *Gothic America: Narrative, History, and Nation*. New York: Columbia University Press.
Godzilla. Directed by Ishirō Honda, Toho Studios, 1954.
Groom, Nick. 2008. "Romantic Poetry and Antiquity." In *Cambridge Companion to British Romantic Poetry*, edited by James Chandler and Maureen N. McLane, 35–52. Cambridge: Cambridge University Press.
Hageman, Andrew. 2019. "Bringing Infrastructural Criticism to Speculative Fiction: China Miéville's 'Covehithe.'" *C21 Literature: Journal of 21st-Century Writings* 7(1). https://doi.org/10.16995/c21.594
Herold, Emanuel, Sina Farzin, and Susan M. Gaines. 2016. "Between Fact and Fiction: Climate Change Fiction." *Science Fiction Studies* 43 (3): 609–611.
Huang, Michelle. 2017. "Rematerializations of Race." *Lateral* 6 (1). https://doi.org/10.25158/L6.1.11
Jameson, Fredric. 2005. *Archaeologies of the Future: The Desire Called Utopia and Other Science Fictions*. London: Verso.
Kalaidjian, Andrew. 2017. "The Spectacular Anthropocene." *Angelaki* 22 (4): 19–34.
Luckhurst, Roger. 2017. "The Weird: A Dis/orientation." *Textual Practice* 31 (6): 1041–1061.
Macdonald, Graeme. 2014. "Improbability Drives: The Energy of Sf." *Paradoxa* 26: 111–144.
Margonelli, Lisa. 2013. "How the New Mad Max Movie Messed With Namibia." *Slate*. February 19. https://slate.com/technology/2013/02/mad-max-fury-road-filming-and-environmental-damage-in-namibia.html.
Mattes, Ari. 2019. "Friday Essay: Eco-disaster Films in the 21st Century – Helpful or Harmful?" *The Conversation*, December 12. https://theconversation.com/friday-essay-eco-disaster-films-in-the-21st-century-helpful-or-harmful-127097.
Miéville, China. 2015. *Three Moments of an Explosion: Stories*. New York: Del Rey.
Miéville, China. "China Miéville Talk at KU, pt. 5," presented 24 Sept. 2009, *YouTube*, last modified 1 Nov. 2011, http://www.youtube.com/watch?v=EPABFo9cg2U/. Last accessed 07 Nov. 2020.
Morton, Timothy. 2016. *Dark Ecology: For a Logic of Future Coexistence*. Columbia University Press.
Mulvey, Laura. 2009. *Visual and Other Pleasures*. 2nd ed. New York: Palgrave Macmillan.
Niblett, Michael. 2017. "Energy Regimes." In *Fueling Culture: 101 Words for Energy and Environment*, edited by Jennifer Wenzel and Patricia Yaeger, 136–139. New York: Fordham University Press.
Pacific Rim. Directed by Guillermo del Toro, Warner Brothers, 2013.
Polack, Fiona. 2019. "Corporate and Worker Photographs of the Offshore Oil Industry: The Case of the Ocean Ranger." *Journal of Canadian Studies* 53 (1): 152–177.

Rancière, Jacques. 2009. *The Emancipated Spectator.* Translated by Gregory Elliott. London: Verso.

Ritson, Katie. 2019. *The Shifting Sands of the North Sea Lowlands: Literary and Historical Imaginaries.* New York: Routledge.

Savoy, Eric. 1998. "The Face of the Tenant: A Theory of the American Gothic." In *American Gothic: New Interventions in a National Narrative*, edited by Robert. K. Martin and Eric Savoy, 3–19. Iowa City: University of Iowa Press.

Shelley, Mary. 2003. *Frankenstein, or, The Modern Prometheus.* 1931. Edited by Maurice Hindle. Penguin.

Shomura, Chad. 2017. "Exploring the Promise of New Materialisms." *Lateral* 6 (1). https://doi.org/10.25158/L6.1.10

Stoker, Bram. 1987. *Dracula.* Edited by Glennis Byron, Broadview Press, 1998.

Strombeck, Andrew. 2019. "The Weird, the Ontological, and the Normal." Review of *None of this is Normal: The Fiction of Jeff VanderMeer* by Benjamin J. Robertson. *American Literary History* 31(2): 347–355.

Terminator 2: Judgment Day. Directed by James Cameron, Carolco Pictures, 1991.

The Day After. Written by Edward Hume, directed by Nicholas Meyer. ABC, 20 November 1983.

Tompkins, Kyla Wazana. 2016. "On the Limits and Promise of New Materialist Philosophy." *Lateral* 5 (1). https://doi.org/10.25158/L5.1.8

Transformers. Directed by Michael Bay, Dreamworks, 2007.

Wells, H. G. 2003. *The War of the Worlds.* 1897, Edited by Martin A. Danahay, Broadview Press.

6

CONTESTED AND EMERGENT FUTURES

Film and energy regimes of the Newfoundland offshore

Rachel Webb Jekanowski

In October 2019, as I walked my usual route to work at Memorial University in St. John's – the capital city of the Canadian province of Newfoundland and Labrador, settled on the homelands of the Beothuk, Mi'kmaq, Inuit, and Innu Nations[1] – I noticed a sign perched in the window of one of the houses. Hand painted on white poster board, it read: "what future am I planning for?" This sign was one of hundreds carried at the city's Fridays for Future Global Climate Strike a few weeks earlier. The protest, organized by the St. John's chapter of the international climate movement Fridays for Future (founded in August 2018 by Swedish high school student Greta Thunberg) and other community organizers, broke provincial records for political gatherings ("Climate Strike Draws Thousands to March on Confederation Building in St. John's," 2019). That day, over 4,000 people marched from Memorial University to the Confederation Building, the seat of provincial power.[2] Stopping to snap a photo (Figure 6.1), I was struck again by the urgency of this question. Faced with global warming, species loss, a melting Arctic, rising sea levels, and other environmental and social impacts of carbon-fuelled climate change, many are asking variations on this question: *what future am I planning for?*

The offshore oil and gas sector has become profoundly embedded in the province since 1979, with the discovery of the first major oilfield. The Hibernia field is located approximately 300 kilometres off the east coast of the island of Newfoundland, in the rich fishing grounds of the Grand Banks. Since the production of first oil in 1997, this resource has radically reshaped the province's social landscape and economic forecast, altering, in turn, the ways in which residents imagine individual and collective futures. As Angela Carter and Leah Fusco note, "the promise of oil" was particularly powerful in Newfoundland and Labrador as a "have-not" province, especially following the collapse of the cod fishery and the federal government's 1992 cod moratorium, triggering further socio-economic crisis in rural and outport communities (Carter and Fusco 2017, 104).

DOI: 10.4324/9781003024125-6

FIGURE 6.1 Climate strike sign reading "What future am I planning for?"
Source: photo by author, 2019

Today, Newfoundland and Labrador is the third largest oil producer in Canada and until the oil market crashes of mid-2014 and 2016, the industry provided up to 30% of the province's total revenues (Carter and Fusco 2017, 104; "Provincial and Territorial Energy Profiles" 2020).

For residents and visitors, offshore oil's imprints onshore are both ubiquitous and subtle. A trip to Walmart or other big box stores in the east end of St. John's requires drivers to navigate streets named for North Sea oil capitals: Aberdeen Avenue and Stavanger Drive. Similarly, offshore oil platforms appear on street signs like that of White Rose Drive. Prior to the 2020 coronavirus pandemic, tourists to downtown St. John's could encounter prominent displays of scale models of oil tanker ships and drilling platforms at the Visitor Information Centre. Petroleum, and its iconography, also seeps into the campus infrastructure and finances of the province's sole university, Memorial University. Suncor Energy Hall (previously, Petro-Canada Hall) houses the School of Music, and the Faculty of Business Administration hosts the Husky Centre for Excellence in Sales and Supply Chain Management, established in 2019 with a contribution of over $1 million CAD from Husky Energy (White 2019).

In what follows, I examine these historical, and highly geographically specific, associations between offshore oil and future imaginaries through a collection of films produced during the early days of exploratory drilling. In the early 1980s, as Newfoundland and Labrador contemplated developing a commercially viable oil industry, Memorial University's Extension Service (MES) produced a collection of films about this new resource in collaboration with the National Film Board of Canada

(NFB)'s Atlantic Studio.[3] This chapter focuses on three films to emerge from this partnership, which imagine, model, and question how oil production in the North Atlantic might shape the province's economy, social fabric, and environments. As future-oriented texts, *Offshore Oil: Are We Ready?* (dir. Paul MacLeod, 1981), *Oil Means Trouble* (dir. Bruce MacKay, 1985), and the uncredited video *Speculation* (1980) adopt different narrative and formal techniques to image and communicate the impacts of cold water oil during the intervening years between the identification of oil reserves (1979) and the start of commercial production at Hibernia (1997).[4] It must be noted, however, that all three films concentrate predominately on the *island* of Newfoundland rather than the province as a whole despite the implications of provincial jurisdiction over newly discovered undersea oil fields.[5]

Informed by fieldwork in St. John's and archival research, I analyse how these media forms act as sites of future-making on a textual level. In other words, how do the films model or communicate knowledge about potential times-to-come for the island of Newfoundland and particularly the capital city of St. John's? Next, I examine how the gendered and racial dynamics of these imagined oil futures frame social relations with energy, oceans, and more-than-human life; whose economic and social futures are imagined through cold water oil? Finally, I turn to the modelled ecological impacts of these imagined futures, and the ways the films help produce the oil fields in the Grand Banks as an *offshore*, an extractive zone and new form of nature.

Offshore Oil: Are We Ready?, *Oil Means Trouble*, and *Speculation* represent one localized example of a global phenomenon of petro-cinema: nontheatrical, sponsored, and nonfiction films made about petroleum and the industries that drill for it. Emerging as a mode of sponsored filmmaking primarily in the decades following World War II, petro-films developed a shared lexicon of aesthetic and narrative conventions, including voice-over (male) narration; animated sequences to depict subterranean industrial operations; aerial footage of remote drilling sites and landscapes; talking-head interviews; fictionalized narratives featuring oil workers, their families, and consumers; and an overall narrative didacticism. The majority of these films were financed and released by petroleum companies – including British Petroleum, Iraq Petroleum Company, Imperial Oil, and Royal Dutch Shell (which had its own film production unit) – across a range of colonial and national contexts, from Iraq and the Middle East, to colonial-era Italy, to post-Leduc Canada, and Cold War America (Banita 2014; Canjels 2009; Damluji 2015; Jekanowski 2018; Tyagi 2020). Throughout this geographically and ideologically diverse corpus of petro-films runs a strong association between what Mona Damluji calls the 20th-century "story of oil" and "the experience of modernity" (Damluji 2015, 147). As I will show, these filmic conventions continue to influence cinematic depictions of offshore oil, and even override the ideological premises of non-corporate films like *Offshore Oil: Are We Ready?* and *Speculation*, which adopt more critical stances towards the industry's financial and ecological excesses.

Strong parallels exist between our current moment – defined by intersecting crises of global warming and racial injustice – and the pivotal period in which these films were made. By the late 1960s, scientists and oil executives alike were already aware of the growing body of scientific evidence about the harmful impacts of carbon emissions and industrial waste on waterways, air quality, and climate as corporate environmentalist films like *The River Must Live* (dir. Alan Pendry, 1966, Shell Film Unit), *Clean Water* (director uncredited, 1967, Shell Film Unit), and *The Shadow of Progress* (dir. Derek Williams, 1970, British Petroleum and Greenpark Productions) attest (Jacobson 2019, 166). The conflict between scientific concerns about the ecological impacts of oil pollution, especially on coastal ecosystems, with the economic prospects of a new offshore energy sector in the late 1970s and early 1980s strongly parallels contemporary concerns about economic development and the climate crisis. I conclude this chapter by asking how these archival films and futures past might be used to locate hope for present-day climate justice movements.

In their introduction to a special issue on environmental futures, Andrew S. Mathews and Jessica Barnes explore how future imaginaries are continuously remade in relation to place and the forms of knowledge they facilitate. Visions of the future are charted – and contested – from within government offices, scientific laboratories, and by activists in the street (Mathews and Barnes 2016, 11). Tracing historical speculations about future times enables energy scholars to ask new questions about the ways in which oil futures are remade in relation to media and geography as well as the emergent politics that future imaginaries can produce. This chapter therefore contributes to the turn towards theorizing relationships between time (pasts and futures), environments, and media within cultural studies and anthropology (Kristoffersen, Bridge, and Steinberg, this volume; Mathews and Barnes 2016; Weszkalnys 2014). At the same time, it participates in the political reorientation of environmental film and media studies by bringing historical visions of energy futures to bear upon ways of thinking about energy transition in the present.

Communicating contested futures

With its seemingly limitless abundance of sea life and geological resources, generations of settler communities have made their fortunes, lost their lives, and struggled to survive off the cold waters of the North Atlantic Ocean. In the mid-to-late 20th century, residents of Newfoundland and Labrador faced a depressed economy and rural, resource-dependent communities became hollowed out by out-migration and dwindling fishing stocks (Delisle 2013, 2). Today, archives offer contrasting and, at times, conflicting narratives about what the future was thought to hold for this struggling province. Many of these texts, from films to government reports, tether the province's economic and cultural survival to either the development of *new* resources or the wiser management of pre-existing ones. The 1974 NFB documentary *Tomorrow is Too Late* (dir. Bané Jovanovic, Douglas Keifer, and Don Virgo), for instance, argues that the decline of inshore fishing jeopardized a whole way of life for outport communities. Over footage of cod fishers hauling in their lines and men processing catch on the wharf, the narrator states that overfishing by industrial trawlers is exhausting cod stocks and pushing small-scale fishers out of business. Only a modern, well-regulated fishery can ensure sustainable yields and prevent the "thoughtless waste of yesterday." Throughout the film, future-oriented language is used to link fishing practices of the past with the implementation of new federal regulation and conservation practices. While the hammy delivery of lines like "tomorrow must be cared for now, today" might elicit groans, the documentary provides insight into the ways that federal agencies – here, Environment Canada and Fisheries and Marine Service, which sponsored the film – modelled the future of this resource sector. In this regard, "care" for the future becomes synonymous with, and limited to, maintaining fish stocks through Western scientific and technological means, from commercial fish farming to the use of electronic sonar devices.

Crucially, the "tomorrow" imagined by this documentary is a white one, shaped by what Eve Tuck and K. Wayne Yang call a "settler futurity" (Tuck and Yang 2012, 1). Here, the future is a racialized, economic category, which can be rescued through the careful preservation of the fisheries – a resource sought by Europeans for centuries, bringing waves of immigrants to Newfoundland and Labrador's shores. The racial make-up of this "tomorrow" is visually conveyed in the documentary by the sea of white faces charactering the industry as well as the narrative silence about Indigenous fishing practices on these same waters. Similarly, a 1978 report published by the Department of Fisheries, *Fish is the Future*, assesses the status of Newfoundland and Labrador's fishing industry and offers a development program with possible avenues for future expansion. As the title suggests, *Fish is the Future* promotes the fisheries as the "hope" for the province's economic future (Bonnell Public Relations 1978, 1). It characterizes fish as "the original Newfoundland resource" as well as the "mainstay" of white Newfoundlanders' cultural heritage (Bonnell Public Relations 1978, 2). The photograph on the report's back cover reaffirms the racial undertones of this future. It depicts a young child with boyish clothing and short blond hair, leaning against the gunnel of a boat in the foreground and watching a group of white fishermen towing their nets from another dingy in the image's centre. A quotation from the report floats above the scene: "[...] the future is ours at last." Rebekah Sheldon, theorizing the figure of the child within climate films and fiction, argues that the image of the child serves as an embodiment of the future. As a mode of timekeeping, the child can be used to track the reproductive potential of white settlers and the social reproduction of their way of life (Sheldon 2016, 3). The future of *this* child, the report infers, depends upon the management of North Atlantic fish stocks. Moreover, the fishers' repossession of "the future" as expressed in the quotation positions times-to-come as contested territory, to be controlled by some to the exclusion of others.

In contrast to these fishy future visions, the MES-NFB productions later in the decade probed a rival future imaginary for the island of Newfoundland:

that of offshore oil. *Speculation* (1980), *Offshore Oil: Are We Ready?* (1981), and *Oil Means Trouble* (1985) were created during a period in which oil's future possibility was on the horizon, offering hope in a province mired in economic underdevelopment and rural out-migration. However, as these films evidence, the years following the Hibernia oilfield's discovery saw public debate and research about the potential negative consequences of development, especially on fishers' livelihoods and ocean environments.

A 1979 project proposal compiled by the Extension Service provides insight into the motivations for producing what would become *Offshore Oil: Are We Ready?* The document, plainly entitled "Oil and Gas Proposal," identifies a need for more knowledge about the industry's impacts and greater understanding amongst the public and decision-makers. The solution it recommends is an "educational project to disseminate information and facilitate discussion and participation" about the development of offshore petroleum resources in Atlantic Canada ("Oil and Gas Proposal" 1979, 1). The unnamed authors contend that "[e]xperience in other parts of the world notably in the North Sea suggests that the impact of such development on economics, societies and cultures such as ours can be considerable;" it is therefore "vital that those who are affected understand the developments that are taking place and plan appropriate action" ("Oil and Gas Proposal" 1979, 1). This educational project, pitched to commence in winter 1979 and continue through fall 1981, included the creation of film and video productions at each stage – including the production of film modules about various aspects of the oil industry, local television programming, and community screenings of films to solicit feedback. The Extension Service planned to lean on the expertise of government agencies and organizations familiar with the industry – including the Centre for Cold Ocean Engineering (C-Core), Northern Oceans Research and Development Corporation (NORDCO), and the federal Departments of the Environment and of Industrial Development – as well as the resources of Memorial University ("Oil and Gas Proposal" 1979, 6–7).

Two phases of film and video production were to be undertaken jointly by the Extension Service and the NFB Atlantic Studio, including a 30-minute "introductory film" intended to provide an "overview" of offshore oil and gas developments in the North Atlantic, for distribution on television and by the NFB ("Oil and Gas Proposal" 1979, 41). This film would help to inform the public and empower more positive gains, by attempting to "maximize the positive effects of oil development and keep the negative effects to a minimum" ("Oil and Gas Proposal" 1979, 1). In other words, the proposed "introductory film" would be both a modelling and pedagogical tool, helping viewers to imagine what oil could mean for the province's finances and environments, while also teaching Newfoundlanders and Labradorians how they might work with this powerful industry. The National Film Board committed to financing the co-production beginning in winter 1979, and the Canadian Broadcasting Corporation also expressed interest in providing television time and other resources to help increase the project's reach ("Oil and Gas Proposal" 1979, 15–16).

The proposal's underlying assumptions that oil development is at once controllable and inevitable translate in ambiguous ways into the resulting films. All three contain footage shot and collected as part of the proposed project, including site visits to Scotland and Norway. The two documentaries, *Offshore Oil: Are We Ready?* (1981) and its shorter classroom adaptation *Oil Means Trouble* (1985), also include footage repurposed from corporate oil films produced by BP and Chevron, including *The Shetland Experience* (dir. Derek Williams, 1977) sponsored by BP about the industry's environmental clean-up efforts at the Sullom Voe Terminal in Shetland. Released four years after its predecessor, *Oil Means Trouble* uses the same footage, edited down to half the original running time, under a more overtly critical title. The NFB intended this shorter documentary to be a classroom adaption, according to the one-page educator's guide designed to accompany high school screenings ("Oil Means Trouble" 1986). Like *Offshore Oil: Are We Ready?*, the narrative of *Oil Means Trouble* is organized around four locations (the island of Newfoundland, Norway, northern Scotland, and the Shetland Islands), featuring cross-cutting between talking-head interviews, establishing shots and press briefings in Newfoundland, and sweeping footage of North Sea drilling platforms and building yards. Notably, *Oil Means Trouble* switches to a female voice-over, performed by Linda Lee Tracy. Tracy's narration serves as the connective tissue between the interview subjects and conveys a more modern aesthetic than the conventional "voice-of-God" narration characteristic of the mid-century NFB documentary.[6] The decision to cast a woman as the narrator could also reflect the amplified pedagogical intent of this classroom adaption, replicating the gendered labour of childhood education.

Unlike the other two documentaries, *Speculation* was not produced by the NFB although it reuses some of the same footage originally shot for *Offshore Oil: Are We Ready?* (Paul MacLeod, director of *Offshore Oil: Are We Ready?*, also worked on this production and provided its voiceover narration.) Produced by Memorial's Extension Service for a public town meeting in January 1980, *Speculation* raises concerns about the influx of oil wealth and speculative investors in St. John's on residents of the Goulds, a rural community southwest of the capital city zoned for agricultural purposes at the time. By focusing on how oil development threatens to reshape the geography of land ownership in the Goulds through real estate speculation, the video speaks specifically to local audiences, many of who may have been farmers or homeowners.

While created for different purposes, all three films speculate upon the future of oil resources through a shared technique: comparisons to Northern European oil-producing nations with lucrative offshore oil and gas sectors built over a decade earlier. Scotland, the Shetland Islands, and Norway are presented in the films as comparable models for successive development and object lessons for Newfoundland and Labrador, based on their shared cultural connections with the sea, fishing heritage, and experiences with uneven development of rural areas. In other words, as the *Oil Means Trouble* educator's guide explains, viewers are transported to European "communities whose geography and way of life, based largely on fishing industries, most resemble those of Atlantic Canada" ("Oil Means Trouble" 1986). The MES films did not invent this comparative technique; rather, it reflects a broader interest in the Norwegian and Scottish responses to offshore oil within publishing and scholarly circles in St. John's during the 1970s and 1980s, as evidenced by Maurice Scarlett's edited collection *Consequences of Offshore Oil and Gas: Norway, Scotland, and Newfoundland* (1977) published by Memorial University's Institute of Social and Economic Research (ISER) and issues of the popular interest magazine *Decks Awash*, published by the Extension Service.

While never explicitly stated, the racial composition of these majority white countries also emphasizes a settler imaginary of Newfoundland as an island of European immigrants, descendants of these same Northern European fishing communities, set against the "disappearance" of the Beothuk.[7] Through these comparisons, the films evaluate environmental risks, speculate upon the social impacts of rapid development, and model positive and negative outcomes for Newfoundland. In so doing, the films perform a sophisticated means of producing knowledge about the future through documentary storytelling about what the films call the "North Sea experience." Each of the comparable communities becomes like case studies or data points, used to model some of the different approaches local governments and community organizations have taken elsewhere in response to offshore oil development. In effect, they afford the viewer both a means of evidence-based speculation and an outlet to explore the uncertainty facing Newfoundland and Labrador at that time, when "it is still unclear what the consequences of this discovery will be" ("Oil Means Trouble" 1986).

Offshore Oil: Are We Ready? begins by laying this central question out for the viewer. Setting the scene, the narrator states that the global demand for energy has brought the oil industry to the province's shores, turning "its attention westwards to our side of the Atlantic Ocean." On an animated map of Newfoundland and Labrador's eastern coast, white dots representing offshore exploration sites proliferate across the Grand Banks. The Hibernia field appears marked with a triangle, before the camera zooms in to provide a closer view of the prospective tanker routes that could connect the oil platform to St. John's. Narratively, the documentary rotates around a slew of speculative questions. Early on, the narrator asks what might come to pass if "Hibernia's reserves are proven in commercial quantities." Likewise, the title of the film itself provocatively summarizes its angle of inquiry; are Newfoundlanders "ready" for the massive changes that would inescapably accompany offshore oil? Apart from several establishing shots of the city of St. John's and animated maps of exploratory drilling sites in the Grand Banks, both *Offshore Oil: Are We Ready?* and *Oil Means Trouble* consist almost entirely of footage shot in Shetland, Aberdeen, and Stavanger during the late 1970s. In other words, the futurity of Newfoundland and Labrador's offshore industry is conveyed quite literally by the absence of documentary imagery, and the filmmakers' resort to animation to depict offshore exploratory permits. In contrast, the sweeping footage of North Sea offshore oilrigs, towering transport ships, and busy harbours prominently announces the material success of this industry on the other side of the Atlantic, while transferring the documentary value of these images from one geography and decade to another.

The more-than-human scale of petroleum development and striking visual presence of the North Sea oil infrastructures in several scenes convey a textual ambivalence towards oil's promise and the darker sides of heavy industry. A montage depicting the construction of the *Ninian Central Platform* (completed in 1978 in Kishorn, in the northwest Scottish Highlands) in the opening sequence of *Offshore Oil: Are We Ready?* celebrates the project as a technological wonder and feat of engineering. Employing rapid zooms into and away from contractors and other employees labouring within dry-dock construction sites, the camera leaps from shots of the human to wide shots of the offshore platform under construction, finally zooming out to reveal the enormous scale of the infrastructural project. As the film explains, the *Ninian* platform was the largest project of its kind in the world at that time. Like corporate-sponsored photography that emphasizes the "massive scale of offshore rigs and dwell[s] on the infrastructure's technological sophistication" (Polack 2019, 154), this sequence creates an operatic dance of infrastructural and human bodies to awe the viewer at the material scope, ingenuity, and apparent permanence of these developments. Yet, even as the *Ninian* platform appears to materialize the "promise" of infrastructure to enact a new social or economic future for Scotland (Anand, Gupta, and Appel 2018, 27), the documentary's attention to the workers who built the rig and the surrounding community of Kishorn undercuts the technological utopianism of this swirling montage. In a later scene, a Kishorn resident voices regret about the *Ninian* project, saying that if the village had known the dockyard would bring an influx of almost 3,000 men to their village of several hundred residents, they would never have supported it. Presented with few options and little information by the project stakeholders, the villagers' experiences offer a compelling counter-narrative to the industry's promises of prosperity and jobs for local workers for Canadian viewers.

Whose futures? Race, gender, and temporalities of development

In the process of modelling the ways in which the Newfoundland public might respond to offshore oil development, the films also reproduce dominant Canadian ideologies about whiteness, masculinity, and the preservation of settler culture in their depictions of oil futures. In their focus on the North Sea experience, *Offshore Oil: Are We Ready?* and *Oil Means Trouble* both reinforce a geographical imaginary of Newfoundland as part of a North Atlantic archipelago of predominately white coastal and fishing communities. Interviews with Norwegian and Scottish fishermen also cast the tensions between fishers and "oil men" as a masculine affair, negatively impacting the social reproduction of men's livelihoods first and foremost.

Both stories and scientific models of the future always include implicit and explicit decisions about the geographic scope and time frames in question as well as the populations included in (or excluded from) such times-to-come. In writing on the temporalities of global development, Cheryl Lousley argues for an attentiveness to the distribution of power and resources when imagining the passage of time and the trajectory of progress; asking "*whose* futures are envisioned and enabled to flourish" (Lousley 2015, 246)? Within the MES films, oil futures predominately focus on the threats and opportunities presented to men, particularly those who work at sea already as inshore fishers. In *Oil Means Trouble*, for example, the viewer is first introduced to some of oil's detrimental impacts on the fisheries by Rob McColl, secretary of the Fishermen's Federation, in Scotland. McColl explains that Scottish fishers learned the hard way about the loss of access to fishing grounds and damage to fishing vessels caused by oil rigs and pipelines, and "the ever present risk of pollution from blow-outs, accidental spills, and what have you." As he lists off these conflicts, the film depicts a towering platform, a diver inspecting undersea pipelines, and oil slicks on the ocean's surface. Across the many interviews with Scottish and Norwegian fishers, the industry – and the waters into which oil extraction encroaches – is represented as an entirely white, male enterprise. Depictions focus predominately on small-scale and inshore fishing, shot outdoors at the harbour or aboard the boats. It should be noted that other cinematic depictions of Canadian offshore oil development from the period share this attention to the labouring male body. In *Ressources sous-marines* (dir. John Ralph, 1973), a contemporaneous documentary about exploratory drilling off the coast of Halifax, Nova Scotia, depictions of the *Sedco-H* platform emphasize discipline and the security of the vessel. Part of the safety protocols governing the all-male workers, according to the narrator, is its exclusion of women. Regulated as a homosocial space, the rig doubles as a work site and a man camp at sea. Close-ups of workers, like a scene of two white men operating the drill, visually reaffirm the sea as a space of white masculinity and arduous physical labour.

This spatial conflict between the fisheries and oil development operating in the same waters, as Berit Kristoffersen, Gavin Bridge, and Philip Steinberg argue in their study of offshore oil production in Norway's Lofoten/Vesterålen/Senja (LoVeSe) region, is accompanied by a temporal conflict of "*differentiated times*" between the two industries (Kristoffersen, Bridge, and Steinberg, this volume). Oil contains a cluster of geological and financial temporalities including the deep-time of fossil fuel formation, the anticipatory futures of administrative planning, and the temporality of capitalization and investment (Kristoffersen, Bridge, and Steinberg, this volume). In contrast, the films' textual emphasis on white working men at sea communicates a struggle to preserve the futurity, or social reproduction, of the men's livelihoods. At stake is the inter-generational continuity of the fishers' identities, culture, and source of employment. The subject of the reproduction of fishers' futures comes up in later interviews as well. Contrasting oil and fish, Norwegian sociologist Cato Wadel points out that fish when properly conserved "is a renewable resource," whereas "oil might be a twenty-year [...] fifty-year thing." Invoking the next generation, Wadel calls for an "economy of permanence" based on sustainable resource use, which would provide stability when "our grandchildren [...] take over our society." Notably, the future potential of the fisheries is represented by footage of Norwegian school kids, ice skating, and playing outdoors. As the camera breaks away from the crowd to follow a young boy, gleefully skating across the rink on his own, the universality of the white child as a stand-in for the future reappears in force. "Saving the child," according to Sheldon, "appears tantamount to saving the future, just as saving the future is done in the service of protecting the children" (Sheldon 2016, 25). Like the youngster on the back cover of *Fish is the Future*, this child represents both the concept of the future *and* the preservation of a specific future for white fishers.

In Canada, oil films perform settler colonial imaginaries about landscapes, identity, and race, and the Memorial Extension films are no exception. As I have argued elsewhere in relation to the racial formations of Canadian popular science films, the erasure of Indigenous Peoples from documentaries and corporate films about extractive industries in Canada upholds settler colonial myths of *terra nullius* and white possession of First Nations, Inuit, and Métis lands (Jekanowski 2019). In her analysis of white settler colonialism across the former British Empire, for instance, Aileen Moreton-Robinson asserts the connections between the possessive and racial logics of settler colonialism. Ownership of nation, she argues, stems from the accumulation by dispossession of Indigenous Peoples, supported by legal systems that support patriarchal and racial hierarchies within society. Moreton-Robinson argues that "the logics of white possession and the disavowal of Indigenous sovereignty are materially and discursively linked" (Moreton-Robinson 2015, xiii). This maps onto these imperial logics of white men's legal and "natural" right to access and remake the physical world – including the world's oceans and seabeds – to ensure settler cultural futurity.

This logic of white possession manifests in the framing of the fisheries-oil conflict through the use of language and the aforementioned representations of the North Atlantic waters as geographies for white men. In an interview that appears in both *Offshore Oil: Are We Ready?* and *Oil Means Trouble*, Wadel draws an analogy between "oil and fish" and "the white Americans and the Indians." Norwegian fishermen, he argues, worked the coastal waters prior to the incursion of multinational oil corporations, so their economic rights should be legally protected. Wadel was very familiar with the Atlantic Canadian context having conducted field research on unemployment in the Newfoundland fisheries while teaching at Memorial's Department of Sociology and Anthropology in the late 1960s.[8] Although this provocative comparison acknowledges the continuity between racial and extractive capitalism that underpins land and resource theft, it also reflects a lack of awareness about the existence of Indigenous Nations in Canada. Norwegian fisher Olav Landsnes of the Fishermen's Association echoes Wadel's language in his own reference to Norwegian fishers' "traditional rights." Fishing communities, he believes, whose heritage and livelihoods depend on the ocean have traditional rights that oil companies should respect.

When transported to the Canadian context, Landsnes' language both parallels and obscures longstanding Indigenous struggles to uphold treaty rights to fishing and hunting. The period's settler amnesia towards the Mi'kmaq and Beothuk seeps into moments such as these, causing frictions to arise within what the films otherwise present as comparable coastal communities.

Another metaphorical guardrail imposed on the scope of the films' future imaginary is the geographical focus on the island of Newfoundland and the capital region of St. John's. Labrador, despite constituting 70% of the province's landmass, is barely mentioned within any of the three films. To some extent, this creative choice could be justified by the proximity of the Hibernia oilfield to St. John's, being only 300 kilometres off the coast of the Avalon Peninsula.[9] The capital is also the province's largest city and the seat of legislative and financial power. However, the marginalization of Labrador – and its majority Indigenous population – from the films' concerns exposes the settler lens through which the texts both imagine and critique the making of oil futures. Labrador has historically been treated as an infinitely exploitable source of natural resources – rare minerals, seals, salmon, hydroelectricity – by Newfoundland and white mainlanders alike. As survey records and other documents from the period attest, exploratory activities extended into the coastal shelf of Labrador, with federal oil and gas exploratory permits granted in the Labrador Sea as well as the Grand Banks by 1976 ("Labrador Sea" 1977, 12). Both culturally and geographically removed from the capital region, Labrador inhabits a separate imaginary in the films, with little thought given to how employment in a new oil sector might benefit workers from Labrador, how drilling may impact coastlines, or how investments from oil revenues might be used to address housing, education, or food security in Northern communities.

The films' technique of using comparisons between Atlantic Canada and oil-producing nations of Northern Europe to probe the island of Newfoundland's oil future exemplifies the settler colonial tendency to treat Europe as more socially and culturally advanced, a model to be reproduced or outdone. At the same time, however, this technique echoes Smallwood-era tensions around modernization and settler anxieties about a development gap between the former Dominion and mainland Canada. In effect, Newfoundland and Labrador's future is relegated visually and narratively to Scotland and Norway's recent pasts. Predicated upon a linear model of progress, the island of Newfoundland is depicted as inhabiting an earlier, less advanced stage of development than the more technologically and culturally advanced nations of Northern Europe. In *Time and the Other*, anthropologist Johannes Fabian theorizes this temporal distancing between the Western subject and Indigenous, racialized, rural, and non-Western populations. Fabian characterizes this "denial of coevalness" as a move to dehumanize and render "primitive" non-Western peoples (Fabian 2014, 25). This temporal distancing serves to justify the objectification of peoples and cultures by Western science and the seizure of their lands and resources. While all the subjects in the films are white, the denial of coevalness between Newfoundland

and Europe nevertheless has racial implications for the depicted rural and fishing communities. Early in *Offshore Oil: Are We Ready?*, the narrator describes the rural fishing villages and outport communities of Canada, Norway, and the United Kingdom as places where "life has always been lived on a less frantic scale." Associating rural life with slowness and pre-industrial stages of development places these communities as out-of-step with urban populations. By inhabiting a different stage of development, rural Newfoundlanders are *less white* according to the colonial racial hierarchy than white Europeans or white Canadians living in more modernized society. Indigenous communities, mentioned only in passing such as by Wadel, are implied to be even more primitive and relegated to history rather than inhabiting the present.

Producing the offshore and environmental futures

In speculating upon the emergence of cold water oil in Newfoundland, the films also provide a lens on the process by which capitalism produces resource horizons and future extractive zones. As other contributors to this collection have observed, the *offshore* transforms the ocean in its all biological diversity and human-sea relations into a site of extraction, a spatial zone delineated by the limits of international and maritime law. The offshore is at once central to contemporary global capitalism and dependent on the illusion of being peripheral to it (Potts 2019, 198). It is a space of international finance and tax havens, military interventions, special economic zones, and outsourced production (Urry 2014) – practices that "rely on boundaries that separate and define" to be seen as distinct from "'onshore' spaces where 'normal' business occurs" (Potts 2019, 198). Yet, the offshore is always already onshore as well (Potts 2019, 200). As the MES films demonstrate, offshore oil's operations, those who benefit from development, and the industry's cascading impacts are enacted and felt on land. Even the resource itself must be shipped from sea to shore for refining, transportation to market, and consumption.

In contemporary Newfoundland and Labrador, the offshore is synonymous with oil and gas development, an industrial practice that is itself integrated within global finance and transnational supply lines. The translation of underwater petroleum reserves into valuable natural resources, in turn, creates new temporalities for these waters, inscribing "the offshore" within a capitalist "temporality that is linear, continuous, measurable, and finite" (Mathews and Barnes 2016, 21). In other words, capitalist accumulation organizes social relations and people's management of the physical world both on the ground through practices of environmental use and within cultural ideas about the human and nonhuman world (Moore 2015, 2). This historical transformation of the North Atlantic from an expanse of seawater to an offshore happened across different spheres: the production of new scientific knowledge through seismic testing and exploratory drilling, new public policy, changes in language use, and, of course, in media representations like these documentary films.

Practices of future-making, like that within these films, rework the natural world alongside spatial and temporal categories (Mathews and Barnes 2016, 21). A key scene in *Oil Means Trouble* recounting Shetlanders' experiences of oil pollution demonstrates this reworking of nature. The scene opens with footage of two men in orange coats walking along the Shetland coast, flanked by an Esso tanker ship and other industry transport vessels on the water behind them. The British environmentalists from the Sullom Voe Authority (an environmental advisory group, with representatives of the Shetland Islands Council and oil companies) and the Nature Conservancy Council speak about a damaged coastline, the challenges of remediation, and the speed at which disaster struck. Shortly after the *Ninian* platform was installed off the coast of Shetland in 1978, Martin Heubeck of the Sullom Voe Authority explains that there was a large spill from one of the tanker ships. As Heubeck speaks, edited shots of oil-soaked bird carcasses, seaweed blackened with crude, and skeleton remains of a ram on the Shetland coastline bring home the dire ecological consequences of this industrial accident. Ruefully, Dr. Mike Richardson, a scientist with the Nature Conservancy Council, remarks that his agency should be renamed "Rent-a-Kill," since he spends more time picking up carcasses and euthanizing oil-slicked seabirds than conservation work.

This footage, shot years earlier in Scotland, once again uses the temporal slippage of editing to communicate a conservationist message. These scenes of oil damage from Scotland's past are used to model potential disasters in the near future for Newfoundland fishers. In her analysis of oil exploration in the African Atlantic island state of São Tomé and Príncipe, Gisa Weszkalnys identifies this anticipation of the detrimental impacts of oil development as "the temporal politics of a disaster yet to come" (Weszkalnys 2014, 212). While Weszkalnys focuses on efforts to anticipate economic disasters of oil's "resource curse," her reading of anticipation offers lessons for environmental impacts as well. The films' suspicions that offshore oil will likely result in greater environmental risks for Newfoundland's coastal inhabitants – as well as economic risks for those livelihoods depend upon the fisheries – serve to draw the effects of future oil disaster into the present, asking the viewer to think about ways these events might be forestalled. While Hibernia remains in the realm of "not yet," "a state of liminality in which potential futures radically define the present" but remain uncertain (Kristoffersen, Bridge, and Steinberg, this volume), the images of Shetland's beaches slick with crude reiterate the disaster always already contained within an oil future.

Framed another way, by posing increased future risks to lives and livelihoods in extraction zones, cold water oil produces a new form of nature defined by both short and long-term temporalities of disaster. In the short term, offshore drilling (particularly without proper regulatory oversight) poses risks to workers' health and safety as well as risks to nonhuman sea life through blow-outs and spills (see Polack, this volume). Longer term, global warming from the consumption of fossil fuels will threaten these same communities through sea-level rise, warming seas, more extreme weather events, and changes to fish populations. Petroleum extraction, as with the globalized economic system of late capitalism in which it operates, renders all forms of life precarious, but in ways distributed unevenly across geographies (Tsing 2015).

Conclusion: locating hope in emergent futures

Possible futures jostle for space in the collective imagination. The three Memorial Extension Service films examined here offer important insight into the ways that offshore developments were imagined as seeding grounds for new societal relations, economic growth, and the production of nature during the early years of oil exploration in Newfoundland and Labrador. By documenting the material ways in which corporate and government fantasies of oil wealth had already started to reshape the social fabric and environments in North Sea petro-states, *Offshore Oil: Are We Ready?*, *Speculation*, and *Oil Means Trouble* affords a glimpse into some of the concerns, anxieties, and hopes held about offshore oil development. At the same time, as artifacts of imagined futures past, these films reveal how environmental damage and threats to people's livelihoods in oil-producing countries rupture assumptions about "the future" and relationships to time.

Returning to the climate strike sign with which I opened this chapter, and turning this question back upon the offshore oil films, the sign's question *what future am I planning for?* resonates as strongly then as it does now. While the contours of these imagined futures have changed over the decades, the documentaries' subjects voiced concerns about the survival of their livelihoods and the consequences of progress, which many contemporary residents of the province would well recognize. Following the interventions of Lousley, Fabian, and Tuck and Yang, however, this question must be unpacked further to ask *which* futures are being planned – and for whom are they designed? "The future," as these films show, is a moving target; an historically contingent imaginary of times-to-come, a goal post to aim for when writing environmental policy, an economic framework for estimating financial risk and investment, and a racialized futurity invested in the survival of certain populations at the expense of others. Projecting into future times through data gathering and long-term planning – as these films do through their interviews with academics, government officials, workers, local residents, and scientists who have already experienced oil development – affords opportunities to reflect on how decision-making in the present brings these futures into fruition.

While all three films focus primarily on the *already lived* impacts of oil development on workers, economies, and coastlines, they each contain moments that prompt the viewer to critically reflect on the structures of power creating these conditions and, in so doing, imagine alternative ways that Newfoundlanders might organize and respond to development. *Speculation*, for instance, uses Aberdeen's skyrocketing real-estate prices in the 1970s to illustrate how local residents could be priced out of their own communities by oil wealth. After setting up this comparison between the island of Newfoundland and Scotland, however, the video abruptly interrupts itself with a sharp referee whistle and a black slide reading: "St. John's is not Aberdeen." Against a montage of St. John's featuring shots of the Confederation Building, the downtown core, the Harbour, and icebergs at sea, the narrator suggests that "[w]hat may happen here, may not be like anything that ever happened over there." The subsequent two-thirds of the video addresses the viewer directly, seeking to mobilize citizens in response to the threat of land speculation. Footage of St. John's, Shetland, and Aberdeen give way to shots of residential homes, roadways, and farmland in the Goulds, telescoping the arc of the video's narrative back to the local. Emphasizing that whatever happens to the community as a whole "will affect our lives and the lives of our children," the narrator directs viewers to "think carefully and move cautiously," to know their rights and protect themselves with strong legal counsel. *Speculation* advocates for residents of the Goulds to become active citizens within their community, encouraging them to work with their elected leaders to rein in the volatility of the real estate market. This appeal to citizen action through local governance is reiterated in the video's conclusion, with remarks by Ian Clark, Chief Executive of the Shetland Islands Council. "The developments will either be controlled by you," Clark intones, or they will determine "your community's future."[10] Acknowledging the limits of citizens' ability to prevent development, *Speculation* nevertheless articulates citizen engagement as a possible solution to the ways in which offshore oil entrenches economic inequalities and unequal terrains of power within communities onshore.

Two interviews with Scottish fishers in *Oil Means Trouble* similarly address the viewer, instructing Atlantic Canadians to learn from their experiences. The documentary's final interview with Raymond Cowie, a fisherman from Scalloway, Shetland, encapsulates both the enormous power of the oil companies and the disempowerment experienced by many local residents and workers. Speaking directly to fishers in Newfoundland, Nova Scotia, and Labrador, Cowie advises that if "the oil men come," make sure to have "a proper agreement with the oil people because they might destroy your [fishing] rights, and your fisheries, and also your agriculture." He continues: "Those oil people have been all over the world, they have great experience from the capitalist side and they know what's best for their own interests." But even contracts offer no guarantees, he suggests, recommending that workers "read the agreement a few times over, and also watch the small print. Because they're up to all the crafty moves." Rob McColl says even more bluntly: "If I were a fisherman in Canada, I would make sure I would learn from the other fishing industries in the world where there has been oil and fish interaction."

These scenes point to the ways that peoples' experiences of offshore oil in the North Sea can create space for an emergent politics to reimagine what is possible (Mathews and Barnes 2016, 21) – even if these possibilities did not, in fact, germinate into a full-fledged social movement in Newfoundland at the time. As Angela Carter and Leah Fusco observe in their study of anti-fracking campaigns in Newfoundland and Labrador, historically there has been limited public opposition to oil extraction in the province despite concerns raised by environmentalists, scientists, and non-governmental organizations (Carter and Fusco 2017, 99). Organized community mobilization against hydraulic fracturing ("fracking") for petroleum in western Newfoundland started to emerge in the fall of 2012, and the province has witnessed growing decarbonization and climate justice movements in subsequent years. Nevertheless, cinematic comparisons between Newfoundland, Scotland, and Norway afford contemporary viewers the opportunity to revisit these moments of potentiality and anticipation, where emergent energy imaginaries could produce new forms of political assemblages and identities.

Crucially, these associations between the future and ocean resources cut across political identities, framing people's personal and collective investments in oil as a livelihood as well as critiques of offshore oil's impacts on ecosystems. In April 2020, as the price of crude plummeted and the province's oil and gas industry faced layoffs and work stoppages due to the COVID-19 pandemic (Gushue 2020), the Newfoundland and Labrador Oil and Gas Industries Association (NOIA) launched a pro-oil social media campaign. Promoting the hash tag #MyOffshoreMyFuture through their Twitter account, NOIA asked people to "spread messages of hope" and say "why it is important to show continued support for oil and gas" ("Messages of Support Shared Online for Oil and Gas Industry," 2020). Workers, their family members, local politicians, and industry representatives responded to NOIA's call, circulating photographs, videos, anecdotes about working in the industry, cries for government bailouts, and worries about unemployment. Sprinkled throughout these tweets were ones from local activists, environmental scientists, and other campaigners repurposing the hash tag to share messages about a post-carbon future and the urgency of a just energy transition. While relatively small in scale, these social media skirmishes reaffirm the contestations around offshore oil as a basis for identity- and future-making – and the production of the *offshore* as a geography and concept to be possessed.

As much as these films reveal about the embeddedness of oil and extractive uses of the sea within visions of future times, they also say a great deal about hope. Hope is, fundamentally, future-oriented, marrying energy to action within "an embrace of the unknown and the unknowable" (Solnit 2016). According to Rebecca Solnit, hope locates itself within the "spaciousness of uncertainty" (Solnit 2016), emerging from fissures in our thinking about the ways things are now and could instead be. However, hope can be appropriated by corporate and government interests, like in the example of NOIA's twitter campaign #MyOffshoreMyFuture that capitalized upon the hopes of workers and their families amidst a public health crisis. The MES films also contain diverging filaments of hope in the face of oil: hopes that offshore oil would not decimate fishing groups, hopes from working-class white communities that they might harness those economic developments to their benefit, hopes that viewers of these documentaries might be swayed by the science of petroleum's ecological hazards. Collectively, these media artifacts share an investment in exploring how offshore oil might open up emergent pathways to other futures, amidst financial and environmental crises. Our challenge is to learn from imagined futures past and work towards collective survival in the here and now.

Acknowledgments

This research was supported by the Social Sciences and Humanities Research Council (SSHRC) through a Banting Postdoctoral Fellowship.

Notes

1 The province of Newfoundland and Labrador comprises two separate landmasses: the island of Newfoundland and the eastern portion of the Labrador Peninsula on the North American continent. The province's name was changed in 2001 to acknowledge both entities. Labrador is the traditional homeland of the Inuit of Nunatsiavut and NunatuKavut and the Innu of Nitassinan. The island of Newfoundland is the ancestral lands of the Mi'kmaq and Beothuk Nations.

2 While estimates of the size of the protest vary, CBC coverage of the climate strike on September 27 puts the number at around 4,000 people ("Climate Strike Draws Thousands to March on Confederation Building in St. John's," 2019). For the sake of transparency, I would like to acknowledge my own involvement in the climate strike as a participant and fellow organizer. I collaborated with Fridays for Future St. John's and gave interviews with local media about the event.

3 Memorial University's Extension Service was founded in 1959 to provide adult education programs and serve rural communities, as part of broader drive to modernize Newfoundland after confederation in 1949. The Extension Service first partnered with the National Film Board in 1967 on the Fogo Process films, shot by Colin Low on Fogo Island, as part of the now famous *Challenge for Change/Société Nouvelle* program. The Extension Service established its own Film Unit the following year, with the NFB's support. For a concise history of Memorial's Extension Service, see Webb 2014. For a contextual understanding of filmmaking in Atlantic Canada and changing practices within the NFB, see Druick 2007 and Varga 2015.

4 The first seismic surveys were conducted in the Grand Banks in the early 1950s. Mobil Oil Canada acquired the first federal exploratory permits for the Grand Banks in January 1965 and exploratory drilling in the area commenced the next year. Although oil was first identified at the Hibernia P-15 discovery well in September 1979, the death of 84 men aboard the *Ocean Ranger* on February 15, 1982 dramatically impacted the Newfoundland public's belief in "promise of oil" (Dodd 2012, 5). The provincial government banned offshore drilling in 1983, a decision later reversed by the federal government citing the limits of the province's jurisdiction. The 1980s witnessed the creation of management bodies and regulatory regimes for the Atlantic offshore area. Commercial oil production began in November 1997, following the installation of the Hibernia platform in June of that year. For a timeline of corporate exploratory activities and development of the Hibernia oilfield, see "Milestones" (*Hibernia* n.d., https://www.hibernia.ca/milestones.html).

5 For the sake of clarity, in this chapter, I will refer to the island of Newfoundland when addressing the island portion of the province, Labrador in reference to the mainland section of the province, and Newfoundland and Labrador when referring to the province as a whole.

6 The original narration in *Offshore Oil: Are We Ready?* is voiced by Cyril Byrne.

7 The Beothuk People resided on the island of Newfoundland. Their culture came under particular stress in the 18th century after Europeans began establishing permanent settlements on the island, in contrast to earlier temporary visits during the summer fishing season. European histories have claimed that the Beothuk disappeared by 1829, with the death of Shanawdithit, a Beothuk woman captured in 1823 along with her mother and sister. However, Mi'kmaw and Innu oral histories and scholars have contested this settler narrative of Beothuk erasure. For more on the changing and contentious historical perspectives on the Beothuk, see Polack 2018.

8 Cato Wadel also appears in the Fogo Process film *Thoughts on Fogo and Norway* (dir. Colin Low, 1967), produced by the National Film Board and the Extension Service.

9 The naming of the Hibernia oilfield and platform likewise evidences the settler cultural fabric that envelops these depictions of Newfoundland. "Hibernia" comes from Latin, meaning Ireland. As the corporate website for the Hibernia development notes, Hibernia has been a popular name for ships in the North Atlantic Ocean, given by mariners of Irish descent ("About Hibernia," n.d.). While there is some debate as to whether the Hibernia oilfield was named for a steamship or a shipwreck ("About Hibernia," n.d.), this naming convention nevertheless draws upon histories of immigration and settler imaginaries of shared heritage between Ireland and Newfoundland to naturalize parallels between these islands.

10 This footage was recorded in St. John's at an oil and gas conference organized by Memorial University's Extension Service in October 1979. Coverage of the conference events is featured in the "Energy" issue of *Decks Awash* Vol. 9, no. 1 (February 1980).

References

"About Hibernia." N.d. *Hibernia.* Accessed 27 November 2020. https://www.hibernia.ca/about.html.

Anand, Nikhil, Akhil Gupta, and Hannah Appel, eds. 2018. *The Promise of Infrastructure.* Durham and London: Duke University Press.

Banita, Georgiana. 2014. "From Isfahan to Ingolstadt: Bertolucci's *La via del petrolio* and the global culture of neorealism." In *Oil Culture*, edited by Ross Barrett and Daniel Worden, 145–168. Minneapolis: University of Minnesota Press.

Bonnell Public Relations. 1978. "Fish Is the Future: The Development Program for the Newfoundland and Labrador Fishing Industry to 1985." St. John's: Department of Fisheries.

Canjels, Rudmer. 2009. "Films from beyond the well." In *Films that Work: Industrial Film and the Productivity of Media*, edited by Vinzenz Hediger and Patrick Vonderau, 243–255. Amsterdam: Amsterdam University Press.

Carter, Angela V. and Leah M. Fusco. 2017. "Western Newfoundland's Anti-Fracking Campaign: Exploring the Rise of Unexpected Community Mobilization." *Journal of Rural and Community Development* 12 (1): 98–120.

Clean Water. 1967. Director uncredited. Produced by the Shell Film Unit.

"Climate Strike Draws Thousands to March on Confederation Building in St. John's." 2019. CBC News, 27 September. https://www.cbc.ca/news/canada/newfoundland-labrador/friday-climate-change-rally-1.5299400.

Damluji, Mona. 2015. "The Image World of Middle Eastern Oil." In *Subterranean Estates: Life Worlds of Oil and Gas*, edited by Hannah Appel, Arthur Mason, and Michael Watts, 147–164. Ithaca and London: Cornell University Press.

Decks Awash. 1980. "Energy" issue, 9.1 (February). St. John's: Memorial University Extension Service.

Delisle, Jennifer Bowering. 2013. *The Newfoundland Diaspora: Mapping the Literature of Out-Migration.* Waterloo: Wilfrid Laurier University Press.

Dodd, Susan. 2012. *The Ocean Ranger: Remaking the Promise of Oil*. Halifax and Winnipeg: Fernwood Publishing.

Druick, Zoë. 2007. *Projecting Canada: Government Policy and Documentary Film at the National Film Board*. Montreal & Kingston: McGill-Queen's University Press.

Fabian, Johannes. 2014. *Time and the Other: How Anthropology Makes Its Object*. New York: Columbia University Press.

Gushue, John. 2020. "N.L. Already had Multiple Crises. The Pandemic Just Pushed Them All Into High Relief." *CBC News*, April 18. https://www.cbc.ca/news/canada/newfoundland-labrador/covid-economy-nl-crises-1.5535567.

Jacobson, Brian R. 2019. "The Shadow of Progress and the Cultural Markers of the Anthropocene." *Environmental History* 24: 158–172.

Jekanowski, Rachel Webb. 2018. "Fuelling the Nation: Imaginaries of Western Oil in Canadian Nontheatrical Film." *Canadian Journal of Communication* 43 (1): 111–125.

———. 2019. "Scientific Visions: Resource Extraction and the Colonial Impulse in Canadian Popular Science Films." *Canadian Journal of Film Studies/Revue Canadienne d'études Cinématographiques* 28 (1): 1–24.

"Labrador Sea." 1977. *Decks Awash*, "Off-shore Oil & Gas" issue, 6.1 (February): 12. St. John's: Memorial University Extension Service.

Lousley, Cheryl. 2015. "Narrating a Global Future: *Our Common Future* and the Public Hearings of the World Commission on Environment and Development." In *Global Ecologies and the Environmental Humanities: Postcolonial Approaches*, edited by Elizabeth DeLoughrey, Jill Didur, and Anthony Carrigan, 245–267. New York and London: Routledge.

Mathews, Andrew S. and Jessica Barnes. 2016. "Prognosis: Visions of Environmental Futures." *Journal of the Royal Anthropological Institute*, Special Issue: Environmental Futures, 22 (S1): 9–26.

"Messages of Support Shared Online for Oil and Gas Industry." 2020. VOCM Local News, April 25. https://vocm.com/2020/04/25/noia-oil-and-gas-support/.

"Milestones." N.d. *Hibernia*. Accessed 27 November 2020, https://www.hibernia.ca/milestones.html.

Moore, Jason W. 2015. *Capitalism in the Web of Life: Ecology and the Accumulation of Capital*. London and New York: Verso.

Moreton-Robinson, Aileen. 2015. *The White Possessive: Property, Power, and Indigenous Sovereignty*. Minneapolis and London: University of Minnesota Press.

Offshore Oil: Are We Ready? 1981. Directed by Paul MacLeod. 37 minutes. Produced by Memorial University of Newfoundland Extension Service and the National Film Board of Canada. Memorial University of Newfoundland – Digital Archives Initiative. http://collections.mun.ca/cdm/ref/collection/extension/id/551.

"Oil and Gas Proposal." 1979. St. John's, Memorial University of Newfoundland Extension Service.

Oil Means Trouble. 1985. Directed by Bruce MacKay. 15 minutes, 27 seconds. Produced by the National Film Board of Canada. DVD.

"Oil Means Trouble [Study Guide]." 1986. Montreal: National Film Board of Canada. http://www3.nfb.ca/sg/66271.pdf.

Polack, Fiona, ed. 2018. *Tracing Ochre: Changing Perspectives on the Beothuk*. Toronto, Buffalo, and London: University of Toronto Press.

———. 2019. "Corporate and Worker Photographs of the Offshore Oil Industry: The Case of the Ocean Ranger." *Journal of Canadian Studies* 53 (1): 152–77.

Potts, Shaina. 2019. "Offshore." In *Keywords in Radical Geography: Antipode at 50*, edited by The *Antipode* Editorial Collective, 198–201. Hoboken and Oxford: Wiley Blackwell.

"Provincial and Territorial Energy Profiles – Newfoundland and Labrador." 2020. Canada Energy Regulator, Government of Canada. June 24. https://www.cer-rec.gc.ca/nrg/ntgrtd/mrkt/nrgsstmprfls/nl-eng.html?=undefined&wbdisable=false.

Ressources sous-marines (*Offshore*). 1973. Directed by John Ralph. 19 minutes, 46 seconds. Produced by the National Film Board of Canada for Energy, Mines and Resources Canada. DVD.

River Must Live, The. 1966. Directed by Alan Pendry. 20 minutes. Produced by the Shell Film Unit.

Scarlett, Maurice, ed. 1977. *Consequences of Offshore Oil and Gas: Norway, Scotland, and Newfoundland*. St. John's: ISER Books.

Shadow of Progress, The. 1970. Directed by Derek Williams. 24 minutes, 30 seconds. Produced by British Petroleum and Greenpark Productions. https://www.bpvideolibrary.com/record/481.

Sheldon, Rebekah. 2016. *The Child to Come: Life after the Human Catastrophe*. Minneapolis and London: University of Minnesota Press.

Shetland Experience, The. 1977. Directed by Derek Williams. 25 minutes. Produced by British Petroleum. https://www.bpvideolibrary.com/record/465.

Solnit, Rebecca. 2016. "'Hope Is an Embrace of the Unknown': Rebecca Solnit on Living in Dark Times." *The Guardian*, July 15. https://www.theguardian.com/books/2016/jul/15/rebecca-solnit-hope-in-the-dark-new-essay-embrace-unknown.

Speculation. 1980. Director uncredited. 22 minutes, 10 seconds. Produced by Memorial University of Newfoundland Extension Service and the National Film Board of Canada. Memorial University of Newfoundland – Digital Archives Initiative. http://collections.mun.ca/cdm/ref/collection/extension/id/546.

Thoughts on Fogo and Norway. 1967. Directed by Colin Low. 16 minutes. Produced by the National Film Board of Canada in collaboration with Memorial University of Newfoundland. http://www.nfb.ca/film/thoughts_on_fogo_and_norway/.

Tomorrow Is Too Late. 1974. Directed by Bané Jovanovic, Douglas Kiefer, and Don Virgo. 28 minutes. Produced by the National Film Board of Canada for Environment Canada and Fisheries and Marine Service. https://www.nfb.ca/film/tomorrow_is_too_late/.

Tsing, Anna Lowenhaupt. 2015. *The Mushroom at the End of the World: On the Possibility of Life in Capitalist Ruins*. Princeton and Oxford: Princeton University Press.

Tuck, Eve, and K. Wayne Yang. 2012. "Decolonization Is Not a Metaphor." *Decolonization: Indigeneity, Education & Society* 1 (1): 1–40.

Tyagi, Ila. 2020. "Inscribing Interiority and Ideology: Representing the Visually Elusive in the American Petroleum Institute's Cold War Films." *MediaTropes* VII (2): 1–18.

Urry, John. 2014. *Offshoring*. Cambridge and Malden: Polity Press.

Varga, Darrell. 2015. *Shooting from the East: Filmmaking on the Canadian Atlantic*. Montreal and Kingston: McGill-Queen's University Press.

Webb, Jeff A. 2014. "The Rise and Fall of Memorial University's Extension Service, 1959–91." *Newfoundland and Labrador Studies* 29 (1): 1719–1726.

Weszkalnys, Gisa. 2014. "Anticipating Oil: The Temporal Politics of a Disaster Yet to Come." *The Sociological Review* 62 (S1): 211–235.

White, Susan. 2019. "Preparing to Compete." Faculty of Business Administration, Memorial University. December 6. https://www.business.mun.ca/news.php?id=12846&type=news.

7

DISPATCHES FROM TWO COLD WATER OIL CULTURES

Norway and Newfoundland and Labrador

Brigt Dale and Danine Farquharson

Our chapter offers a series of snapshots describing contemporary realities of cold water oil in Norway and Newfoundland and Labrador (Canada). This work is a result of Brigt Dale (BD)'s research on energy futures in Norway that prompted responses from Danine Farquharson (DF) regarding the similarities and differences with Newfoundland and Labrador (NL).[1] As early as the 1970s, NL looked to Norway as a role model for offshore oil and gas development, and comparisons of the two places continue through the decades (Reid and Collins 2012, 7). However, connections between Norway and NL go beyond our petrocultures; arguably, the earliest being histories of trans-Atlantic voyages to, and landings in, NL by Vikings (Fein 2017, 27). The North Atlantic cod fishery – an industry based on a migratory species – is another link that continues in the present (Arnason et al. 2000, 233), but the relationship between the two places deepened with the initiation of offshore oil exploration in NL.[2] All this despite factors that make the two places distinct – the difference in their status as sovereign nation (Norway) and provincial jurisdiction (NL) being the most obvious and significant because Norway also has more protectionist oil and gas policies than NL (Carter 2007, 18). Other differences, such as offshore expenditures, number of producing fields, and amount of oil and gas produced, are profound. Norway's fossil fuel industry is more mature: it spends more, has more fields, and produces more oil and gas than NL.[3] One way to describe the cold water oil relationship between the two places is that NL looks to Norway as iconic inspiration to develop offshore oil strategies. There is even a "Stavanger Drive" street sign in NL's capital city (Figure 7.1).

We deliberately use the word "dispatches" in our title first to indicate the idea of messages sent back and forth; but also, second, to describe the uneven nature of the exchange to follow. We corresponded with each other between 2018 and 2021 and that dialogue (artificially reproduced here) was open-ended and unstructured, yet always bound by a working through of some, but by no means all,

DOI: 10.4324/9781003024125-7

FIGURE 7.1 Stavanger Drive street sign in St. John's
Source: Photo by author, 2021

of the similarities and differences between our two cold water oil locations. Our conversation evolves from different scholarly frames of reference and different types of source material (social sciences and humanities; industry publications and social media), with the goal of offering petroculture snapshots – sometimes descriptive, sometimes analytical, sometimes activist, sometimes all three.

The dialogue to follow takes up three themes to ground our contemporary comparisons of these two cold water oil cultures, all of which are based in a deeply rooted sense of ontological security that oil-dependency previously provided for both places and peoples. "Oil Matters" unpacks some communications through different media about the exploration, development, and government support for oil and gas. "Shocks to the System" further contextualizes our dispatches within recent global crises such as the COVID-19 pandemic and the rollercoaster of oil and gas prices. "Tales of Offshore Oil" interrogates our ontological security. We conclude in "Hopeful Futures?" with some of the challenges facing Norway and NL as we contemplate post-oil futures.

Oil matters

BD: Almost every day I receive news in my inbox from the Norwegian Petroleum Directorate (NPD) about activities on the Norwegian offshore shelf. These notices include information about drilling licences awarded in both new and mature licencing areas, the initiation of test drillings, the expectations operators have about the assets below the sea bottom, and the results of such drillings, positive or negative. Mere minutes after I receive these messages, some are picked up by various media outlets: business news outlets first and then, if the content is juicy enough, by Norwegian mainstream media channels. News about oil *matters* in oil-dependent Norway and, for those who know what to look for, the bureaucratic, somewhat uninspired language of these messages hides dramatic tales well worth telling.

One example is a message that ticked into my email about a drilling licence being awarded for well 7132/2-1 in licence 857 in December 2018 (NPD 2018). As usual, I opened the message and then the link to an online map to see where the test drilling is to take place. I found out that this specific well was situated close to the fairly recently drawn Norwegian–Russian separation line in the Barents Sea, based on the 2011 agreement signed by Foreign Ministers Sergei Lavrov of Russia and Jonas Gahr Støre of Norway (see Ryggvik in this volume).[4] Further, the message said that the drilling licence was awarded in the 23rd licencing round in 2015, and that this was the first well to be drilled in this particular licence. The message revealed nothing about all the work put into the process leading to the decision to test drill right there, nor did it say anything about either what it would mean for the future of Norwegian petroleum if they found hydrocarbons there, or if the well ended up being dry.

The importance and relevance of these specific drilling sites for Norwegian society were thus hidden, yet such messages are important snapshots from processes that last for years and involve a number of specialists and decision-makers, both within and beyond the petroleum industry. The brevity of the messages hides the chain of events that precede the initiation of test drilling, the years of preparing for such moments, and are therefore a sombre reminder of the influence of bureaucracy and objectivist principles on an elaborate process; the enthusiasm and excitement of exploration and "black gold digging"[5] is wrapped in a rhetoric suitable for decision-making in a knowledge-based society.

This rhetoric of oil exploration should be about science, not passion, the argument goes: "doing oil" should be knowledge-based and *rational*. It is "*what we Norwegians do*" (Dale 2011, 3) and is thus the basis for Norway's collective ontological security.[6] Information about a new well like 7132/2-1 is taken up by journalists and reported in hopeful terms both by opponents and proponents of extending Norway's petroleum era. Proponents hope that a new finding will prove commercially viable; opponents hope is that the well is dry. The results of test drillings are presented in sober language, completely free from expressions of joy or disappointment, and promise nothing more than objective results. In the case of well 7132/2-1, the opponents of Arctic oil extraction could breathe a sigh of relief. In April 2019, the NPD reported that the licence owner Equinor declared this well to be dry: no hydrocarbons were found. The message thus ends, as they always do in these instances with: "the well will be permanently plugged and abandoned" (NPD 2019).

The steady stream of messages about the state of the petroleum resources on the Norwegian shelf are a glimpse into the relentless process of maintaining the status quo (or prolonging its demise), but they are also a testament to both the resilience of the petroleum nexus in Norwegian society and to the anonymous and uncounted workers who makes up the system and whose lives are intertwined with and dependent on the processes and their outcome (Haarstad and Wanvik 2016). Thus, the neutral and bureaucratic language of these messages hides a much more dramatic and fundamental, even existential, question: Is there any more to find? Can we find enough to still make a living of this? *Are we safe in our oil dependency?*

DF: Media reactions to and coverage of offshore oil exploration and development in NL is remarkably similar to what Brigt describes in Norway, mostly in the sense that oil news *matters*, but also because industry reports frequently initiate public discussions and debate. My personal newsfeeds are more attuned to social media platforms, such as a Facebook group called "Newfoundland Offshore" which, while ostensibly dedicated to celebrating offshore vessels and rigs, is also a powerful oil industry advocacy group. Other social media sources include the Newfoundland and Labrador Offshore Industries Association (NOIA)'s twitter feed. The Canadian Broadcasting Corporation (CBC) reports regularly on licences granted, and productions started or stopped. Oil industry news is always a leading news story in NL because the province is so heavily dependent on the one resource.

One recent example that illustrates the importance of oil news – and its embeddedness in climate crisis realities – is the myriad, immediate, and oppositional reactions to a CBC interview with Dr. Angela Carter, author of the recently published *Fossilized: Environmental Policy in Canada's Petro-Provinces* (UBC 2020). On November 16, 2020, CBC's host of the local 'drive-home' show, *On the Go*, spoke with Carter about her book, which argues persuasively that NL, and other Canadian provinces, face enormous economic and environmental risk in continuing offshore oil and gas developments.[7] A brief but critical context for this interview is that it occurred just before the Canadian government announced (after months of lobbying from industry) that it would be further subsidizing the oil industry during the pandemic (Bakx 2020). *On the Go*'s host, Ted Blades, posted on Twitter mere minutes after the interview aired that "We've already gotten a complaint from a local oil advocate basically saying we shouldn't have talked to [Carter] because she's biased against our industry" (Blades 2020). The reaction to Blades' interview with Carter didn't stop with one complaint, there was enough audience feedback – on Twitter, Facebook, over email and from telephone callers – that *On the Go* ended up broadcasting four more segments (between December 2 and 8, 2020) on the topic of energy transition away from fossil fuels and the province's oil industry. In the first of these, Blades had Dr. Carter back on the line, joined with a member of Iron & Earth (an oil industry worker-led organization dedicated to global carbon neutrality by 2050) in order to respond to questions that were notably submitted to *On the Go* by the past

President and CEO of NOIA.[8] The CBC is Canada's national broadcaster, and these few radio conversations are symptomatic of how oil news plays out in NL, on air, in print, and online: it is omnipresent. As Brigt rightly notes, oil matters are presented by various media sources as concomitant with ontological security.

Shocks to the system

BD: Norway's oil reserves rest below offshore cold waters, and the industry's future is also in deep trouble, if not immediately so, then at least in the not-so-distant future. There have been two global shocks to the oil industry since the 2008 financial crisis, and both have deepened or broadened awareness that our petrocultures are changing. The first shock, the plummeting of oil prices in 2014–2015, was a consequence of the combined effects of international reactions to the Russian annexation of Crimea, the Syrian War, and an OPEC move to bring American fracking under control (Baffes et al. 2015 and Dale et al. 2018). The second shock, in 2020, came from the combination of tumbling oil demand (due to varying lock downs/shut downs in order control the spread of COVID-19) and a lack of agreement between Russia and OPEC on limiting production (IEA 2020). The combined effect of these events, a one-two punch to the gut of global economies, left Norwegian oil executives and political leaders with the grim prediction that their future might be beyond their control. Being a relatively small player in the raging battle for who gets to produce and sell the remaining oil, Norway is currently stuck between opting for the well-known but increasingly unstable path of continuing into a petroleum future, or (as promoted by the Executive Director of IEA) to opt for "green energy at the heart of stimulus plans to counter the coronavirus crisis" (Birol 2020). This choice is before decision-makers at a time when oil revenues are down and peak earnings are a thing of the past, even though there is an estimated 15.6 billion standard cubic meters of oil equivalents remaining on the Norwegian shelf (see Figure 7.2). Another variable at play in forecasts of future oil revenues is when peak *demand* will arrive. Oil price variation is likely more important than how much is found/produced, in estimating future earnings from the remaining revenues. Another complicating factor is that the more petroleum that is found, the higher the risk for the Norwegian state, as Norwegian petroleum taxation regulations allow for tax exemptions on produced petroleum covering the initial investment costs (including exploration drilling). In short, betting on revenues, perhaps decades into the future, is an increasingly risky bet.[9] At the time I write this, in autumn 2021, the center-right Cabinet of Erna Solberg has in fact suggested that amendments to the tax scheme that has made investments in Norwegian oil less risky than in all other sectors should be considered, with the aim of adapting it to the new reality that oil just might not be that profitable for very much longer. As the Minister of Finance said during the press conference: "The changes introduce stricter tax rules with a more neutral effect on investments" (Government.no 2021). Even though the sitting government is by no means prepared to politically slow down the industry due to its contribution to the burning of fossil fuels, the September parliamentary election campaign has for the first time brought the issue of climate change to the fore of the agenda, and the generally agreed upon perception in Norway is that further offshore oil exploration can only serve to slow the industry's decline not reverse its fall (NPD 2020).

DF: Contrary to Norway, the end to offshore oil exploration and extraction is not central to the political conversation in NL. Or, if it is agreed upon by oil industry advocates and elected officials, then they are not saying so in any public way. There is not enough space in this chapter to historicize the current economic situation in NL, but it is no exaggeration to write that the province was in financial trouble before the two shocks to the system described above. Currently, one outrageous and devastating blow to the provincial economy – and one that will continue to harm the province's well-being – is the debacle of the Muskrat Falls Hydro-electric project in Labrador, a project managed by people with oil industry experience. Cost overruns, which continue to grow, and public inquiries do not nearly encompass the full scope of the harm the project is already wreaking for Indigenous lands and communities.[10] *The Commission of Inquiry Respecting the Muskrat Falls Project*, subtitled "A Misguided Project," lists many problems, not the least of which are the suppression of Indigenous community concerns and environmental damage (LeBlanc 2020).[11] These few details about a hydro-electric "boondoggle" – a project designed to transition away from fossil fuels – are offered to establish not only that the province was already in shock over energy projects and our energy futures but also to argue that any post-petroleum transition in NL needs to be one deeply committed to social justice.

Nonetheless, there is one revealing snapshot from 2020 and 2021 that illustrates how quickly the situation of oil production can change in NL and, in particular, since the COVID-19 pandemic. In September 2020, Husky Oil asked the provincial and federal government for a 2.2 billion dollar bailout to rescue its West White Rose development, "an oil and gas drilling venture off the Newfoundland coast that it had planned to expand until the COVID-19 pandemic sent global fossil markets into a tailspin" ("Husky" 2020). On September 15, 2020, the premier of NL, Andrew Furey, told reporters that the province could not financially help Husky, and his Finance Minister stated that "There will be a lot of work that has to be done post-COVID to consider how we may ensure our financial stewardship and our financial success as a province is maintained [...]. But now is not the time" (Smellie 2020). Very shortly afterwards, by December 2020, not only was a new Oil and Gas Recovery Task Force established but there was also direct financial support distributed to oil companies through the Oil and Gas Industry Recovery Assistance Fund (Newfoundland and Labrador 2020). These bailouts of the oil industry were nation-wide, and by the end of March 2021 – six months after the province said no bailouts were coming – over 18 billion dollars in subsidies to Canadian operators in the oil and gas industry were distributed. The precarity of the industry is obvious if federal and provincial money is required to keep production running. Equally in no doubt is that NL is doubling down on oil supports.[12] The shocks to the system are being absorbed by taxpayers and paid out to multi-national corporations. And those corporations, along with powerful industry advocates like NOIA, are constructing compelling narratives about local, regional, and global need for such oil and gas operations as well as conscious or unconscious erasure of the potential for oil to lose its monetary and cultural value.

Tales of offshore oil

BD: There are at least two dominant but contradictory narratives about energy futures in general and, more specifically, the future of fossil fuel extraction in Norway. One narrative is that the role of fossil fuels in the future energy mix is *unpredictable* but ends with a post-petroleum future. A second and opposing narrative is that the energy future can be imagined, assessed, and assumed; it is *predictable and knowable.* The narrative of unpredictability can be frightening (as all uncertainty can be), but it is also used by oil pundits and politicians to argue that oil must remain as an inevitable part of the future's energy mix. In such views, no alternative is plausible and the weakening of the oil and gas sector, as proactive climate politics demand, would lead to global energy shortages.[13] The second narrative of knowability asserts that the future must be planned for through the use of scenarios and predictions based on statistics, historic developments, and (first and foremost) economic evaluations – in other words, on *what is known today to be true.* This narrative is appealing to politicians, business executives, and much of the public because it satisfies a desire that everything will stay the same or, that change will be what our resilient present can handle: that *we're still safe.* In short, the status quo feels less threatening than the transformative efforts required to transition beyond fossil fuel dependency.

Continuing oil production and use relies on imaginaries in which oil is part of the solution to the challenges facing the world in the 21st century; a myriad of challenges of which the climate crisis is by far the most urgent. The presence of a relatively large potential fortune of oil and gas revenues on the shelf north of the Polar Circle has presented a new opportunity for the industry. Simultaneously, that potential presents a number of new challenges, of which environmental concerns and regional tensions are prominent. As such, a new – a third – narrative has developed wherein Norwegian technological know-how and environmental management principles are presented as producing *greener oil* than the rest of the world, and that production from the Norwegian shelf should therefore be supported as global initiatives such as carbon pricing should make Norwegian oil competitive. The proponents of this greener/cleaner narrative argue that the remaining and necessary oil should be developed by those producing with less emissions. Cold water oil from Norway is thus be presented as being a part of the solution to the climate change challenge.[14] Because these three narrative threads are woven together and mutually supportive of continuing oil production and use, the tangled web merits further discussion.

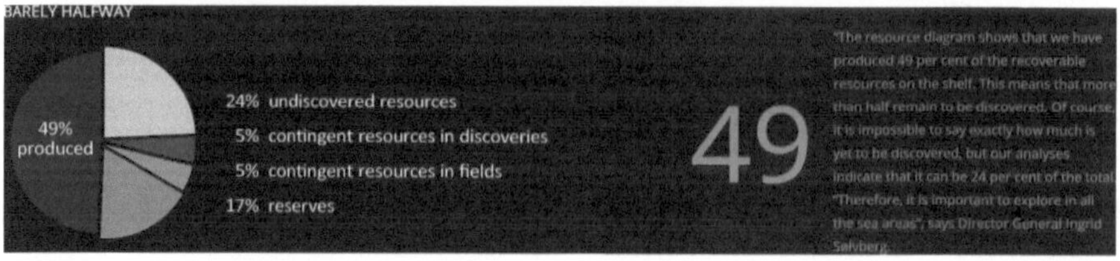

FIGURE 7.2 Barely Halfway
Source: Norwegian Petroleum Directorate

For any of the 'keep drilling' narratives to work, there must be resources left to deplete. The NPD provides regular updates about the state of affairs on the Norwegian shelf, which are then translated into infographics like "Barely Halfway."

"Barely Halfway" challenges the assumption that the oil age is over; in fact, what is stated in the text accompanying this figure is that "more than half remains to be discovered" (NPD 2018). From a resource availability perspective, the oil era is far from over, the resources are available, and we should go get it. The most politically potent, but also controversial argument, is that Norwegian oil and gas should be produced for two reasons: first, its production is 'greener' than the world average; and, second that a relatively large proportion of the energy mix under the Paris Agreement's 1.5° scenario should be oil and gas from Norway, as – again – we produce it greener. The rationale for this argument deploys assessments of both future demand and potential supply to present a window of opportunity and a *reason to drill.*

The International Energy Agency's World Energy Outlook shows that there will, based on an assessment of the development of currently known sources of energy, be an energy cap in 2040 that, in the words of the IEA in the illustration, "must be filled by production from conventional crude oil projects yet-to-be approved" (IEA 2016). Combine the IEA outlook with assertions from the NPD that this gap should – and can – be filled with Norwegian crude yet-to-be approved (and found). That crude, according to the NPD, exists almost entirely in the speculative: not yet producing, still to be confirmed, still to be discovered. To be clear, the argument asserts that drilling for oil and gas on the Norwegian shelf, no matter its material reality, is good for the climate. In fact, the argument might seem an oversimplification or exaggeration, but it has nonetheless been presented throughout the last decade by parliamentarians, ministers, and lobbyists in Norway as fair and relevant arguments for the basis of political policy (Jensen 2012).

Returning to my opening comment about ontological security, these tales pay heed to the obvious: the future will look different from anything we have experienced, and the highly uncertain status of oil in the future is destabilizing. Recent developments show that future oil wealth is far from as certain as it has been presented. Global investments in oil exploration fell sharply during 2014 and 2015, a fall that is reflected in scenario predictions pointing towards the need for a substantial boost of explorations efforts, as the IEA report of 2016 also indicated. The 2014 oil price shock was also a watershed moment for Norwegian debates about the future of oil. From 2014 to the present (the crisis following the oil price wars and the COVID-19 lock-down), has seen a gradual change in the way oil is debated and narrated. The chain of events essentially starts with the 2014 plummeting of the oil price, moves through the signing of the Paris agreement the year after, is followed by a gradual development of more complicated debates that gained new traction due, in no small part, due to Greta Thunberg's now globally active school strikes. Importantly, a 2017 to 2020 court case followed in which a consortium of plaintiffs sued the Norwegian government for breaching the constitutional rights of future generations when they opened for a 23rd licensing round for petroleum exploration in the Barents Sea – a round followed up just two years later by a 24th round (NPD 2019).

It is also in the years following the oil price collapse and subsequent redefinition of the public discussion over oil in Norway that the most politically controversial of all debates about the development of the Norwegian shelf to date was, for all practical purposes, decided. That decision provides important insight into how an oil-dependent country starts grappling with its inevitable post-petroleum future and showcases a fourth narrative: saying no to oil.

Historically, the practice has been that, when there has been a need for new areas, they have eventually been opened. Not so though with regard to the sea areas outside the Lofoten, Vesterålen and Senja regions (see Kristoffersen et al. in this volume). Multiple objections from the people who live there became compelling challenges for an industry that had become used to getting access to any zones it needed on the Norwegian shelf.[15] At their annual convention in 2019, the largest political party in Norway (the Labour Party) made the controversial decision for the first time to say "no" to opening an offshore region for exploration of oil and gas reserves. This decision ended, for all practical purposes, an almost two-decade long debate about access to the area referred to in 2006 as "the jewel" of the Arctic petroscape (Kristoffersen and Dale 2014, 212). In 2001, a licence to drill in immediate proximity to the world's largest cold-water coral reef outside the island of Røst in Lofoten was revoked by the Stoltenberg Cabinet after massive protests from fishers and environmental organizations (Dale 2011, Kristoffersen 2014, and Andersen 2017). This revocation initiated postponements of test drilling activities before the matter was put to rest with the 2019 decision from the Labour party, ensuring that there was no longer a parliamentary majority for oil extraction from the area. The issue pits fishers against oil executives, party bosses against their youth organizations, environmentalists against industrialists, and engages previously apolitical citizens in a people's movement. The 2001 temporary ban would be re-imposed by all new Cabinets up to the present, notwithstanding that they have to partner with smaller political parties who have all declared oil as *activity non grata* in LoVeSe (Dale and Kristoffersen 2018, 253).

What tipped the scale in the direction of not drilling for oil in Lofoten?[16] Together with colleagues, I have argued that the answers involve environmentalist concerns backed by a national culturally-based resistance movement rooted in a place-based understanding of both nature and natural resources. These forces are central to how people in Lofoten imagine a desirable, sustainable, and meaningful future beyond oil. The question in the region of whether or not to be included in the oil era has been largely tied to the most vital and symbolic resource upon which society in these islands has been sustained for thousands of years: the safety of migrating cod or *skrei*, from the potential environmental consequences of oil exploration and drilling. This concern, combined with the realization that alternatives to a petroleum-driven local economy exist, has changed Lofoten from being a potential carbonscape to a post-petroleum region. A striking example of how the discourse around oil has changed from "yes, more oil" to "no more oil" lies in how future imaginaries are created and by whom.

Prior to the fall of oil prices in 2014–2015, large national industrialization actors such as Statoil (now Equinor), The Norwegian Oil and Gas Association, and the Confederation of Norwegian Enterprise (Norwegian abbr. NHO) produced reports that included scenarios for a future in which oil and gas development was necessary for economic and societal development. In fact, one of the most influential scenario reports was produced in 2010 by Statoil depicting what they called "industry sketches Nordland," in which the LoVeSe regions feature prominently (Statoil 2010). The report envisioned three phases of development and production that suggested impressive figures on local, regional, and national income and job creation potential. Industry protagonists would argue that the regions could not pass over this possibility for development. However, the report said nothing about future quality of life or how that was perceived locally. Further, the figures presented have little basis in what geological data actually existed about the potential petroleum resources in the region (Dale 2011; Kristoffersen and Dale 2014). Such assessments pay little to no heed to the concerns and wants of people in Lofoten, neither do they succeed in the strategy of establishing the centrality of the area to the future of North Norwegian oil development.

Interestingly, a recent futuring effort is a report from consultancy firm SALT called *Jakten på Lofoten* ("The Search for Lofoten"), financed by Innovation Norway and regional banks. This report used ethnographic and co-production methods (together with local stakeholders and communities) to outline a desirable future based on the needs, desires, and hopes of locals (Ørnes et al. 2019). "The Search for Lofoten" is full of stories about lives and imaginations about possible – and desirable – futures anchored in local, lived experiences and stands in stark contrast to the oil-infused future imaginations produced by either the oil companies themselves or hired consultancy firms focussing on economic and industrialization potential.[17] The report is an example of how successful alternative stories about the future can be produced in oil-dependent Norway where the ontological security of communities and individuals is taken care of. The Lofoten case shows that despite the seemingly hard-coded embeddedness of Norwegian petroleum in the very fabric of Norwegian identity and society, there are developments that point in the direction of disruptions to that system:

towards opposition to oil and alternative future narratives. Disruptions like the Lofoten case can be just the thing needed to break out of the construct that what has provided security in the past and present also, inevitably, provides security in the future.

As described earlier, well 7132/2-1 in the Barents Sea was found to be dry. That well is but one of many disappointments in the oil push to the north (the 23rd and 24th licensing rounds have not as of yet led to any significant findings of oil or gas). From an oil exploitation perspective, the dry well supports the desire to gain access to new unexplored areas where potential for oil or gas exists. However, recent developments, culminating in the IEA assessment of 2021, show that no new areas of oil and gas are needed if the 1.5° scenario is to be met. In other words, continuing to rely on more oil and gas to secure the future welfare of Norway could very well turn out to be a deeply flawed strategy. What might, in fact, be most threatening to the future security of the Norwegian population is the lack of transformational capacity; ontological security needs paradoxically becomes a future ontological security *threat* if we miss out on opportunities to prepare for a future different from the present. The easiest way of restoring a sense of ontological security in a time of upheaval, which highlights the contingent nature of the future, is to refer back to what was secured before: oil. That eternal return to the known, however, requires that the public be persuaded that the status quo is *secure*. That persuasion entails a persistence in the order of things that, many would argue, necessitates overlooking the risks we are all facing, and thus the need for transformative change.

DF: I want to hone in on and unpack two of these narratives and, in the context of the "time of upheaval" Brigt mentions, at least how I experience them in NL. The first narrative relates to "futures" and the second considers "clean" or "green" oil. On Wednesday, September 16, 2020, at a rally to support the oil and gas industry in the province, Premier Andrew Furey said: "there is no future without oil and gas." This particular rhetorical statement traffics in a tried-and-true strategy that Mark Fisher succinctly describes in *Capitalist Realism: Is there No Alternative?* (2009). Fisher writes that capitalist realism can be broadly understood as not only hegemonic but also presented as the *only* viable economic system; thus, there can be no imaginable alternative (Fisher 2009, 16). Furey's "no future" statement does not stray far from Fisher's contention that capitalist realism creates and sustains a "pervasive atmosphere" that affects areas of cultural production, political-economic activity, and general thought. Nine days after that industry support rally, there was a "Fridays for the Future" march for climate justice, and it is striking that these two rallies happened so close together and at a time when the Canadian federal government was considering major financial bailouts for oil and gas industries. Both public rallies worked with the word and the concept of "future" but in contradictory ways (see Jekanowski in this volume).

While I am unable to quantify which of the public events was more successful, what is interesting in terms of the crafting of campaigns is that the offshore oil industry was able to pivot their messages quickly and effectively during the

COVID-19 pandemic months of anxiety and uncertainty. One example is how NOIA translated a widely distributed national image of general support for oil and gas into a regionally resonant one. CanadaAction.ca produces merchandise (t-shirts, hats, and stickers) that proclaim: I [heart] Oil & Gas, all including the iconic Canadian maple leaf. Signs used at a protest in support of NL government continued financing of the fossil fuel industry adapted CanadaAction's main logo/promotional image; first, the tag line is WE [heart] Offshore Workers, and second, the maple leaf is replaced with images of the NL provincial flag and an outline of the province's geographic boundaries.

The revision of CanadaAction's logo is important for at least four reasons: it makes clear the geographic region of concern, it shifts the focus from general support for an industry to support for people (workers), and it moves from the personal "I" to the collective "we" – all to address local concerns over job security in the offshore sector that is threatened by both shocks to the system mentioned earlier in this chapter. The last significant change, and this is related to what follows about the greening of oil narratives, is the elimination of the words "oil" and "gas" all together. The cumulative effect of the newer campaign image is an articulation of a province-wide effort (and one imbued with regional identity via the provincial flag on display) to come together (as we have done during the pandemic) to support our own communities. The voice is no longer a single person articulating a belief; it is now a collective statement of solidarity with workers who should be deemed "essential." Premier Furey's statement about there being no future in NL without oil tacitly makes the case that offshore workers are essential to *all futures* of the province.

The second idea worth commentary is how industry advocates and governments have absorbed and incorporated some key messages from the climate crisis activist groups. Again, Fisher is helpful in making clear what is going on: "climate change and the threat of resource-depletion are not being repressed so much as incorporated into advertising and marketing" (Fisher 2009, 18). What Brigt notes in Norway ("the argument is that drilling for oil and gas on the Norwegian shelf is good for the climate") is clearly deployed by the government of NL in conjunction with industry advocate groups. As journalist Ashley Fitzpatrick has written and tweeted about, the cleanliness of Canadian offshore oil

> was the cry through 2020, as then-provincial Natural Resources Minister Siobhan Coady lobbied for federal support after the oil price crash and early challenges of COVID-19. "The importance of the offshore industry in Newfoundland and Labrador cannot be overrated. It represents one of the *lowest carbon-per-barrel footprints* in the world, contributes an estimated 30 per cent GDP, and directly employs 6,390 people with thousands more in supporting industries." (Fitzpatrick 2021, emphasis mine)

The statements by Minister Coady noted above, alongside others such as this one from Ken Hanley (the Executive Director of Newfoundland and Labrador

Environmental Industry Association) make clear the rhetorical strategy of greenwashing oil and gas: "With its commitment to achieving net zero by 2050, Canada's offshore industry provides the basis from which enormous *clean growth and economic diversification* is possible for Newfoundland and Labrador" (We Are NL Offshore, no date, emphasis mine). While the narrative is clear, who is producing and driving that narrative is more tangly. The statement from Minister Coady was quickly reproduced on both the We Are NL Offshore website (a joint government, industry, and union initiative) and the twitter hashtag #MyoffshoreMyfuture, among others. The mechanisms at work in such public statements are a form of greenwashing, but also what Jordan Kinder calls "petro-turfing" (Kinder 2020, 168). In his discussion of the Ethical oil campaign in Canada, Kinder develops the idea of petro-turfing as a direct "countermovement" to the dirty oil campaigns of climate and environmental activists of the late 1980s and 1990s. And, as Kinder notes, it is effective. The effectiveness squares perfectly with Fisher's idea of capitalist realism in that the strategy incorporates the most effective critiques of capitalist structures into a defence of that structure itself. This incorporation is arguably best described by Andrew Shmuley in "Totality, Hegemony, Difference: Henri Lefebvre and Raymond Williams" wherein he writes:

> a dominant formation's most successful strategy regarding those who would question or threaten its dominance can be found in the attempt to incorporate their views and practices in a manner that effectively diffuses them of their revolutionary potential: a perverse 'negation of negation' that allows the reigning relations of power to remain fundamentally unaltered (2008, 219).

Finally, on this point, it is worth heralding Lefebvre and Williams' mutual belief that such hegemonic practice (and it is a practice not "real") can never be totally or exclusively quieted (Shmuely 2008, 222). Rather, the answer to the absorption of revolutionary actions is to stop being defensive and to counter attack (Shmuley 2008, 226). Perhaps, these ideas of counter action (or counter-counter action), which Kinder also mentions, can be made manifest in efforts to transition away from fossil fuel dependency.

The capitalist realism of government spokespeople, industry advocates, and unions stands out next to the narratives about energy transitions because the latter are more complicated: there are many paths towards a post-petroleum future, whereas maintaining the status quo is simple to understand. That multiplicity is much harder for policy makers to work with because of the unpredictability and uncertainty. In NL, the political will is clearly to stick to the same thing; indeed, the province has "doubled down" on the oil future, whatever that may be. Brigt's comment that "environmentalist concerns backed by a national culturally-based resistance movement rooted in a place-based understanding of both nature and natural resources [...] has imagined a desirable, sustainable and meaningful future – beyond oil" is not something that is happening, yet, in NL.

Hopeful futures?

BD and DF: The contemplation of oil futures, different from past and current realities, in Norway and NL is ongoing. The challenge for such futures, as Beck and others have argued, is deeply connected to the risks we face, how we understand and interpret them, and how we deal with them (Beck 2009). As a consequence, the matters of what a viable energy transition looks like, who secures what and how, are dependent on a number of factors: the role and courage of the political leaders, the involvement of constituent communities, and, ultimately, how any decisions influence that deeply rooted sense of ontological security that oil-dependency has provided for both our subject populations. Thus, the changing circumstances under which a redefining of the future of oil takes place leave us with the reality that no future *immediately feels safe*, and referring to security produced from past and present dependencies seems the right thing to do for many. As so many of our experiences during the pandemic have proven, the mixed sense of fragility and urgency continues to increase in times of rapid change.[18] The consequences of local and global shut-down responses to the COVID-19 pandemic are shocks to the system that are felt as both threatening but also full of possibilities; it may very well be a tipping (or intervention) point for how the global economy will develop, particularly in terms of tackling the overarching threat of global climate change (Farmer et al. 2019; Galafassi et al. 2018; Tàbara et al 2018). Recent academic work on societal transformation argues that unexpected and unprecedented events may be just what is needed for transformations (while being scary and seemingly impossible when they occur in hindsight): triggering important choices and decisions that secure a desirable and sustainable future, post-petroleum.[19] It remains to be seen if Norway and NL will take advantage of the opportunities that crises create to decrease the insecurity that a petroleum-based path represents and to move into a future beyond oil.

Two things that will benefit a future beyond oil are a collective ownership of oil and all its attendant financial rewards and environmental harms, plus a collective spirit of we're all in this together. Norway's accumulation of wealth based on its ownership of petroleum rent surely has its bizarre flip side; the profits earned as a consequence of the soaring oil price due to the wars and global unrest following 9/11 are far from being the normal upon which future ontological security can be based. But the fact that there is a strong element of collective ownership to oil in Norway, and that the state has met its obligation to secure its population through an active involvement in the way revenues are shared between petroleum companies and the Norwegian public, means that oil is not just one of a number of industries in Norway – it's the industry upon which Norwegians have gradually relied more and more. And it is in this broader, more fundamental, coupling of a sense of ontological security with oil that we find important reasons, even more important than any potential economic short-term losses, for why a transformation away from oil will be difficult in Norway and NL. In this, understanding the importance of *what* offers people a sense of security is important, as it makes clear the rationale behind both the opposition to and initiatives for transformation away from oil. And there is no doubt that replacing oil will be difficult in both our locations; there are very few export goods that can provide the same revenue, and there are thousands of local people employed by the industry. Further, it is not easy to disrupt the assumption that currently underpins our ontological security – that *we do oil, and we do it well.* Still, the inevitability of a future after oil is something that has only been reluctantly and with hesitance introduced into the discussions of how to secure society in the future, not only economically but indeed also ontologically.

The oil economy remains important for Norway and NL, and its effects are far from only negative. The last chapter has yet to be written on how the continued focus on extracting oil might lead us further into a dependency that can be expensive to get away from, and as the recent debates over economic rescue packages to the industry from the government have shown, the undoing of an economic, political, and ontological dependency developed over decades of increased prosperity, and sense of *being secure* will not easily be dismantled. The expense level for maintaining our current lifestyle and a safety net to secure it is so high that alternatives to oil often fall short of expectations of what might secure our future. Therefore, the obligation to transform cannot be left to individuals or communities alone (as campaigns about changing individual lives suggest), but require inspired political and business leadership with a desire to create the fundamentals for a secure future beyond oil. Thus, both Norway and Newfoundland and Labrador must couple their narratives with active polices and bold action to leave the remaining oil assets where they have been for millions of years – underneath cold waters.

Notes

1 As Jekanowski rightly notes in her chapter, the "province of Newfoundland and Labrador comprises two separate landmasses: the island of Newfoundland and the eastern portion of the Labrador Peninsula on the North American continent. The province's name was changed in 2001 to Newfoundland and Labrador to acknowledge both entities."

2 Two examples that discuss and analyse the cold water oil relationship between Norway and NL include Maurice Scarlett's 2017 *Consequences of Offshore Oil and Gas: Norway, Scotland, and Newfoundland,* which is a selection of background papers prepared for a colloquium on the potential impact on the province of future commercial oil/gas discovery offshore, and also Colin S. Doody's 2007 *Offshore Oil and Gas Benefit Regimes of Newfoundland & Labrador and Norway: An Evaluation.*

3 At the end of 2020, Norway had 90 fields in production compared to NL's four fields, and Norway produces an average of well over 10 times more barrels of oil per day than NL. The difference in expenditures is commensurate with these disparities of scale. Facts and figures for both locations can be found at the Ministry of Petroleum and Energy, the Norwegian Petroleum Directorate, and the Canadian Association of Petroleum Producers websites.

4 The delimitation line agreement between Norway and Russia ended a decade-long deliberation process where the two countries' sovereignty over large sea areas was finally agreed upon. The consequence of this agreement was that both countries immediately sought to explore potential oil and gas revenues on their side of the line.

5 A phrase used in personal communication between BD and a geologist employed in the Norwegian oil company Equinor, July 2019.
6 Ontological security is broadly defined as "a person's fundamental sense of safety in the world" (Giddens 1991), and the concept makes for an understanding of the rationale behind people's reactions to threats and concerns, and as an extension of how transformation – potential or real – is met by real concerns about how these changes might interfere with what makes the world make sense. See also (in relation to Lofoten): Karlsson and Dale (2019) and Dale and Kristoffersen (2018).
7 In Chapter 3 of *Fossilized*, titled "Newfoundland and Labrador: Economic Miracle and Environmental Debacle," Carter writes: "It is difficult to overstate the significance of oil to Newfoundland and Labrador [...] undoubtedly, oil brought great benefits" but "expansive risks and impacts of oil development were inadequately regulated in Newfoundland's offshore oil sector during the oil boom due to several outstanding gaps in or concerns regarding environmental policy surrounding the industry" (90, 91).
8 This interview and other show segments are available at https://www.cbc.ca/listen/live-radio/1-47-on-the-go/clip/15812515-otg-asks-oil-advocates-questions-green-economy
9 See Bang and Lahn (2019), Kama (2019), Safari et al. (2019) and Dale and Andersen (2018).
10 Leblanc's *Commission of Inquiry* lists other problems: "strategic misrepresentation and political bias," the concealment of information, the exploitation of public trust and "unrealistic optimism" (3). The report goes on to describe "unachievable" work schedules (16), risk reports that were not made public (18–20), and a deeply flawed environmental impact assessment process that suppressed the voices of NunatuKavut Community Council and Innu of Ekuanitshit.
11 In particular, the Nunatsiavut Government repeatedly articulated concerns (and other community concerns) about methylmercury poisoning that would result of the flooding of the reservoir. And yet, as the Commission report notes, "The reservoir is now fully impounded and none of the promised mitigation measures have been implemented" (37).
12 Lucian Ashworth's 2020 article in *The Independent* does an excellent job of detailing the "Doubling Down of the Fossil Fuel Fairy Tale" and the recent resuscitation of the *Terra Nova* oil platform is further evidence of this provincial mindset.
13 See, for instance, Stoddart et al. (2020) and Dale and Kristoffersen (2018).
14 These claims frequently argued by the Norwegian petroleum sector have never been supported in scientific work, but it's still put forward as a valid argument for why Norway should be the last producer of oil and gas to shut down production.
15 There are obviously state-run and controlled processes instigated and finalized before access is given, processes that assesses both environmental, societal, and economic consequences of drilling, extraction, and industrialization. A thorough description of these processes is beyond the scope of this chapter.
16 Even though all three LoVeSe-regions are included when political decisions are made concerning the potential petroleum development in the area, the Lofoten region is by far the most profiled, researched, and symbolically important of the three. For this reason (and the fact that it is here I have worked and lived for the past 12 years), I focus mainly on how its development has been influenced by the debate – and how Lofoten has become both an important political issue in and of itself.
17 Examples are plentiful, such as NPD (2010), unfortunately, none of these reports have been translated. See Dale (2011), Dale and Kristoffersen (2018), and Kristoffersen and Dale (2014) for descriptions and analysis.
18 Excellent work has been done on risk, vulnerablity, and danger in the context of the COVID-19 pandemic, including Barrios and Hochberg (2020), Giritli Nygren and Olofsson (2020), Karwowski et al. (2020), and Weible et al. (2020).
19 See, for example, White and Roberts (2020) and O'Brien (2016).

References

Andersen, G. 2017. *Parlamentets Natur. Utviklingen av en legitim miljø- og petrokeumspolitikk* (1945–2013). Oslo: Universitetsforlaget.

Arnason, R., R. Hannesson, and Willam E. Schrank. 2000. "Costs of Fisheries Management: The Cases of Iceland, Norway and Newfoundland." *Marine Policy* 24, no.3: 233–243.

Ashworth, Lucian. 2021. "Doubling Down on the Fossil Fuel Fairy Tale." *The Independent*. Web. January 20, 2021.

Baffes, J., et al. 2015. "The Great Plunge in Oil Prices: Causes, Consequences, and Policy Responses." CAMA Working Paper, Crawford School of Public Policy. http://dx.doi.org/10.2139/ssrn.2624398.

Bakx, Kyle. 2020. "How Ottawa Is Providing a Lifeline to the Oilpatch." CBC Online. 8 Dec. 2020. https://www.cbc.ca/news/business/bakx-feds-oilpatch-pandemic-1.5829095

Bang, G. and B. Lahn. 2019. "From Oil as Welfare to Oil as Risk? Norwegian Petroleum Resource Governance and Climate Policy." *Climate Policy* 20, no. 8: 997–1009.

Barrios, J.M. and Y. Hochberg. 2020. "Risk Perception through the Lens of Politics in the Time of the Covid-19 Pandemic." National Bureau of Economic Research. https://www.nber.org/papers/w27008.

Beck, U. 2009. *World at Risk*. Cambridge: Polity.

Birol, Faith. 2020. "Put Clean Energy at the Heart of Stimulus Plans to Counter the Coronavirus Crisis." *IEA*, 14 March 2020. https://www.iea.org/commentaries/put-clean-energy-at-the-heart-of-stimulus-plans-to-counter-the-coronavirus-crisis.

Blades, Ted. @OntheGo., Nov 16, 2020, 5:32 pm. Canadian Broadcasting Corporation. Newfoundland and Labrador, Nov 16, 2020, 5:32 pm.

Carter, Angela V. 2007. *Economic Development Benefits of the Oil and Gas Industry in Newfoundland and Labrador*. Project Report. The Harris Centre. Web. https://www.mun.ca/harriscentre/reports/research/2007/HC_-_OG_Conference_Report.pdf.

———. 2020. *Fossilized: Environmental Policy in Canada's Petro-Provinces*. Vancouver: UBC Press.

Dale, B. 2011. "Securing a Contingent Future: How Threats, Risks and Identity Matter in the Debate Over Petroleum Development in Lofoten, Norway." PhD Dissertation, University of Tromsø.

Dale, B. and B. Kristoffersen. 2018. "Post-Petroleum Security in a Changing Arctic: Narratives and Trajectories towards Viable Futures." *Arctic Review of Law and Politics* 9: 244–261.

Dale, B. and G. Andersen. 2018. *Til Dovre Faller? Norsk Olje og Grønn Omstilling*. Grønn Omstilling: Norske Veivalg.

Dale, B., et al. 2018. "Petroleum as a Challenge to Arctic Societies: Ontological Security and the Oil-Driven 'push to the north'." *Journal of Extractive Industries and Society* 6, no. 2: 367–377. https://doi.org/10.1016/j.exis.2018.10.002.

Farmer, J.D. et al. 2019. "Sensitive Intervention Points in the Post-Carbon Transition." *Science* 364, no. 6436: 132–134.

Fein, Judith. 2017. "Return of the Raiders: Sagas, Swords, and Surprises on the Trail of All Things Viking from Newfoundland to Norway (OBSESSIONS VIKINGS)." *National Geographic Traveler* 34, no. 1: 27.

Fisher, M., 2009. *Capitalist Realism: Is there No Alternative?* John Hunt Publishing.

Fitzpatrick, Ashley. 2021. "Greenwashing Our Hands of it – N.L.'s Oil Challenge." *Atlantic Business*, 2 March 2021, https://atlanticbusinessmagazine.net/article/greenwashing-our-hands-of-it/

Galafassi, D. et al. 2018. "Restoring Our Senses, Restoring the Earth. Fostering Imaginative Capacities through the Arts for Envisioning Climate Transformations." *Elementa: Science of the Anthropocene* 6, no: 69. https://doi.org/10.1525/elementa.330

Giddens, A. 1991. *Modernity and Self-Identity. Self and Society in the Late Modern Age.* Cambridge: Polity Press.

Giritli Nygren, K. and A. Olofsson. 2020. "Managing the Covid-19 pandemic through individual responsibility: the consequences of a world risk society and enhanced ethnopolitics." *Journal of Risk Research* 23, no. 7–8: 1031–1035.

Government.no. 2021. "Proposed amendment to the petroleum taxation system." https://www.regjeringen.no/en/aktuelt/proposed-amendment-to-the-petroleum-taxation-system/id2869684/

Haarstad, H. and T.I. Wanvik. 2016. "Carbonscapes and Beyond: Conceptualizing the Instability of Oil Landscapes." *Progress in Human Geography* 41, no. 4. https://doi.org/10.1177/0309132516648007.

"Husky Seeks Government Bailout for $2.2 Billion Fossil Project Off Newfoundland Coast." 2020. *The Energy Mix.* September 10, 2020. https://theenergymix.com/2020/09/10/husky-seeks-government-bailout-for-2-2-billion-fossil-project-off-newfoundland-coast/.

IEA. 2016. "World Energy Outlook." Paris.

———. 2020. "Global Energy Review 2020: The Impacts of the Covid-19 Crisis on Global Energy Demand and CO2 Emissions." Paris.

———. 2021. "Net Zero by 2050. A Roadmap for the Global Energy Sector." Special Report, F. Birol. Paris.

Jensen, L.C. 2012. "Norwegian Petroleum Extraction in Arctic Waters to Save the Environment: Introducing 'Discourse Co-optation' as a New Analytical Term." *Critical Discourse Studies* 9, no. 1: 29–38.

Kama, K. 2019. "Resource-making Controversies: Knowledge, Anticipatory Politics and Economization of Unconventional Fossil Fuels." *Progress in Human Geography* 44, no. 2: 333–356.

Karlsson, M. and Dale, B. 2019. "It belongs to the world." Oil, conservation and futures in the making in Lofoten, Norway. *Environment and Planning C: Politics and Space* 37, no. 8: 1452–1470. doi:10.1177/2399654419835828

Karwowski, M. et al. 2020. "When in Danger, Turn Right: Covid-19 Threat Promotes Social Conservatism and Right-Wing Presidential Candidates." *PsyArXiv* 31. doi:10.31234/osf.io/pjfhs

Kinder, J.B. 2020. "From Dirty Oil to Ethical Oil: Petroturfing and the Cultural Politics of Canadian Oil after Social Media." *Journal of Environmental Media* 1, no. 2: 167–183.

Kristoffersen, B. 2014. "Drilling Oil into Arctic Minds? State Security, Industry Consensus and Local Contestation." PhD Dissertation, UiT The Arctic University of Norway.

Kristoffersen, B. and Dale, B. 2014. Post-Petroleum Security in Lofoten: How identity matters. *Arctic Review of Law and Politics* 5, no. 2: 201–226.

LeBlanc, Justice Richard. 2020. *The Commission of Inquiry Respecting the Muskrat Falls Project.* Web. https://www.muskratfallsinquiry.ca/.

Newfoundland and Labrador Department of Industry, Energy, and Technology. 2020. "NL Offshore Oil and Gas Industry Recovery Assistance Fund." N.d. https://www.gov.nl.ca/iet/nl-offshore-oil-and-gas-industry-recovery-assistance-fund/.

NPD. 2018. "Barely Halfway." Accessed 5 May, 2021. https://www.npd.no/en/facts/publications/reports2/resource-report/resource-report-2019/summary/more-than-half-remains2/.

NPD. 2019. "Dry Well in the Southeastern Barents Sea – 7132/2-2." Last modified 2 April, 2019. https://www.npd.no/en/facts/news/Exploration-drilling-results/2019/71322-2/.
NPD. 2019. "Resource Accounts as of 31 December 2019." Accessed 5 May 2020.
NPD. 2018. "Drilling Permits." Accessed 20 June 2021.
O'Brien, K.L. 2016. "Climate Change and Social Transformations: Is it Time for a Quantum Leap?" *Wiley Interdisciplinary Reviews: Climate Change* 7, no. 5: 6218–6626.
Ørnes, R. et al. 2019. "Vi er Lofoten." Svolvær, SALT.
Reid, Scott and Jeff Collins. 2012. "Oil, Politics and Public Policy: A Case Study of Newfoundland and Labrador." Canadian Political Science Association. Accessed 28 June 2021. https://cpsa-acsp.ca/papers-2012/Reid-Collins.pdf.
Safari, A. et al. 2019. "Natural Gas: A Transition Fuel for Sustainable Energy System Transformation?" *Energy Science and Engineering* 7, no. 4: 1075–1094.
Shmuely, Raymond Williams Andrew. 2008. "Totality, Hegemony, Difference: Henri Lefebvre and Raymond Williams." In *Space, Difference, Everyday Life*. Routledge, 226–244.
Smellie, Sarah. 2020. "Newfoundland and Labrador Can't Afford Buy-in for Husky-led Oil Project: Premier." *The Toronto Star*. Web. 14 September 2020. https://www.thestar.com/business/2020/09/14/nls-struggling-oil-sector-dominates-debate-as-legislature-reopens.html.
Statoil. 2010. "Industriskisser Nordland VI/VII."
Stoddart, M.C.J. et al. 2020. "Envisioning Energy Futures in the North Atlantic Oil Industry: Avoidance, Persistence, and Transformation as Responses to Climate Change." *Energy Research & Social Science* 69. https://doi.org/10.1016/j.erss.2020.101662.
Tàbara, David. J. et al. 2018. "Positive tipping points in a rapidly warming world." *Current Opinion in Environmental Sustainability* 31: 120–129. doi:10.1016/j.cosust.2018.01.012
We Are NL Offshore. "Support Newfoundland and Labrador's Offshore." https://wearenloffshore.ca/support/supportnl/.
Weible, C.M. et al. 2020. "Covid-19 and the Policy Sciences: Initial Reactions and Perspectives." *Policy Sciences* 53: 225–241.
White, D. and T. Roberts. 2020. "Post Carbon Transition Futuring: For a Reconstructive Turn in the Environmental Social Sciences?" *The Cambridge Handbook in Environmental Sociology*, Volume 1: 223–242.

8

ART AND THE OFFSHORE

Sue Jane Taylor

Introduction

In December 1984, on a dark December night, as I boarded the *Stirling Teal* supply vessel in Aberdeen Harbour, I never imagined that this was going to be a life-changing artistic experience for me. I was one of 12 recent art graduates commissioned by Stirling Shipping Company to depict and celebrate their ten-year involvement in the North Sea oil and gas industry (Taylor 2005, 1). Sailing out on the high seas opened up a visually exciting new world. I was entering a world of extremes, a tough, brutal environment in every way, dominated by men.

This oil and gas industry has framed my life. It was part of my contemporary Highland culture when I was growing up. In the 1970s, the oil boom hit the North like a gold rush and the region where I lived was geographically and socially changed in perpetuity. It became very much part of my contemporary landscape (Taylor 2005, 5). My approach in recording the industry is from an individual visual artist's perspective. I view these industrial sites with different eyes to most people employed there. We work in different contexts. For offshore platform personnel, theirs is a place of work, purely functional. For me, these are inspirational places; they represent an aspect of our contemporary age that many people do not think about or try to ignore. I am fascinated by the industry's work-gear and artefacts which, when taken out of their environment and placed in a museum, become powerful symbolic statements. As an illustration of this, Aberdeen Archives, Gallery & Museums has in its collection two original Murchison Field drill-floor safety mats which once surrounded the drill hole/wellhead. To me, these are imposing industrial objects of our time. As I describe in my diary: "Over the years these mats had borne the brunt of tough drillers' laboured treads and ancient Jurassic-Triassic age crude oil slops" (Taylor 2017, 35).

DOI: 10.4324/9781003024125-8

Since oil was discovered in the North Sea 50 years ago, the industry's public identity has been predominantly portrayed in cold, hard economic terms and its image has been very tightly controlled in photography and film. Its production platforms are typically located in remote offshore environments, and this remoteness is reflected in its somewhat secretive and hidden corporate identity. For an individual outside the industry, it is extremely difficult to gain access offshore, almost like winning "the golden ticket."[1] That said, most people would view these places as tickets to hell; terrifying, dangerous, and certainly not on their bucket list. I am often asked how I manage to gain access to these very private environments. The answer is that I don't know. It has certainly not been easy. Few company managers have an understanding or appreciation of the arts, viewing them as a pastime rather than a profession. Oil companies also seem to have been born suspicious; it is in the nature of the beast. Their sites are potentially dangerous workplaces housing highly volatile hydrocarbons and chemicals. However, if company managers can be persuaded to see an advantage in it for them – in terms of public relations and support for a local community – then at times they are happy to help. A generally benevolent attitude towards the arts could act as a powerful counterbalance to a widely held view of oil companies as corporate giants riding roughshod over community concerns. Indeed, I have been fortunate through the years in collaborating with some individuals in the industry who were instrumental in getting me offshore. Even achieving all my self-sponsored offshore survival certificates, though, and having a reputation as an established artist on this subject, doesn't guarantee access. From a company's perspective, an extra person on the chopper flight and onboard one of their assets means more expenditure and an extra liability.

Very few artists have experienced and recorded the North Sea oil industry at first hand. Three, interestingly all female, made one-off residency visits offshore: Fay Godwin in the 1970s, Kate Downie in 1988, and Fionna Carlisle in 2005. George Mackie created an important series of poignant paintings depicting Aberdeen Harbour's transition, during the 1970s, from a fishing port into an international oil and gas hub. It is also important to note the artworks and objects that were created by the oil workers whilst they were off-duty on their platforms, from banjos, to steam engines, to cartoon drawings and paintings. These creations constitute a kind of offshore folk art.[2]

In the UK, there is a long tradition of visual artists recording industrial maritime subjects; for example, the Clyde shipyards were depicted in both the First and Second World Wars by Muirhead Bone and Stanley Spencer, respectively. Tackling this North Sea industry is, however, on another level; it is highly controversial, political, and very challenging. Back in the 1980s, when I first began recording offshore and onshore oil-related industrial work in Scotland, I felt it was important to highlight the workers who kept these installations operating safely and who, at that time, were largely ignored, and this has remained the main focus of my work. When the industry was privatized, the global companies operating in the North Sea could not build platforms quickly enough, and they certainly did not identify with their workforce.

This fact came very much to the fore in my association with the *Piper Alpha* platform. This was the first offshore installation on which I stayed – for a week – almost exactly a year before its catastrophic accident when 167 men died. On my visit, I had met and drawn some of the men who perished that night (Taylor 2005, 99–114). This event, still the world's worst oil disaster, changed the mind-set in the offshore North Sea industry for ever. It changed my life as an artist and as a person, too, in the way I portrayed this industry, making me much more wary in dealing and working with oil companies. In the aftermath of this disaster, I experienced the wrath of the platform owners, the Occidental Oil Company (Taylor 2005, 149–168). Shortly after the disaster took place, I was called into their London headquarters, as they had heard I was going to have an exhibition, *Oil Worker Scotland*,[3] which included images from my *Piper Alpha* visit. As I wrote in my diaries at that time: "I could name a price. I was to wrap them [artworks] all up and bring them to head office. The company had never before shown the slightest interest in my work, but I had now become a minor potential threat in that I might keep the name of *Piper* alive in the public eye" (Taylor 2005, 91).

In 1990, I was chosen to be the artist to sculpt the memorial organized by the Piper Alpha Families and Survivors Association and Aberdeen City Council (Figure 8.1). Occidental was very much against a lasting physical memorial; they deemed a book of condolence, signed by dignitaries and the public, to be adequate. It was a real struggle for the Association to raise the funds and to save embarrassment the Scottish Office had to step in and meet the £100,000 target. Only £14,000 was donated by all the oil companies operating in the North Sea at that time. I found the whole experience very stressful as I witnessed the political and corporate game-playing. It seemed to me to be summed up by the fact that in her memoirs Mrs Thatcher made no mention of the disaster, despite it being a very significant event in 1988 when she was Prime Minister. She was, of course, a good friend of Armand Hammer, the head of Occidental.

What keeps me returning to these liminal places that take such a toll on our natural environment is an unanswerable question in my own creative mind. Primarily, it is the people and their work locations that are set in a time warp with no real visual time gauge, just sea, horizon, and sky. These man-made metal islands are in parallel, artificial worlds, comparable perhaps to confined futuristic space stations. The workforce's expansive and colourful PPE (Personal Protection Equipment), the clothes that they wear to work in, capture my imagination – from deep sea divers' gear to survival suits. As an artist, I am also searching for "the source," in a sense the opposite to Gustave Courbet's *La Source de La Loue* river painting series in 19th-century France.[4] One could describe looking down at the massive well risers, soaring out from the sea and up into the platform's belly, as "the source" of our Age of Oil; sooking up the fossil-fuel devil's water from hell; it is our addiction.

FIGURE 8.1 Survival Suit Figure Head, *Piper Alpha* Memorial
Source: Author's original art, Sue Jane Taylor

As we move into the mature years of the North Sea, with the eventual removal of almost all these 400 offshore installations, the decommissioning program has begun. This represents a shift change in the industry: existing oil wells are running dry, which makes them uneconomic, leading to new methods of oil and gas extraction, new attitudes towards the natural and social environments, and an emerging large-scale offshore renewable energy industry.[5] My art practice is tracking this change and, to this end, I recently recorded the *Murchison* platform's last months of production, its removal, and break-up. I was one of the last 17 onboard, experiencing its power shutdown and its death.[6] My lasting *Murchison* memories are of silence; the only noises came from the sea and the wind, a magical moment in time when nature was given back its space. The removal of these fields in recent years is creating nostalgia, reminiscence, and heartfelt sadness amongst the offshore workers; these giant fossil fuel-producing metal beasts were their second homes.

Since 2007, my work has also been involved with offshore renewable energy.[7] These clean energy infrastructures are very different and challenging to draw. They have an air of present-day towering white goddesses and gods of the seas.

Yet, the renewable wind industry is also corporately controlled in the UK. It, too, is a private industry and owes much of its development offshore to the oil and gas industry. It is also, especially onshore in rural areas, a very visible industry because of the large scale of its installations, and thus has received much negative publicity. The artist's portrayal of wind energy is not a new phenomenon; Rembrandt, his Dutch painting compatriots, and Van Gogh have all included windmills in their artworks as important landscape landmarks.

As an artist using traditional methods such as plaster, paints, pencils, and paper, these are fascinating if troubling times to be following the energy industry. I think it is important to record these pioneering projects using old-fashioned pencils and sketch pads, in parallel with new media technology. Artists see and observe the world differently, creating different outlooks and personal impressions.[8] At times, I am a fly on the wall in these remote extreme environments, privileged to be writing my on-site diaries and drawing my visual experiences. My diaries are an extension of my sketch books and visual research contributing towards my final artworks, reminding me of the on-site details, environments and instances with people who I have met and drawn. I hope that, over the years, these documents will become important public social-history references.

During my 30 years recording this industry, I have been privileged to visit many North Sea offshore installations. I have witnessed and drawn the Brent Field's giant platform topsides being lifted from their perches offshore by the largest construction vessel in the world, and I have documented pioneering offshore windfarms and new wave/tidal prototypes in the Orkney Isles' coastal water test sites.

I have now come full circle, returning in 2020 as artist in residence at Nigg Energy Park in the eastern Highlands of Scotland. This area is full of my childhood memories. It was here that, as a young artist in the 1980s/1990s, I drew massive oil platform jackets for the North Sea oil fields being fabricated by a workforce of up to 4,500 men (at present, only several hundred people, men and women, are employed here). During my residency, among other things I will be following the refit of the *Well-Safe Guardian*, an exploration drilling rig being converted into one that decommissions oil and gas wells in offshore installations, and the servicing of Atlantis's giant underwater tidal turbine, currently being tested off the Pentland Firth. I shall also be recording the arrival of components for the new massive Moray East Offshore Wind Farm: 100 turbines to be sited just north of the Beatrice Field, nearby in the Moray Firth. Before this fabrication yard was established in 1972, it was a landscape of undulating sand dunes and beach. Today, I observe the remnants of this natural past, the visiting seabirds returning to nest on its edge; notably, Eider Ducks, Arctic Terns, Oyster Catchers and Ringed Plovers, summertime's contrasting frantic energies – nature and industry.[9]

The following onsite diary excerpts focus on the first slow transitional steps offshore in the North Sea – the death of the old ways with the removal of oil and gas installations and the decommissioning program, leading to the establishment of the new with the emergence of renewable energy.

Brent Delta platform[10]

Excerpts from Artist's Diary: June 2016 (Brent Crude $48.25 per Barrel)

The Brent Field was one of the most lucrative oil discoveries in the North Sea. For 40 years, its four platforms, *Alpha, Bravo, Charlie,* and *Delta* produced over three billion barrels of sweet light crude oil. This oil, evolved and formed from the Jurassic period, was considered the crème de la crème of oils, the equivalent of the perfect malt and a distiller's dream, Brent Blend. By perfect happenstance, Shell initially named all its UK oil fields after water birds in alphabetical order by discovery: Brent refers to the Brent goose, the golden-black goose. It is the price of Brent crude that is used as the norm in global reportage.

In June 2016, I was given permission by Shell UK to visit Brent Field (Figure 8.2). *Brent Delta* was weeks away from final abandonment and personnel evacuation. This platform was very different to any other I had been on. Apart from *Alpha*, all three Brent platforms had vertical flare towers and three giant subsea concrete legs supporting their topsides. Surrounding the base of these legs, sitting on the sea bed, were massive reinforced concrete tanks, subsea storage cells that were once used to

FIGURE 8.2 The Brent Field
Source: Author's original art, Sue Jane Taylor

store and separate crude oil before export. *Bravo* and *Delta* each had 16 reinforced concrete tanks that still contained layers of crude oil, called attic oil.

The Final Farewell, Able UK Seaton Park Yard, Durham County

In April 2017, *Brent Delta*'s 24,200 tonne topside was lifted by the *Pioneering Spirit* from the Brent Field in the northern North Sea. This marked the world's heaviest offshore lift, which took just over 10 seconds. It was transported to Able UK Seaton Park yard in Hartlepool where it was fully dismantled over a 12–18 month period; 98% of the *Brent Delta* platform material was recycled.

This was the first time in UK waters that a complete platform topside on this scale had been brought to shore.

Shell UK granted me permission to visit Able UK yard during the week's events marking the Final Farewell to Delta.

Excerpts from Artist's Diary: June 2017 (Brent Crude $46.37 per Barrel)

The road to the yard ran in parallel with the long stretch of Seaton Carew beach front, and as we came round the corner from the long coastal road stretch, it hit you with a bang: *Brent Delta* topside, now docked on land with its flare tower soaring up to the sky, extinguished and no longer burning off years of unwanted oil-well gases. A beached alien beast of gigantic proportions had reached its final resting place. It stood out from an industrial horizon; to its left, a nuclear power station with huge pylons and wires branching out, burdened with carrying national grid electricity, and hot and cold-stacked jack-up rigs, and to its right, massive oil storage tanks.

In the heat of the day, outside the marquee and away from all the bustle of activity inside, I chose a drawing spot in the shade looking straight up at *Brent Delta*. The only other people who would have viewed it from this angle in its operational offshore lifetime would have been the platform's rope access technicians (RATs), scaffolders, and cargo supply vessel seafarers. This now defunct topside giant had stumped elephant feet, amputated from its three massive concrete legs, which had been left behind in the northern North Sea.

I also drew *Brent Delta* from the outside of its graveyard perimeter. In the early morning, I walked from my hotel to Teesmouth National Nature Reserve, a small patch of undulating sand dunes leading to the sandy shore estuary, the mouth of the river Tees. On a dune hillock perch-top, I set up my art camp to draw the incongruous site before me. There were birds singing all around me, oblivious to the constant flow of dog walkers, cyclists, and runners who were following the reserve's path networks below me. Skylarks were hovering high above; Sandwich Terns, with their dagger-like wings, were swooping back and forth along the shoreline behind me. A hawk even rested for a while on a rock close by, busily preening itself.

The small geographic area of Hartlepool, Middlesbrough, and Redcar reads like a visual textbook. It was here that the British nation's old and contemporary industries were concentrated: an offshore windfarm out at sea, a constant flow of maritime freight vessels, an abandoned and redundant steel works at Redcar. *Brent Delta* stood out amongst these muted grey tones, although its own painted colours were now faded; changed to shades of delicate pinks and rusted yellow ochres after a 40-year battering from its maritime habitat.

For the week's farewell event, every day Shell hosted different parties of invited guests to the site: VIPs, school children, bereaved families who had lost a relative on the platform, and the local community. I was allowed to be present for the Shell pensioners, ex-employees, and contractors' day; people who had worked on *Delta* from its birth right up to its death in 2016.

The exciting part for everyone was a tour of *Delta*, and they got kitted out with safety boots, hard hats, and high-viz jackets from the makeshift stores inside the marquee. As the groups stood under *Delta*, by its enormous elephant feet, they looked like stick insects. An eight-storey scaffolding stairway was constructed for people to get up onto the topside. One lady returned to the marquee saying she had vertigo going up the stairs. St Johns Ambulance officers were on standby.

When people returned, I could see that it had been very emotional for some of the pensioners who had not been on *Delta* since they had retired many years ago. After the day's event, I was given the chance to go up onto *Delta*. From the ground looking up at its torso's underbelly, my guide pointed out a bent metal section, the result of "the 100-year wave," which had hit *Delta*. I could see the massive metal bar frame support system up close. This held up and supported the weight of the topside welding reinforcements above.

Climbing the scaffolding stairway up onto the platform was a very strange and odd sensation. On my previous visit, almost exactly a year before, I had arrived by chopper far out at sea. It was all a bit unsettling walking around and climbing up and down the different levels; looking into where once the wellheads and Christmas trees were located, now empty spaces, ghost-like. All the exterior module doors were welded up in preparation for the lift and sailing. It was even weirder entering the living quarters by torch-light, all areas in total darkness. The control room was just as it was left before final abandonment with work folders all lined up above the monitor screens and old photos still on the walls. Outside on deck the view was of the estuary, bay, jack-up rigs, an oil refinery, and the gentle hill slopes of Durham County in the distance. It did not seem real.

In passing, I overheard a conversation, "In the canteen, tables were lined up in one long row, decorated with plastic flowers. The last personnel onboard had held a special dinner to mark their final night on *Delta* – with the table lay-out resembling the painting of the Last Supper."

FIGURE 8.3 Oilman and Stag
Source: Author's original art, Sue Jane Taylor

The Beatrice Field, Moray Firth

For nearly 40 years, the Beatrice Field, located in the Moray Firth, has been part of the eastern Highlands' contemporary maritime landscape. Unlike most oil and gas installations located far out to sea, Beatrice Field, consisting of *Alpha, Bravo,* and *Charlie* platforms, is visible and close to land. When the oil boom hit the North, Beatrice played a part in that big shift; Nigg Oil Terminal in Easter Ross was constructed on top of sand dunes and beach shore line to accommodate huge storage tanks containing the black-tarry crude oil from Beatrice's pipeline (Figure 8.4).[11]

Since childhood, when travelling up the A9 road to Caithness or Orkney, I would look out onto the Beatrice Field in the firth, with its platforms' bright, burning, orange flares and wonder what it must be like to be on these platforms looking out to land. These installations are clearly visible from the Caithness, Moray, and north-east Sutherland coasts and on my many walks throughout the years along that coast line, the platforms were like tiny mechanical seals, always following me, and they caused me to reflect on their impact and place in this environment.

It was only when I passed close to the Beatrice oilfield in a boat that I could really appreciate the scale and enormity of the *Alpha* and *Bravo* platforms and the

FIGURE 8.4 Looking out to *Beatrice* from Caithness Coast
Source: Author's original art, Sue Jane Taylor

visual maritime footprint that they had made over the years. This was in 2006 and then again in 2018. The first time I had been commissioned to follow the Beatrice Demonstrator Offshore Wind Turbine project from its construction to its installation. The second occasion was as a passenger on a rib-boat observing the construction of BOWL (Beatrice Offshore Windfarm Ltd).

In 2018, I realized that there was an urgency to visit the Beatrice Field as it was about to be downgraded to a Not Normally Attended (NNA) status. *Alpha* was the last remaining platform with workers onboard, and I was curious to find out what it was like for the people working in this offshore location so close to land, and how they were affected by living in this surrounding land-environment.

Excerpts from the Artist's Diaries: September 14th–19th, 2018 (Brent Crude $78.89 per Barrel)
Friday

Once the chopper headed out into the Moray Firth basin, it was a delight to my artist's eyes: the watercolour-effect rain clouds were interspersed amongst the firth's soft mottled blues. As we neared the Beatrice Field I could clearly see, mapped out in 3-D to the north, the BOWL development nearing its completion; coming into view *Beatrice Bravo* and *Charlie*, now decommissioned and abandoned; then our destination *Beatrice Alpha*. South of *Alpha*, five miles away, standing lifeless, were the two decommissioned Beatrice demonstrator offshore wind prototypes. These were the precursors to all deep-water offshore wind turbines in the world, a Talisman Energy project that I had documented from construction to installation (Taylor 2013).

Alpha consists of two jacket installations, AD (Alpha Drilling) and AP (Alpha Production), joined together by an enclosed walkway bridge. Walking on the 100-metre-long see-through metal grid, looking down at the water below, was mesmerizing. Production from *Alpha* stopped in 2015, and its recent decommissioning was now complete, so the noise levels and activity in most modules were very low and, in most cases, silent. The only systems operating were its life support: power, heating, and water supply. Whilst walking and climbing endless stairways, I caught glimpses of the panoramic views. Unlike most platforms that are located far out at sea, *Beatrice* is sheltered in a firth, surrounded by land on three sides with the new addition of the BOWL wind farm on its west side (Figure 8.5).

The Offshore Installation Manager (OIM) talked about the characters who used to work on *Beatrice*. On one occasion, during a safety training exercise, one of the rescue crafts had gone missing. Its crew had sailed up to Weatherspoons in Wick for a jolly. On their return, they were reported for drinking on duty, and they lost their jobs. The close proximity of the pub had been all too tempting. On another occasion, *Beatrice*'s power had gone down with no provision to feed the hungry onboard. So they had phoned the nearest chip shop, *La Mirage*, onshore in Helmsdale, to order fish suppers. Unfortunately, they had to wait a long time for their order, as when they got there, the shop thought that someone had been playing a prank phoning from *Beatrice*.

FIGURE 8.5 The Launch, Cromarty Firth
Source: Author's original art, Sue Jane Taylor

There were only 62 people on board, but I had been warned to avoid certain rush hours in the small canteen. When I went for my supper, I met Steven Dalgetty, a relative of the Dalgettys of Bayview farm at Munlochy, who lived on the opposite hill to where I grew up. His first job was as a welder at *Nigg*, and he had spent 22 years working on *Beatrice*. He also grew up on Nigg farm in the 1970s, experiencing the oil terminal and fabrication yard being built nearby. As an adventurous teenager, when the Beatrice pipeline was being buried through his family farm, he would go out in the evenings to try out the huge operational diggers left abandoned from the day's work.

Saturday

James, my platform "buddie" guide checked his gas monitor before we stepped outside into the bright sunshine. We climbed up the endless stairways to the helideck, the most easily accessible highest platform point. The views from this flat deck were splendid. All looked out onto land. Out west, I could see the red *Pacific Orca* vessel at BOWL site unloading its precious cargo: the towers, generators, cones, and turbine blades being lifted onto the sea-bed-secured, bright acid-yellow sub-sea jackets. The newly assembled turbines all stood in rows of grid-lined symmetry.

From our platform point, the views towards the Caithness coast were magical, and I could make out familiar landmarks, caves, and cliff faces, visually retracing the coastal walks I had done from Berridale up to Wick some years back. This platform is well known as being one of the North Sea's coldest installations; when the snow settles on the nearby hills, the cold chill winds blast off from these hills onto the firth. I was told how once, on crew change day in winter, they had to dig away two feet of snow on the helideck before the chopper's arrival.

We wandered through all the redundant modules. The upper and lower 231 AP levels, where the two main export lines, separator tanks, crude and water injection pumps, sand filters, and de-aerators were all now standing in perpetual silence, and rusting away, were decidedly eerie. During its production years, these places would have been deaf-defyingly noisy and smelly. We passed Steven and his welding mate and stopped for a chat. Steven was concerned about the rapid metal fatigue and deterioration since the plant was shut down two years ago. He explained that when moisture gets into carbon steel it rusts very quickly, and it was therefore worrying to leave these platforms standing for so long before their final removal five years down the line.

Out at sea on the horizon, a giant cruise liner was heading north after stopping at Invergordon port. In 2014, the Beatrice Alpha platform had to be evacuated as a precaution because a ship carrying radioactive concrete waste material from Dounreay had caught fire and was drifting seven miles away towards this platform.

After supper, I went up onto the helideck again to capture the fading lights and to experiment with my new infrared camera attachment. As the light faded in the distance, the shore lights grew stronger, and I watched the A9 traffic light-trails. Looking out in the firth, the reassuring standby vessel lights flickered, and beyond that were the windfarm's red nav-aid lights. I looked into the distance where somewhere, 32 nautical miles east, was my home near Dornoch, but it felt very distant; the land was like a maritime mirage playing tricks on your eye. I recalled one of the worker's comments: "A lot of the guys like the close proximity to land, it's something to look at compared to other North Sea assets where you can't see anything apart from the other offshore installation lights in the distance at night time. Seeing the land, there's a bit of comfort and the thought that you are not far from home. But a lot of guys don't like it either, especially on special events, such as bonfire night in November where they see the bonfire and fireworks ashore. They are so close but so far from home."

Monday
It was high tide, and some seals were basking on the AP leg-corner landing areas below. This was an everyday sight here, but for me, it was quite magical to witness nature's ingenuity and adaptability where humans have intruded into their habitats; such a fine fragile balance. This area is officially designated as a Nature Conservation Marine Protected Area, a Special Area of Conservation, and a Special Protection Area.

Wednesday
I thought about what the OIM told me when he worked on another offshore platform close to shore in Morecombe Bay: "You could see the Blackpool Tower and the Big Dipper all lit up at night. At the weekends you could also hear the noisy music from these amusements. Although we are closer to land up here, you certainly don't hear any music from Buckie or Helmsdale!"

Thursday
I spoke to the guy who had been operating the crane during the night shift. He said it was dead flat calm out there all night then suddenly, out of nowhere, in the early morning the wind blew up – Storm Ali was coming. There would be no abseilors dangling from heights today, 27 knots of direct wind was their cut-off point.

Despite the weather, I was determined to go out and draw. Davey was my buddie. We walked around to find the most sheltered areas. The stand-by vessel was dodging the weather by going into the wind. Gannets were enjoying the exciting wind currents, and a gathering of razor bills were bobbing on the water, washing themselves and taking storm shelter by staying close by the platform. Davey told me that there had been a few small crude oil spillages over the years.[12] Thanks to decommissioning, the diverse wildlife was now spared from devastating fossil-fuel spill hazards; this area harbours abundant bird life and is also home to dolphins, porpoises, and transient visiting whales. The rain started to really drive in as we went inside. The platform's foghorn started to blast out ominous sounds.

That night when I had returned home, I looked out to the firth. In the blackness, I could just see *Beatrice A*'s lights, a flickering fossil-fuel dead star, soon to be extinguished. It was hard to believe that I had been there; it could have been light years away in time. I remembered one of the guys' comments: "Again just another platform, a lump of metal in the sea, not supposed to get emotional about that sort of thing."

The *Pioneering Spirit*: the *Brent Bravo* topside heavy-lift

Excerpts from the Artist's Diaries: June 16–21, 2019 (Brent Crude $64.22 per Barrel)

Monday

Onboard the biggest construction vessel in the world, I was more than ready to witness and record, in a painterly way, the lifting of the 25,000-tonne topside from the *Brent Bravo* platform and the journey to its graveyard in Hartlepool.

I was nursed to sleep by the constant heart-beat drumming, humming, and thudding of the mighty engines deep below deck, as we headed up to the northern North Sea. In the morning, I got up at 06.30 hours and went up to the Bridge where Oliver, the second mate, showed me around, explaining the various departmental stations: the Bridge-centre, where the captain and his mates were positioned, contained the Dynamic Positioning Control, the tension machine control, and the navigation centre. To the port-side was the ballast control area and behind that, the Topside Lifting System (TLS) station; to the starboard-side were the chief mate's desk, the survey station, the Heli-control and the communication sections.

Oliver showed me a computer-generated film about the complex lift procedures and the dynamic positioning that I was about to witness for real in the next couple of days. I chose a spot to start sketching the highly sophisticated scene in front of me; not quite the *Starship Enterprise* bridge, but certainly, the *Earthship Enterprise* bridge. I had brought along some folded landscape concertina drawing paper which fitted this scene perfectly, emphasising this interior's scale.

Although the outlook was foggy with slightly choppy seas, there was hardly a tilt or sway onboard; it was as smooth as a Rolls Royce on land. We were making good headway and, in the distance, I could just make out two platform silhouettes: *Charlie* and *Bravo* platforms in the Brent Field. We were now in the Brent Field, and it was quite a sight from the sailor's viewpoint; the three remaining platforms all running in a regimented north to south line. To our south was *Alpha*, the only jacket-supported platform in this field; the three other platforms were concrete gravity-base structures. Directly north was *Charlie*, its bright orange/red flare-flag burning, the last producing platform in this field. In the sea mist, north of *Charlie*, the three stumped capped legs of *Delta* were sticking out of the sea.[13] I had mixed emotions as I looked out from the comfort of my bridge drawing spot. It was good to witness the removal of these platforms, which for the past 40 years had "sooked up" billions of crude oil and gas for our hungry, fossil-fuel guzzling nations. But now, *Bravo*'s life was over, her remains waiting to be taken away. She was about to be this field's second defunct, hydrocarbon, metal star (Figure 8.6).

To the south, I spotted a chopper landing on *Alpha* platform. Brightly coloured little stick people, in their survival suits stood on the helideck, all lined up for

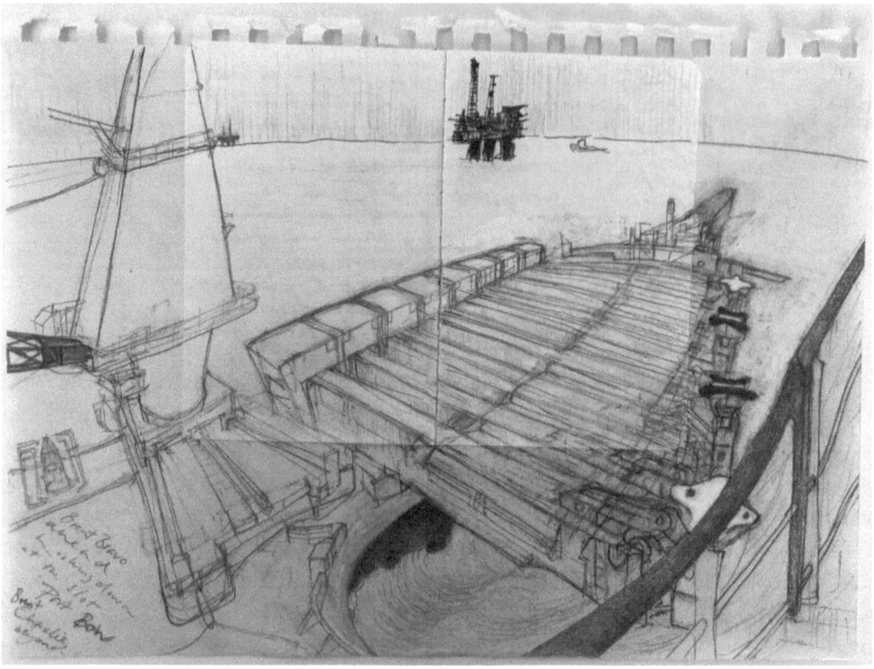

FIGURE 8.6 Looking Out to *Brent Bravo*
Source: Author's original art, Sue Jane Taylor

boarding. On take-off, the pilots did a fly-by around our vessel and *Bravo* before landing, due north, on *Charlie*. This was an opportunity for the chopper's passengers to view our spectacular vessel and to bid their personal farewells to Bravo; some of them would have worked on her.

People were gathered on E-deck to watch the action. The Allseas film crew were busy with their video camera filming peoples' reactions, and everyone had their phone cameras out, snapping away. As we moved closer to the giant three-legged metal beast that was *Bravo*, I could distinctly make out the platform levels and the accommodation block; the rust now dominated over its lemon yellow all-weather protective paint. Our vessel was now in position to adjust the final ballast levels in readiness for the lift. We were all issued with safety glasses to wear while outside as a precaution against any loose metal particles in the air from the platform.

I reflected on my previous offshore *Bravo* trip (Taylor 2017, 66). I had been on that platform back in November 2016, and now I was staring across at it from this huge vessel. After two years of abandonment, there might have been evidence of roosting sea birds, but there was nothing, no sign of life. To many, this hunk of metal was simply the end of a filthy, fossil-fuel polluting monster, but to thousands of her former workers, it had been a second home, a financial saviour with regular and steady work; for its shareholders, owners, and the UK Government, it had represented lucrative investments and tax revenues.

There were remotely placed Leica time-lapse cameras positioned all over the vessel; no visual angle should be missed, there could be no retakes by the Allseas film crew who were now up on the bridge, taking close-up shots of people working intently at their monitor screen stations. Meanwhile, I, the archaic-traditional artist, was storing and absorbing all this action and visual information mostly in my head, but also constantly drawing fleeting little sketches and scribbling in my drawing pad – with photos as back-up.

Tuesday

At 01:30 hours, I wandered down to C-deck and positioned myself at the best viewpoint. More and more people were gathering there; the top "high-heidians" started to assemble, so I knew something was about to happen. We were all huddled together in a corner, the atmosphere tense. The wait seemed endless. The vessel's tungsten spotlights were all pointed towards this crucial scene: *Bravo*'s three legs highlighted like never before. I could clearly see the slim cut line on all of the three legs. The cuts between the topside and the supporting legs were made during the final months of the platform preparations. Until the lift, the structure was held in place by purpose-built shear restraints that prevented lateral movement.

Suddenly, creaking, echoing, haunting metallic sounds, created by *Bravo*'s 25,000-tonne topside being lifted up and away from her three concrete legs, resonated all around the vessel. For the nine second lift duration, there was an audible unified chorus of breath intake and awe-inspired gasps, and then came the fortissimo finale of clapping, cheering and whistling. This mighty weight-lifting champion had, once more, shown its prowess as the industrial goddess of the seas, breaking all records.

From a high vantage point on the top accommodation deck, I looked down into the utility shaft – a 39-foot diameter, steel-reinforced, concrete leg, standing in nearly 160 metres of water – recalling the three lives that had been lost in that space. The topside and the vessel gently swayed together as one, on the relatively calm waters. The bridge crew were outside on deck taking celebratory selfies, with *Bravo*'s giant nameplate as a backdrop, totally oblivious to this platform's tragic loss of life.[14]

As I looked up at the topside, I noticed that there was a ghostly optical illusion, created by the setting sun's reflection on the dirty accommodation cabin windows; it was as if the lights were still switched on. I was half expecting the OIM, Alan Stenhouse, and his crew to come out and wave down at us; flash-backs of that earlier trip sped through my mind. I experimented with the infrared camera attachment on my iPhone, pointing the camera to the topside. This magical gadget picked up my heat-silhouette reflection from the bridge's window, looking onto the cold, dead *Bravo* – interesting shots.

Eventually, people dispersed and I retired to my cabin for a couple of hours sleep. Some months earlier I had bought a book, *The Sea Journal: Seafarers'*

Sketchbooks, which illustrated some of the most fascinating mariners' and artists' experiences on sea voyages throughout the ages (Lewis-Jones 2019). While drawing on board, I would reflect on some of these. By comparison with these historic voyages, my experiences were "space age" and luxurious.

Although there was a biting westerly wind, it was a bright day, and the oil- and gas-producing metal islands were clear and sharp. *Charlie*'s fossil-fuel flag burned brightly, and *Delta*'s stunted legs were very distinct. In the far distance, to the north-east, the *Thistle* platform was belching out its smoky, smutty flag. To the west, I could just make out three little dots on the horizon: the *Statfjord* platforms in the Norwegian sector. I reflected on my time on *Murchison*, 18 miles north, beyond *Delta*'s legs, now all gone and past history.

Amidst this industrial menagerie, a company of gannets were fishing, graceful and acrobatic, diving into the sea, oblivious to all our "he-man" operations. Their bodies and wing-span were black and white colour flashes set against the backdrop of the man-made islands and the blue-grey seascape.

I returned to the bridge, to my original drawing spot, and worked on my concertina sketch. We were now sailing south – to Able UK, the dismantling yard in Hartlepool. This would be *Bravo*'s final resting place where, like *Delta*, she would be broken up and 98% of her materials recycled. *Bravo* was now at the mercy of *Pioneering Spirit*, which was cradling it to its grave. We were her initial funeral procession and this was her wake.

When I looked down, I was transfixed by the waves being buffeted against the slot between the bows as we sailed along at around 12 knots, with perpetual sea spray in the air. It was now much more subdued and relaxed up on the bridge and I enjoyed my drawing session. Then, looming into sight I noticed a massive concrete, single-base operational platform and another jacket-supported platform, to its south: *Ninian Central* and *Ninian South* platforms. I sketched these quickly from the comfort of the bridge and thought of Peter Stuart and Brian Taylor, the former Murchison Field's OIMs, who had been seconded to the Ninian Field. Very sadly, Brian died early this year, just before his retirement. The other *Murchison* crew members whom I had met had also gone on to work on these platforms. I thought of them watching us as we sailed past some nautical miles west of their field.

I did not want to leave the bridge as there was so much of interest at sea to draw. It was a beautiful evening so the clarity was very sharp and, as it was nearly mid-summer, there was almost continuous natural light. The occasional gannet flew past, reminding me of the fragility of it all. The sheer magnitude of oil production in all these fields we were passing prompted me to wonder how we would ever get out of this fossil-fuel trap.

Wednesday

I went up on the bridge again to draw. We were 65 nautical miles west of Mullen Head. To our starboard was a fishing fleet of six vessels, surrounded by gannets and fulmars looking for scraps. I was told that sometimes small boats got curious and came dangerously close to *Pioneering Spirit*. They often had to be warned off because they did not realize that, because of her size, she was not quick to manoeuvre. I could see why the fishing boats might be curious, and we were not exactly your everyday maritime freight traffic.

The bridge was again busy and I noticed Venetia, the naval architect, working out some ballast calculations. She was the only female among the Allseas office-based engineers and a pioneer in her professional field, working in this nearly all-male environment.

Throughout my short time onboard, people had been curious about my role. Some thought I was "quality control," making notes in my black-bound sketch books. When I explained what I was doing, they were fascinated that an artist should be recording this event.

Thursday

At 01:45 hours, we were anchored five nautical miles off Hartlepool and preparing the ballast height for the *Iron Lady*'s barge arrival. Just before 04:00 hours, the *Iron Lady* was slowly backing into the slot, underneath the topside, with *Bravo*'s elephant feet, its coffin feet, onboard. Sailors were scurrying about throwing, tying, and securing tow ropes as the *Iron Lady* docked. I marvelled again at the scale of everything: the three giant feet were 20 × 20 × 8 metres in height. The R3 crane carefully lowered the gangway into place, allowing access onto the barge. The sun was rising as I watched the whole spectacle. The ballast was now being adjusted so that the mighty beast could be secured onto her awaiting elephant feet and her funeral barge.

I checked in at 05:45 hours for my return to shore and handed over my ID card in exchange for my passport. There was a delay in the launch boat docking; the sea-swell was too rough for people to board safely. They had to get one of the *Iron Ladys*' tugs to be a buffer for sea-swell protection.

The boat was swaying around in the choppy waves as I managed to squeeze into a spot on its small outer deck amongst impatient men, wanting home. I stood up, holding onto a side rail and tried to take some film footage, as we moved away from the giant vessel. What a sight, its scale and presence, from our sea-level angle. As we sped closer and closer to land, *Pioneering Spirit* got increasingly distant, like a mirage on the horizon. I silently bade farewell to both vessel and topside. We were stepping out from the maritime world, back into the world of the shore. We passed some small fishing boats going out to sea, and I recognized some landmarks nearby from my *Brent Delta*'s drawing session at Able UK yard in 2017: the Teesside Offshore Windfarm, its 27 turbines slowly turning as if they were just waking up, and behind these was the redundant massive and decaying Redcar steelworks.

I half wished that I could stay to watch the spectacle of *Bravo*'s funeral barge procession; the second-captured Brent beast brought ashore to Hartlepool. On the train home, I was not too sure what time of day or night it was. I had returned from another world, and I still had my safety boots on.

Notes

1 *Willy Wonka and the Golden Ticket: Charlie and the Chocolate Factory* is a 1964 children's novel by British author Roald Dahl. The story features the adventures of young Charlie Bucket inside the chocolate factory of eccentric chocolatier Willy Wonka.
2 Some examples of North Sea offshore folk art are held in the Aberdeen Archives, Gallery & Museums collection, but most items are kept by the workers themselves. There has been very little research and documentation into these offshore artworks.
3 My *Oil Worker Scotland* exhibition (1989–1991) premiered at the City of Edinburgh Art Centre.
4 Gustave Courbet (1819–1877), the French 19th-century painter, often painted the rocky grotto at the source of the Loue, the river that flows through his native village, Ornans, in the Franche-Comté region of eastern France. Other contemporary artists, for example Tasmanian artist Raymond Arnold, have also reflected on and been influenced by Courbet's painting (Kelly 2008, 32–33).
5 As the UK moves towards a green future, many ask whether there is still a place for the aging UK oil industry. "If you were to ask a dyed-in-the-wool Aberdonian who has worked in the industry for the last 20, 30 years whether they are interested in climate action they would probably only go along with it as it didn't impact their livelihoods," the oil veteran says, "These same people are realising renewables might just save them" (Julian Ambrose, *The Guardian*, March 30, 2020).
6 The Golden Chopper is a term used for the final helicopter flight out after Total Platform Shutdown (TPS).
7 See James Hunter's introductory essay in the *Beatrice Works* exhibition catalogue (Hunter 2013, 4–8) and pages 71–92 of my *Age of Oil: Artwork* (Taylor 2017).
8 For example, see the excerpt published in *The Guardian* from Olivia Laing's novel *Art in Emergency*: "Art can't forcibly induce a change in behaviour. It's not a re-education pill. Empathy is not something that happens when we read *War and Peace*. It's work, for which art can, however, provide us with radiant materials. Art can't win an election or bring down a president. It can't stop the climate change, cure a virus or raise the dead. What it can do is serve as an antidote to times of chaos. It can be a route to clarity and it can be a force of resistance and repair, providing new registers, new languages in which to think" (March 21, 2020).
9 I was allowed back onsite at the beginning of May after a six-week COVID-19 lockdown period for non-essential personnel.
10 A version of *Brent Delta* platform entries appeared in blog form on the NMS website during *Age of Oil*'s exhibition tour (2017–2019). For further reading on my *Brent Delta* visit diary entries, see *Age of Oil Artwork* (2017), 62.
11 "The crude from Beatrice is unlike any other in the North Sea: it has a high content of wax (17 per cent) and a high pour point (about 20° C). In other words, it is fluid only at temperatures above 18° C: below that level it resembles black shoe polish. Once it is reduced to sea temperature after leaving the 80° C environment in the reservoir it becomes viscous. In that condition, a shut-down in the 60-kilometre seabed pipeline could leave BNOC as the proud owner of the longest candle in the world. To avoid this horror, the corporation searched for an additive to keep the oil fluid at low temperature. It settled for a toluene-based chemical that Shell produced. Unfortunately, the additive ceases to be effective if the wax in the crude has already begun to crystallise, so it has to be injected as soon as the oil leaves the reservoir – at a depth of 1000 metres" (Hall 1982, 776).
12 "The Beatrice Pollution Control Plan was the first developed for a near-shore oil field in the UK. The use of conventional dispersant methods proved impracticable when dealing with the Beatrice crude spill, it required the development of innovative dispersal techniques" (Hall 1982, 776).
13 In October 2019, two Greenpeace campaigners boarded Shell's *Brent Alpha* platform in the North Sea. A protester from the environmental group also climbed on to a

concrete leg of the decommissioned *Brent Bravo* platform and placed a banner, "Clean up your mess Shell!" Greenpeace was protesting against the method of decommissioning platforms in the Brent oil field, which is about 116 miles (186 km) north-east of Shetland. The protestors boarded in protest at Shell's decision to leave 11,000 tonnes of oil sediment in the legs of three of its four platforms. In December 2019, Shell obtained a court order banning Greenpeace from carrying out climate protests on their North Sea oil platforms.

14 In September 2003, oil workers Sean McCue, 22, from Kennoway, Fife, and Keith Moncrieff, 45, from Invergowrie, Tayside, died after a sudden escape of gas when they were working inside the utility shaft. When the liquid drain line valve patch gave way, a significant volume of rich hydrocarbons flooded into the enclosed space causing their deaths from asphyxiation. Electrician Graeme Burns, 47, died while carrying out maintenance work with a colleague on *Brent Bravo* on New Year's Day, 2005. Both men evacuated the leg of the oil rig after smelling gas at 02:30, but Graeme Burns collapsed and died.

References

Hall, John. 1982. "BNOC Gets a Taste of the 21st Century." *New Scientist* 17 June, p.776.

Hunter, James. 2013. "The Far North: An Artistic Perspective." In *Beatrice Works* publication, 4–8. Aberdeen: Aberdeen City Council, 2013.

Kelly, Séan. 2008. *Raymond Arnold*. Hobart: Quintus Publication.

Lewis-Jones, Huw. 2019. *The Sea Journal: Seafarers' Sketchbooks*. London: Thames & Hudson.

Taylor, Sue Jane. 2005. *Oilwork: North Sea Diaries*. Edinburgh: Birlinn Ltd.

_____. 2013. *Beatrice Works Publication*. Aberdeen: Aberdeen City Council.

_____. 2017. *Age of Oil: Artwork by Sue Jane Taylor*, edited by Elsa Cox. Edinburgh: NMS Enterprises Ltd.

9

TIME FOR OIL

Competing petrotemporalities in Norway's Lofoten/Vesterålen/Senja Archipelago

Berit Kristoffersen, Gavin Bridge, and Philip Steinberg

Competing temporalities

On a cold winter afternoon in 2010, one of the authors (Kristoffersen) accompanied a local facilitator and an oil industry executive to a public consultation in a small fishing village in Norway's Lofoten Islands. The consultation was one of a series being organized by the Norwegian Oil Industry Association (OLF, now Norwegian Oil and Gas) to generate community support for offshore oil-based development that the industry was hoping to undertake in Nordland VI and VII and Troms II, the group of Norwegian Sea oil licencing areas that abut the Lofoten and Vesterålen archipelagos and the adjacent island of Senja, an area of Arctic Norway collectively abbreviated as LoVeSe.[1] During the drive to the consultation, the local facilitator was prepping the oil executive about the community where she would be speaking. He explained that the community was struggling with outmigration, that it had a "traditional" coastal rural mind-set, and that its economic lifeblood was rooted in fisheries: up to 10,000 tonnes of cod and other fish were landed every year for processing and distribution. In other words, it was a coastal community much like others in the LoVeSe region.

Before the public meeting, we detoured to visit the local dock, as the oil executive announced that she had never visited a fish landing and processing facility. Getting out of the car, we were greeted by a worker who quickly voiced his feelings about the proposed oil development, first by ignoring the oil company representative and then by greeting the local facilitator with the droll compliment: "Congratulations on becoming our local oil minister." The oil company representative interjected to ask the worker what he thought about the proposals to bring the oil industry to Lofoten. "I don't want the oil but I assume that it will come anyways," he replied. "This means continuous seismic shooting air guns to map potential underground resources by the oil companies which means that

DOI: 10.4324/9781003024125-9

the *sei* [*Pollachius virens*][2] will be scared away." For the three previous summers (2007–2009), the Norwegian Petroleum Directorate (NPD) had organized seismic surveys in LoVeSe that had disrupted the fishing industry, and thus people involved in catching and processing fish were keenly aware of seismic mapping's potential impacts.[3]

The oil company representative then leaned over a tub of fish and asked, "Is this *sei*?" The worker confirmed that it was, while winking at me (Kristoffersen) in silent conspiracy because, in fact, the fish were not *sei* but *skrei*, the Arctic cod (*Gadus morhua*) that migrate to the waters off LoVeSe every January, where they are caught in abundance, cured on ubiquitous drying racks that dot the islands, and sold globally as *the* Norwegian delicacy.

Following this diversion, we proceeded to the crowded community meeting, where the oil company representative told the expected story of abundant offshore oil resources that could bring wealth to adjacent communities. Pointing to maps and figures, she informed the gathering that up to three billion barrels[4] could be hidden in the seabed outside LoVeSe, and that working these resources could provide substantial development opportunities to the region. She noted that the community where the meeting was being held might be a particularly good site for an onshore terminal.

Placing the region's opportunities in context, the oil company representative went on to relate the recent history of oil exploration in the region, depicting the industry as one that is both progressive and cautious. She explained how two exploratory licences had been granted in 1996 in waters in the southern reaches of the region: PL 219, which is located just 30 kilometres west of Røst, the island group at the southwestern end of the Lofoten chain, and PL 220, which is slightly further away, to the south of Røst. She also noted that exploration and the issuing of further licences had been halted in 2001 to evaluate how "oil could be done in the North." Brushing over the fact that this moratorium had been enacted in response to strident opposition from environmentalist and fishing interests, the oil executive noted how this reassessment had culminated in the first Lofoten – Barents Sea Management Plan in 2006 (Norwegian Ministry of the Environment 2006). She further explained that, in keeping with that plan, she was here today to discuss what "the oil companies could do" for Lofoten.

The question-and-answer session after the presentation in many ways replayed the conflict between oil and fisheries that had led to the moratorium and Management Plan in the first place. Residents in particular wanted to know more about the industry's plans for seismic testing and how it would impact fisheries. However, as the oil executive revealed in the car driving back from the meeting, the one question that had surprised her was one that added a new measure of *temporality* to the conflict. Under the assumption that there *would* be oil development and a new oil terminal built, one local asked, "What will we live off in this community afterwards?" The question was grounded in the assumption that oil would last only one generation. While the oil executive was prepared to address questions about contemporaneous conflicts between various ocean uses, she had failed to prepare for questions regarding the threat that oil development posed in terms of longer, existential time scales. However, this way of looking beyond the extraction period as a premise for whether or not to develop hydrocarbon resources was typical for debates in Lofoten at the time (Kristoffersen and Dale 2014). Talking further with the oil executive after the meeting, I (Kristoffersen) attempted to elucidate the perspective of the local who asked that question by referring to an interview that I had conducted the previous day in Vesterålen with a 70-year-old fisherman whose son and grandson had both taken up the trade. Referring to the grandfather of this three-generation fishing family, and the likelihood that oil extraction would last a maximum of 40 years, I rhetorically asked the oil executive: "How long is 40 years to him?"

As this series of exchanges between the researcher, the oil executive, and the community demonstrates, what is usually portrayed as a conflict over a *shared space* (the space of fisheries and the space of oil development, and the linked spaces of community that join these and other economies) is also a conflict over *differentiated times*. As the oil executive went about her site visit, for instance, a number of temporalities, and the knowledges associated with them, were placed in conflict with each other: the community's knowledge, accrued across generations, about fish migrations; the oil industry's 40-year projection of the proposed project's lifespan; the community's understanding of seismic testing's impacts gathered over the previous three seasons; the knowledge gathered by the oil executive in one evening of community meetings and site visits. To explore this mosaic of temporalities, the next section of this chapter unpacks the various temporalities imagined and applied by the petroleum industry as it goes through the practice of constructing oil as a "resource." The following section draws on ten years of interviews conducted by Kristoffersen on Røst (from 2010–2020) to uncover the community temporalities (and related spatialities) that are engaged in the region's encounter with an emergent oil economy. In the final section, we explore how the state inserts its own temporalities into the mix as it articulates between industry and community temporalities. The resulting analysis may not give us definitive answers to cold water oil development questions. However, in contrast to the oil executive quoted above, it may prevent us from being blindsided by other perspectives that reflect different temporalities.

Temporalities of oil

For now, the moratorium introduced in 2001 still holds, and no new licenses have been issued in the Nordland VI/Nordland VII/Troms II region. However, if oil production were ever to commence in the waters off LoVeSe, it likely would follow the model of PL 219 and PL 220, and countless other licenses around the world, where exclusive rights to exploration drilling and production are deeded to a private or state-run corporation for a fixed (but extendable) period of time. Since their issuance in 1996, PL 219 and PL 220 have conveyed to

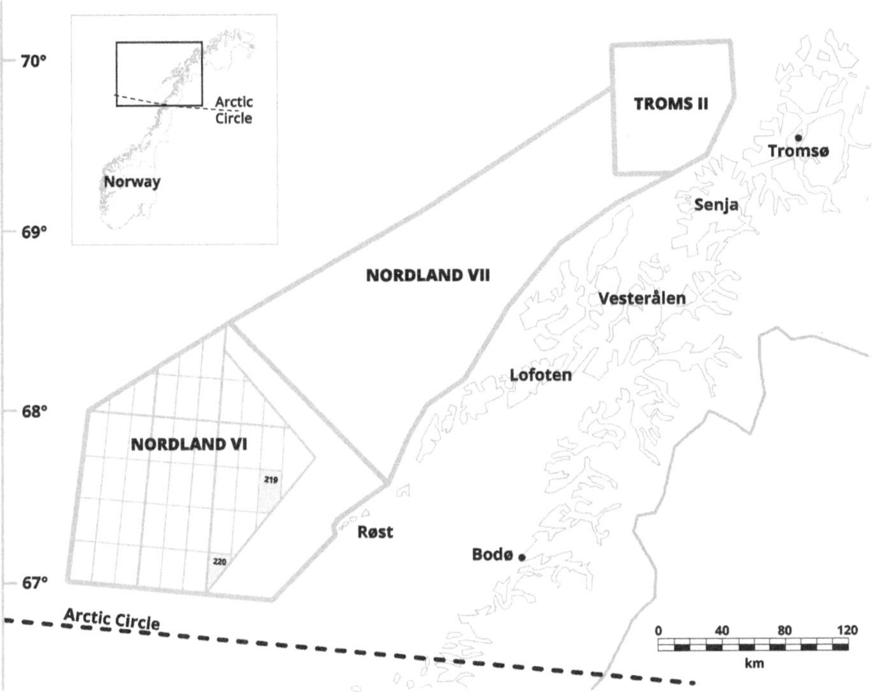

FIGURE 9.1 Offshore petroleum areas in the LoVeSe region
Source: Bjørn Hatteng/UiT

the groups of companies which hold the licenses rights to drill in the designated areas. Although considered the most prospective license areas in Norway (Norwegian Petroleum Directorate 2010, 3), PL 219 and PL 220 are effectively sleeping licenses as there has been no active exploration on them since 2000 due, in part, to the moratorium and subsequent restrictions of the Management Plan. At present, the license to PL 219 is held by a four-company consortium where Vår Energi is the majority (50%) shareholder but Equinor holds the operating license. PL 220 is held by a three-company consortium where Equinor holds the biggest share (50%) and is also the licensed operator.

Such licenses are, in a rather obvious way, spatial instruments of property. However, as we argue below, these licenses also presage future relations – the potential for an extractive landscape – in ways that signal the value of reading such licenses as "timescapes" (Adam 1998). For Adam, time is plural and diverse – an assemblage of temporal layers, tempos, rhythms, and contingencies – that constitute landscape and experience. Oil exhibits this "cluster of temporal features" in often striking ways, and not only because temporalities of future-use are "baked in" to the claims about "resources" and "development" through which oil is often described (Adam 2004, 143). More extensively, as Ferry and Limbert have argued, resources "frame the past, present, and future in certain ways; they propose or preclude certain kinds of time reckoning; they inscribe teleologies; and they are imbued with affects of time, such as nostalgia, hope, dread, and spontaneity" (Ferry and Limbert 2008, 4).

An insight from this perspective is that oil exploration (and the license to cold-water acreage through which it materializes) privileges some temporalities while occluding others: in this regard, it echoes wider western/colonial ontologies of nature and possession that erase or marginalize alternative temporalities. The multiple temporalities (durations, velocities, dynamics, chronologies) aligned with oil form a nexus of "oil-time" (with analogy to Müller's (2019) work on coal), an "articulation of multiple temporalities" (D'Angelo and Pijpers 2018, 216). The remainder of this section disaggregates four of the most significant oil temporalities in relation to PL 219 and PL 220: the deep-time past of geological knowledge; the anticipatory futures of administrative planning; the asynchronous logic of investment and financial return; and the eruptive and stochastic temporalities of oil's environmental legacy.

First, there is the long-term, linear temporality of the deep-time processes of oil formation – a distant past of large deltas stretching over thousands of kilometres, sedimentation and burial processes enduring for millions of years, and then subject to multiple more-million years of pressure and fluid migration – that created the concentrated pools of hydrocarbons that constitute Norway's offshore oil resource. The reserves accessed by PL 219 and PL 220, for example, are hosted by rocks laid down in the Jurassic and Cretaceous period and are, effectively, concentrations of space and time that, if appropriated and brought onshore, would provide an enormous energetic subsidy to modern economic and social life. Drilling through these layers folds space and time – the "rotating drill pushes in an instant from one millennium to the next as it cuts through the sedimentary rocks of the Pliocene, the Cretaceous, the Triassic" – enabling imaginaries of the future that rest, ultimately, on accessing the *longue durée* of deposition, sedimentation, and petroleum formation (Calvino 1974, 171–172). In the process, reserves of energy that formed unimaginably slowly underground flow to the surface, a transfer of geological space and time that (as fuels for transportation and feedstocks for durable plastics) underpins the compression of time and space in modernity (Bridge 2015).

Since Norway first asserted sovereign rights over its continental shelf in 1963, the qualities of this space have increasingly come to be understood by reference to the deep-time of geology. The discovery of gas in the Dutch offshore Groningen Field in 1959, and a subsequent request from Phillips Petroleum for exploration rights to Norway's offshore, initiated an extensive programme of geological mapping and assessment by the Norwegian state and private capital (Noreng 1980). In the process, a space formerly differentiated via maritime registers (e.g. bathymetry and currents) has been reimagined – and increasingly administered – by reference to the qualities of subsurface geology. Here, as in other cold water environments, such as the waters of western Greenland and the Chukchi, Beaufort, Barents, Pechora, and Kara Seas, geological science's chronological "grids of intelligibility" (e.g. the stratigraphic column, regional depositional histories) serve as proxies for probable content and are overlain by legal grids that divide the seabed into blocs to be licensed.

Second, there is the anticipatory temporality of the administrative project cycle – the imagined progression over time from geological assessment through exploration and field development to the first oil of production – that means geological assessments of resource potential usher in an "horizon of expectations" (Weszkalnys 2016, 133). The qualities of territory may be constructed through the repertoire of geological time but, once given objective form (e.g. as quantifiable resources), these qualities then act in the world, generating and demanding orientations that structure political economic relations in the present. This theatre of anticipation plays out in several different ways but revolves, fundamentally, around the tension between current conditions and future prospects – a potential change of state mediated by questions of timing and uncertainty. As Weszkalnys has observed in relation to offshore oil in the Gulf of Guinea, oil's "double obscurity" (uncertain subterranean extension and indeterminate market value) enables an economy of expectations to take hold (Weszkalnys 2015, 625). Such a horizon of expectation is built into the Petroleum Directorate's procedures governing the Norwegian offshore: the Norwegian Petroleum Act requires the Petroleum Directorate to open up the offshore for oil and gas development and to outline and initiate "area opening procedures" of resource and impact assessment prior to licensing (Norwegian Petroleum Act, 1996). These opening procedures, like the subsequent issuance of drilling licenses and publication of seismic data, foster an anticipatory orientation in the present – a sense of imminent acceleration and transformation strongly associated with the time-space compressions of modernity – that is simultaneously affective (feelings of hope, doubt, excitement, and fear) and strategic (expressed through planning procedures and preparatory action).[5] In LoVeSe, these emotions associated with "area opening procedures" were in full display at the community meeting referred to at the beginning of this chapter, as local residents reflected on the seismic assessments that were being carried out in the vicinity of regional fishing grounds.

But, as PL 219 and PL 220 illustrate very well, anticipation does not smoothly fold potential futures into the present: it fundamentally requires this process to be incomplete. Anticipation exists, as Weszkalnys puts it, in the "not yet" – a state of liminality in which potential futures radically define the present (by separating it decisively from the past), even while attaining this future is uncertain (Weszkalnys 2014, 213). From this perspective, the delays and attenuations that so often attend cold water oil and other frontier resource development plans are constitutive of anticipation rather than opposed to it. The status of PL 219 and PL 220 as sleeping licenses – and the "decades of dormancy" in which they are not-yet awake – speaks directly to how delay and indeterminacy can sustain rather than cancel anticipatory economies of offshore oil (Weszkalnys 2015, 616). It also highlights how, in this corner of the Norwegian Sea, the "not yet" has endured for a generation and how temporalities of anticipation can "hav[e] a profound impact on individual and community lives" (Szolucha 2018, 353).

The temporality of anticipation that attends oil development is overdetermined: it is not only structured by administrative and corporate processes that "materialize absent potential" but also by economic considerations (the waxing and waning of price cycles, interest rates, and the availability of debt) and, especially in the context of cold water environments, climate change (Weszkalnys 2015, 616). Although the waters off LoVeSe, warmed by the Gulf Stream, have been ice-free for millennia, speculation about their potential is interlaced with narratives of an impending Arctic resource boom, portended by the opening up of a hostile environment, that for centuries have shaded attitudes towards the region's resource potential.[6]

This speculation, in turn, is stimulated by (and serves to stimulate) the production of vast amounts of information in the form of reports on ecological baselines, impact studies, and disaster planning. In their account of expert knowledge around Arctic oil and gas development in Alaska, Mason and Stoilkova reference the "paper pipelines" of documentation accumulated over decades (Mason and Stoilkova 2012, 84). Such accounts suggest a juddering anticipation, a sense of uncertainty (advances and setbacks) that attends the competition to get in first and seize the prize; where the Arctic's seemingly sinister materialities – ice, darkness, cold, ocean depth – drive a selective "evolutionary" process that determines which techniques, projects, and companies will survive in these Arctic "resource frontiers" (Steinberg et al. 2014). But the anticipation of a melting Arctic, the dream of a future liquid medium in a warming world, also suggests an imminent process of revelation and connection. Imaginaries of a melting and more open Arctic are supported by a series of cold water firsts – the *Noble Discoverer* drill ship in the Chukchi (2012), the *Prirazlomnoye* platform in the Pechora Sea (2013), the *Goliat* oil platform in the Barents (2016). The anticipated opening up of the Arctic as a hostile environment, then, is central to the "not yet" of cold-water resource development.

Third, there is the temporality of capitalization and investment, and the time required between sinking capital in oil development and generating a revenue stream for investors. This asynchronous logic takes us beyond a general interest in "resource-making," centring instead on the assembly of oil resources as a specific form of economic object: the investible proposition. A key window here is the time between initial investment and first oil, a temporal gap between sinking capital and generating cash flow that is spanned by finance. This gap structures the amount of debt a project must bear and, ultimately, it limits the type of economic actors able to operate in the offshore. Cold water oil frequently features sharp temporal disjunctures between production time (when revenue is being generated) and the time-value of money tied up in plant and equipment that are exacerbated, for example, by seasonal ice movements, freeze-thaw cycles, and long periods of darkness. In spanning the gap between investment and production, however, finance sutures the fate of the offshore to the calculus of valuation and risk. These modes of calculation explicitly fold the future into the present: calculations of discounted cash flow (e.g. net present value and internal rate of return), and the interest rate on capital advanced, for example, exert a determining influence on commerciality. Low interest rates make it easier to raise finance to span the gap, while the debt obligations associated with large projects can, in a lower-than-expected oil price environment, lead companies to default.

Calculations of commercial return rest on assessments about the forward state of the oil market and, in particular, the relationship between oil price and costs of production. Long-term projections of demand growth against constrained supply have traditionally created a future "wedge" (i.e. growing shortfall of supply relative to demand) against which many cold-water oil projects have been financed. With costs of production above the industry average, cold-water environments frequently provide the marginal barrel, and their commerciality depends on a relatively high price environment (and cheap debt). Increasingly, however, historic expectations of continuing demand growth for oil have been faltering, and the availability of investment has been shaken by rapid oil price collapses (e.g. 2014) and unprecedented reductions in demand (e.g. due to responses to the COVID-19 pandemic). In this context, a temporally more complex and geographically differentiated narrative about cold water oil environments is emerging. Bolstered by high profile experiences of retreat, such as the decision of Statoil (now Equinor) and partners not to proceed with Shtokman LNG (2012) or Shell's abandonment of its drill programme in the Chukchi Sea (2015), or even Russia's decision to place a moratorium on new offshore exploration licences until (at least) 2020, temporalities of anticipation in reference to cold water environments are now overlain (and undermined) by a more precarious and stuttering temporality of contingency.

Fourth, there are the stochastic and uncertain temporalities of oil associated with the environmental fate of hydrocarbons. By their nature, these defy planning and prediction and so frequently either outstrip or disrupt the planned temporalities of administration or finance. In many instances, it is the persistence of anthropogenically driven environmental processes – i.e. their slowness and brute duration – that challenges an imagined 40-year "life-cycle" of oil development progressing, in a discrete (and ultimately impermanent) way, from exploration and development via maturity and decline to closure and remediation (Gray-Cosgrove et al. 2015). The residence time of carbon in the atmosphere, the duration of plastics in the ocean, or the "zombie" character of abandoned wells exhibit temporalities (and spatialities) that exceed those of planning, capital recovery, and regulatory oversight. Indefinite in duration and uncertain in their effects, these "afterlives" of oil haunt future generations and raise challenging questions about perpetual care and responsibility (Sandlos and Keeling 2013).

Temporalities of community: uncertain entities in meaningful seascapes

To further consider the temporalities of oil, in this section, we zoom in on Røst, the cluster of 365 islands (pop. 498) that extends out from the Lofoten

Archipelago, into the Norwegian Sea and towards Iceland. But even as we zoom in, we also zoom out, in both time and space, because the islands of Røst, like the nearby oil resources of PL 219 and PL 220, exist in a multitude of temporalities and spatial connections.

Archaeological findings suggest that there have been people living off the seas at Røst for 5,000 years, which make it one of the oldest settlements in Northern Europe. Although the Lofoten Islands (including Røst) had been trading dried cod to Europe since the Viking era (ca. 9th-century CE), "for Røst, globalization started in 1432," according to Arnfinn Ellingsen, who served for 16 years as Røst's mayor (Ellingsen 2015)). Ellingsen was referring to the voyage of Venetian merchant Pietro Querini, whose ship left Crete for Flanders in 1431 and was battered by a storm off the Atlantic coast of France. After several weeks drifting northwards, Querini's men abandoned their dilapidated ship for lifeboats, one of which eventually beached on Sandøya, an outer skerry in the Røst group, on 5 January 1432. After surviving on the islet for about a month, Querini and the 10 remaining members of his crew were discovered by local fishermen, who took them into their homes to recuperate. On 15 May, the Venetians began their return journey on a trading ship bound for Bergen, but not before receiving a send-off gift of 60 stockfish (air-dried *skrei*) from their hosts.

After leaving the ship in Trondheim, the men travelled overland across Europe, arriving in Venice on 12 October and still carrying some of the stockfish. Querini (and, later, his two officers) proceeded to publish travel narratives that recounted their adventures and that introduced Venetian readers to the lifeways of the people of Røst. Querini's narrative includes a vivid description of the process by which stockfish are dried on Røst:

> They dry their fish (stockfish) in the wind and sun without salt, and as these fish do not contain high humidity or fat, they become dry as wood. Before cooking the fish, they hit them with a hammer thereby loosening the tendons; then they add butter and spices in order to impart flavour. It is a great and valuable export article from the North Seas. (Humberset 2017)[7]

Like the oil company executive's site visit to Lofoten and the broader arc of the petroleum industry related in the previous two sections, Querini's description of stockfish preparation suggests a number of different temporalities and attendant spatialities. On the one hand, Querini's narrative supports the image of Røst as an isolated outpost frozen in time. His description of the stockfish drying process is as accurate today as it was 600 years ago. On the other hand, Querini's voyage unleashed a history (and geography) that continues to unfold. The stockfish that he brought back to Venice went on to become the central ingredient of the *baccalà* dishes that characterize "traditional" Venetian cuisine, and to this day, 90% of Røst's stockfish exports are sent to Italy. Italy's gratitude for Røst's historic act of hospitality (and its ongoing supply of stockfish) is maintained in cultural and diplomatic exchanges between the two seemingly disparate regions. The Italian

Embassy in Oslo has supported the establishment of a Querini Literary Park on Røst and Røst residents have named an islet Sandrigo after a village in Veneto that holds an annual stockfish festival. According to Mayor Ellingsen, personal friendships, as well as commercial ties, between the regions remain strong. Connections between Italy and this seemingly remote outpost in the Norwegian Sea continue to deepen in unpredictable ways that build upon past entanglements. Vår Energi, the majority shareholder in the consortium that holds the lease to nearby PL 219, is 69% owned by the Italian energy company Eni.

The histories (or temporalities) that connect Røst with Veneto are notable in part because they are so unexpected. However, they are hardly the only connections – historic, ongoing, or anticipated – that make Røst what it is today. There are the migrations of *skrei* between the Norwegian Sea, where they spawn, and the Barents Sea, where they live most of their lives. There are the migrations of puffins and other seabirds, who play an increasing role in Røst's growing tourism economy. There are the migrations of people, as isolated islands like Røst struggle to maintain dwindling populations. And, of course, there is the migration of water, namely the warm waters of the Gulf Stream that are so important for maintaining the region's ecology. Yet, as Querini's narrative so clearly demonstrates, amidst all of these migrations (including his own) and all of the changes that they bring, there is also a sense of *longue durée* continuity that transcends any of the temporalities considered by the petroleum industry. In other words, just as the intersection of multiple temporalities creates disjunctures on land, they also disrupt visions for the offshore.

For Lone Nikolaisen, a young fisher from Røst, the concern about impending oil extraction in nearby waters is as much about what might happen *during* a potential oil boom as it is about what will happen *after* the oil has all been extracted:

> It is estimated that people have lived here for five thousand years. Then you can look around and see that it is actually fisheries that is what we have been doing the whole time. Then you have all the variations that affect this, where people are moving to and from Røst, and then, let's say we get this oil. That we get lots of [oil related] jobs. And if we then take a society that has functioned more or less in the same way for five thousand years, and replace it with something else, [then] the oil will get the recruitment. And if it takes a community that has worked more or less in the same manner for five thousand years, and replaces it with something else, I am anxious to see how it looks after the 30 years, when they leave. How forward looking is their industry? (Nikolaisen 2015)

Rather than focusing specifically on preservation of Røst's economy, or its culture, or its environment, Nikolaisen is expressing concern for the preservation of what Ingold calls a "taskscape" (Ingold 1992). Elaborating on this concept, Ingold writes, "Just as the landscape [or seascape] is an array of related features, so – by analogy – the taskscape is an array of related activities" (Ingold 1992, 44). The concept of a "taskscape" thus adds an emphasis on *practices* and *knowledges* to the concept of the "timescape" previously discussed with reference to Adam. In a taskscape, relations between humans and their interconnected ecologies are reflected in life processes, conditioned by the active engagement and knowledges gathered from continuous interactions. From this, it follows that people who engage directly with their surroundings are dependent on what their surroundings have to offer, which become "familiar affordances." The cod – often referenced in singular (as well as "the oil") – is one of many entities in a taskscape that is networked across not only the marine territories that are the fishing grounds but also the material and cultural landscape of lands as near as Røst and as distant as Venice. These elements are linked through *practices* (or "tasks") and are thus temporal, being remade, and reflected in pasts, presents, and (anticipated) futures. In Ingold's reframing, every task takes "its meaning from its position within an ensemble of tasks, performed in a series or in parallel, and usually by many people working together" (Ingold 1993, 158).

Nikolaisen reports that when she is taking up and gutting fish in the waters around Røst, she often thinks about the fish ending up on Italian families' dinner plates and she thinks of how these connections between distanced livelihoods will be threatened by oil development in the area: "It is a social responsibility to bring food to people," she reflects (Nikolaisen 2015). In particular, Nikolaisen fears that Mortengrunn (Morten's Ground, the submerged bank close to PL 219 where she often goes fishing during the *skrei* season) will be closed to fishing due to the security zones that are established around oil installations. When viewed from such a perspective, the introduction of new temporalities mediated through new tasks (e.g. work at oil landing facilities) or elements (e.g. the oil at PL 219) and practices (the implementation of security zones where fishing is forbidden) can only be disruptive. Røst's future and the *skrei* economy, in particular, thus require the maintenance of a broad set of relations, tasks, and environmental factors, linking elements as diverse as labour relations and weather conditions. As a bookkeeper at one of Røst's fish processing companies noted after a visit to their stockfish storage facilities: "First, you need to get fishermen, especially those you want the most. Then they have to catch it. Then you need to get it hung up at Røst, and then you need good drying conditions. Then, you sell it to Italy, in other words, getting it out to the market" (Kristoffersen 2016). The manager of a landing site added, "We have the best natural conditions in the world for drying fish at Røst. When others in Lofoten are plagued with flies, the brutal truth is that we don't have to deal with that. We get the wind from the north, while the rest of Lofoten lies in the shadow of it" (Kristoffersen 2016). With climate change, as the winter weather gets rougher, the fishermen often have less time to catch the cod when it is at its breeding grounds, making some take even bigger risks, disrupting every element of the taskscape.

Given this multitude of temporalities and spatial connections, it would be shortsighted to interpret debates over the future of Røst (and, more broadly, LoVeSe) as simply a series of competing opinions about the impact of anticipated petroleum development. As we have seen, the islands exist within a complex ecosystem that for millennia has linked human activities, fish migrations, and ocean-atmosphere interactions across both space and time. The way some of these temporalities have been combined to form durable taskscapes is illustrated by Lena Hamnes, author and vice mayor in the Vågan municipality in Lofoten, who paraphrased a Native American saying: "[In] the actions you make today, you have to think seven generations ahead. And when we talk about the potential everlasting value such as the fish resources in Lofoten, this is a criterion" (Hamnes 2017). In Hamnes' appeal to an intergenerational perspective, time is "reappropriated as part of the process of positioning oneself in the field of conflicting interests" (Szolucha 2018, 350), challenging the dominant ontology of oil that pivots around Lofoten as a space of resource abundance.

State temporalities: counting (on) coldwater oil

The timescapes and taskscapes of LoVeSe enrol the Norwegian state in multiple ways, from sovereign resource holder and corporate investor to regulator, rentier, and adjudicator. These multiple roles converge in oil production licenses like PL 219 and PL 220, for example, so that folded into a single spatial instrument of property – and held together by it – are a set of diverse state objectives and temporalities that often are in tension.

PL 219 and 220 are production licenses issued by the Petroleum Directorate, a government agency within the Ministry of Petroleum and Energy, established by the Storting (the Norwegian Parliament) in 1972 to administer petroleum resources on the Norwegian continental shelf. Underpinning the license is the Norwegian state's proclamation (Norwegian Royal Decree 1963) to sovereign rights over the natural resources on the continental shelf and the vesting of these rights in the Norwegian state (Agerup 2013). Together with subsequent legislation governing offshore oil exploration, these acts by the state with regard to the continental shelf have for the past half-century (and since the 1980s for the Norwegian Sea) privileged geological knowledge – and knowledge of the distribution of oil and gas resources in particular – as a primary way in which this part of the state's territory is made legible and actionable. The temporal logic here is one of orderly resource development and resource management to maximize value to the Norwegian state in ways that reproduce the Norwegian nation over time. This involves simultaneous temporalities of permanence (in relation to sovereignty over resources), the punctuated rhythm of licensing rounds in which access to blocks of territory are made available for bid and oversight of the full project cycle from exploration via production to decommissioning. These are the temporalities of planning and administration in relation to offshore oil, and their effect, particularly on communities like Røst, is to generate a sense that the future is owned by the state (Szolucha 2018).

The state is also present in PL 219 and 220 as a corporate operator and investor in the form of the state-owned (67%) Equinor (formerly known as Statoil). The company, which is well-known for being a state-owned firm that operates on commercial principles (Kristoffersen and Young 2010), produces about 1.2 million barrels a day (out of a total 2 million barrels for Norway as a whole) from the 42 fields they operate and is the dominant operator in the Norwegian offshore (Equinor 2020).

Through its operations as Equinor, the state broadly conforms to the temporality of capitalization and investment described in the first section of this chapter: one of financial returns on investment and the time-value of money as measured, for example, by discounted cash flow analysis and calculations of net present value. Significantly, however, transnational firms (including a transnational state firm like Equinor) are weighing up the Norwegian offshore in relation to rates of return on capital sunk elsewhere, setting up a negotiation over the temporalities of administration and those of investment capital (around, for example, the long-term rate of resource development vs the interests of investors). This plays out, for example, in the time limits associated with monopoly licenses like PL 219 and 220, through which the state (in the form of the Petroleum Directorate) attempts to secure its interest in long-term development by not allowing firms (including the state firm Equinor) to lock up acreage without undertaking meaningful exploration work.

Since the 1990s, however, the Norwegian state has contemporaneously been placing oil production within a wider spatio-temporal framework through its Government Pension Fund Global, also known as the Norwegian Oil Fund. In the early 1990s, government planners were anticipating that oil would lose its value in the near future and that, for the medium term, a decline in returns from offshore production could be offset by investing revenues in properties, bonds, stocks, and equities around the world. Established in 1990, and with the first money deposited in 1996, this Fund is now valued at over US$ one trillion (>10 trillion NOK). With holdings in around 9,000 companies worldwide, it is one of the world's largest sovereign wealth funds (Norwegian Bank 2020).

One of the main visionaries behind the establishment of the Oil Fund was Jens Stoltenberg, who went on to become Labour Party Prime Minister (2000–2001 and 2005–2013) and is now general secretary of NATO. In 1993, following the Rio de Janeiro Earth Summit (where Norway's Prime Minister Gro Harlem Brundtland played a key role), Stoltenberg, who was then the Minister of Oil and Industry, warned against leaving oil in the ground because a global climate deal could potentially kill the value of oil over the next decades:

> We cannot exclude that the same will happen with the oil [as with the coal], that new energy sources grow rapidly, reducing the value of the oil dramatically. This is not an argument to pump up all the oil as quickly as possible, but an argument against the claim that future generations will necessarily profit when leaving as much oil in the ground as possible. (*Dagens Næringsliv [Business Daily]*, 1993)

Like PL 219/220 (and other petroleum licenses), the Oil Fund folds and fuses together several different temporalities. Specifically, it sublimates the discrete and disjointed temporalities associated with the development of individual oil fields into a permanent return to the nation. In doing so, it codifies the mutual dependency between the onshore and offshore spaces of Norwegian territory: it establishes, on the one hand, the offshore as a space for hydrocarbon development through which the reproduction of social and economic life onshore can be assured; and on the other, the social welfare goals and aspirations of the population reproduce the offshore as a space of revenue generation via hydrocarbon extraction. Since the late 1990s, less than half of the NOK 10 trillion that now composes the pension fund has been directly supplied from oil revenues (Norwegian Bank 2020). This puts state elites in a favourable financial situation and creates a buffer between the changing level of oil prices and the government's returns. As Ranveig Frøiland, Minister of Oil and Energy from 1996 to 1997, stated, it was more favourable to have oil "on the account" than to have oil "in the ground"[8] during times of low oil prices, as was the case from the mid-1990s to early 2000s. The Storting accepted this argument, as exemplified when, in 1996, the Energy and Environment Committee justified depositing revenues in the fund because the alternative, leaving oil in the ground, was "fraught with great uncertainty." As the Committee reported to the Storting, "The question of the pace of extraction, especially for the oil resources, is first and foremost a discussion of which form of wealth can best ensure the highest possible value creation and help to ensure long-term welfare and employment" (Norwegian Standing Committee on Energy and the Environment 1996). Here, then, it becomes clear that the temporalities of geologized property – i.e. the hydrocarbon potential of territory – do more than simply consolidate the state's claim to territory (cf. Braun 2000): they also secure the conditions for developing, optimizing, and sustaining (national) life. In short, Norwegian biopolitics are geologically subtended and bounded together by the anticipatory logics of oil for development, re-distribution, and welfare (Clark and Yusoff 2017).

Conclusion: time for oil?

In many respects, the various narratives related here concerning oil extraction in the Norwegian Sea affirm a Norwegian version of what Lewis Mumford coined as "carboniferous capitalism," in 1934, which is now reflected in the expansion of the workforce and welfare state as dependent on state-led management of fossil fuel supplies (Dalby 2020, 98–99). Certainly, in Norway, since creation of the Oil Fund, there has been an ongoing process of "mainstreaming" and intensification of dependence upon hydrocarbons, materially, economically, and discursively, from the halls of power in Oslo to the frigid waters off remote outposts like Røst. Like any state-managed process, the Norwegian state's management of the oil economy has required it to broker between competing interests and actors that transcend not just different social groupings (industrialists, environmentalists, fishers, pensioners) but also different spatialities and flows. Just in this chapter, we have seen the movement of knowledge between LoVeSe and Oslo, the movement of live *skrei* between LoVeSe and the Barents Sea, the migration of dried *skrei* between LoVeSe and Italy, and the migration of waters between LoVeSe and the Caribbean.

Here, however, we have stressed that each of these spatialities also encompasses a number of temporalities and that decision-makers, at all levels, are working within multiple timeframes. The oil executive on the site visit is considering the geologic time of petroleum formation, the fluxes of the market, the long-term trend towards decarbonization, and the delays resulting from the Management Plan's approval process with the potential impact on the commercial value of their asset, while also perhaps wondering if weather conditions will delay her flight back to Oslo. The fisher is considering seasonal temporalities of fish migrations, the ways that marketing patterns reflect historic and ongoing interregional connections, the delay in cash flow before investment in fishery infrastructure can generate revenue, and the impact that impending oil development will have on the long-term sustainability of the community. The state is considering the balance between the aspect of its legitimacy that is based on its ability to protect the environment with the prospect of increasing state revenues, each of which, in turn, requires a balancing of short-term and long-term perspectives.

These temporalities advance unevenly: the steady progress towards globalization is disrupted, but ultimately enhanced by Querini's shipwreck. Temporalities also are deployed strategically: An elderly fisherman worries over whether he can pass on his trade to future generations whilst the state affirms its legitimacy by boasting of its ability to provide for tomorrow's pensioners. This chapter has suggested that there is no singular "time for oil" but rather a series of alignments and intersections among different temporal frames. Cold water oil may arrive very quickly or very slowly but, regardless, it materializes by folding together pasts, presents, and futures. As we have shown, the anticipatory temporalities that sustain cold water oil – temporalities of petroleum abundance, technology development, and enduring demand – mean that it has an impact on coastal communities even when it does not flow at all.

Notes

1 Roughly speaking, Troms II lies west of Senja, Nordland VII lies west of Vesterålen and Lofoten, and Nordland VI extends southward and westward from Røst, the island group at the southwestern tip of the Lofoten archipelago (see Figure 9.1).

2 *Sei*, generally known in English as pollock (US) or saith, coalfish, or coley (UK), is a member of the cod family and is fished year-round off the coast of northern Norway.

3 These activities were carried out as part of the assessment mandated by the Management Plan for the Lofoten-Barents Area (Norwegian Ministry of the Environment 2006), first passed in 2006 and updated several times since.

4 The estimate was in reference to an industry-led report from Konkraft (2008). A more conservative estimate by the Norwegian Petroleum Directorate (2010) suggests 1.5 billion barrels, with the greatest potential in Nordland VI.

5 See Abram and Weszkalnys 2011; Weszkalnys 2016.
6 See Avango et al. 2011; Avango, Hacquebord, and Wråkberg 2014.
7 The complete works of Querini and his officers have not been published in English. This quotation is from Humberset's (2017) English translation of excerpts from Sommerfeldt's (1908) translation into Norwegian. The travel narratives have also been published in modern Italian (Nelli 2018).
8 This is translated from the Norwegian wording "bedre med olje på bok, en olje på fot."

References

Adam, Barbara. 1998. *Timescapes of Modernity: The Environment and Invisible Hazards.* New York: Psychology Press.

_______. 2004. *Time.* Cambridge, UK: Polity.

Abram, Simone and Weszkalnys, Gisa. 2011. "Introduction: Anthropologies of Planning-Temporality, Imagination, and Ethnography." *Focaal* 61: 3–18.

Agerup, Mette Karina Gravdahl. 2013. "Petroleum Law Spring 2013: Introduction and Access to Resources." University of Oslo. https://www.uio.no/studier/emner/jus/jus/JUS5411/v13/undervisningsmateriale/ageru p_v13-intro-access.pdf.

Avango, Dag, Louwrens Hacquebord, Ypie Aalders, Hidde De Haas, Ulf Gustafsson, and Frigga Kruse. 2011. "Between Markets and Geo-Politics: Natural Resource Exploitation on Spitsbergen from 1600 to the Present Day." *Polar Record* 47 (1): 29–39.

Avango, Dag, Louwrens Hacquebord, and Urban Wråkberg. 2014. "Industrial Extraction of Arctic Natural Resources since the Sixteenth Century: Technoscience and Geo-Economics in the History of Northern Whaling and Mining." *Journal of Historical Geography* 44: 15–30.

Braun, Bruce. 2000. "Producing Vertical Territory: Geology and Governmentality in Late Victorian Canada." *Ecumene* 7 (1): 7–46.

Bridge, Gavin. 2015. The Hole World: Scales and spaces of extraction. *Scenario Journal* 5: Extraction. https://scenariojournal.com/article/the-hole-world/

Calvino, Italo. 1974. "The Petrol Pump." In *Numbers in the Dark: And Other Stories*, Italo Calvino, 170–175. New York: Vintage.

Clark, Nigel and Kathryn Yusoff. 2017. "Geosocial Formations and the Anthropocene." *Theory, Culture & Society*, 34 (2–3): 3–23.

D'Angelo, Lorenzo and Robert J. Pijpers. 2018. "Mining Temporalities: An Overview." *The Extractive Industries and Society* 5 (2): 215–222.

Dagens Næringsliv [Business Daily]. 1993. *Risikerer at oljen blir verdiløs [Risking that the Oil Will Lose Its Value].* March 28th. https://www.retrievergroup.com/no/.

Dalby, Simon. 2020. *Anthropocene Geopolitics: Globalization, Security, and Sustainability.* Ottawa: University of Ottawa Press.

Ellingsen, Arnfinn. 2015. Interview conducted and recorded by Berit Kristoffersen at Røst, September 8th.

Equinor. 2020. *Norway.* https://www.equinor.com/en/where-we-are/norway.html.

Ferry, Elizabeth Emma and Mandana E. Limbert, eds. 2008. *Timely Assets: The Politics of Resources and Their Temporalities.* Santa Fe, NM: School for Advanced Research Press.

Gray-Cosgrove, Carmella, Max Liboiron, and Josh Lepawsky. 2015. "The Challenges of Temporality to Depollution & Remediation." *SAPIENS: Surveys and Perspectives Integrating Environment and Society* 8 (1). https://journals.openedition.org/sapiens/1740.

Hamnes, Lena. 2017. "Local Perspectives on Oil Extraction in Lofoten." Keynote Lecture at "Supply-side Climate Policy" Conference Organised by UiT, Cicero, and the Stockholm Environment Institute, August 4th, Svolvær, Norway.

Humberset, Bård. 2017. "Pietro Querini's Amazing Journey." *Sandalsand Global Travelling*. https://sandalsand.net/pietro-querinis-amazing-journey/.

Ingold, Tim. 1992. "Culture and the Perception of the Environment." In *Bush Base, Forest Farm*, edited by Elisabeth Croll and David Parkin, 39–56. London: Routledge.

_______. 1993. "The Temporality of the Landscape." *World Archaeology* 25 (2): 152–174.

Konkraft. 2008. "*Olje-og gassvirksomheten i nord.* [The Oil and Gas Industry in the North]." https://konkraft.no/wp-content/uploads/2016/04/KonKraft-rapport-6-Olje-og-gassindustrien-i-nord.pdf

Kristoffersen, Berit. 2016. Fieldnotes from Visit to Fishing Landing Site at Røst, March 12th.

Kristoffersen, Berit and Brigt Dale. 2014. "Post Petroleum Security in Lofoten: How identity Matters." *Arctic Review on Law and Politics* 5 (2): 201–226.

Kristoffersen, Berit and Stephen Young. 2010. "Geographies of Security and Statehood in Norway's 'Battle of the North'." *Geoforum* 41 (4): 577–584.

Mason, Arthur and Maria Stoilkova. 2012. "Corporeality of Consultant Expertise in Arctic Natural Gas Development." *Journal of Northern Studies* 6 (2): 83–96.

Müller, Katja. 2019. "Mining, Time and Protest: Dealing with Waiting in German Coal Mine Planning." *The Extractive Industries and Society* 6 (1): 1–7.

Nelli, Paolo, ed. 2018. *Pietro Querini, Nicolò De Michele, Cristofalo Fioravante: Il naufragio della Querina: Veneziani nel circolo polare artico*, 2nd ed. Roma: Casa editrice Nutrimenti.

Nikolaisen, Lone. 2015. Interview conducted and recorded by Berit Kristoffersen at Røst, September 13th.

Noreng, Øystein. 1980. *Oil Industry and Government Strategy*. Boulder, CO: ICEED

Norwegian Bank. 2020. *About the Fund*. Norwegian Bank Investment Management https://www.nbim.no/en/.

Norwegian Ministry of the Environment. 2006. "Meld. Nr. 8 (2005–2006): Helhetlig forvaltning av det marine miljø i Barentshavet og havområdene utenfor Lofoten (forvaltningsplan). [White Paper #8: Management Plan for the Marine Environment in the Barents Sea and the Ocean Zone Outside Lofoten (The Management Plan)]." https://www.regjeringen.no/no/dokumenter/stmeld-nr-8-2005-2006-/id199809/?ch=1.

Norwegian Petroleum Act. 1996. "Act No. 72 from 29 November 1996 Relating to Petroleum Activities." https://www.npd.no/en/regulations/acts/act-29-november-1996-no2.-72- relating-to-petroleum-activities/.

Norwegian Petroleum Directorate. 2010. "Petroleumsressursene i havområdene utenfor Lofoten, Vesterålen og Senja [Petroleum Resources in the Offshore Areas of Lofoten, Vesterålen, and Senja]." https://www.npd.no/globalassets/1- npd/publikasjoner/rapporter/od_seismikk_rapport_15-4.pdf.

Norwegian Royal Decree. 1963. "Royal Decree of 31 May 1963 Relating to the Sovereignty of Norway over the Sea-Bed and Subsoil outside the Norwegian Coast." http://www.un.org/Depts/los/LEGISLATIONANDTREATIES/PDFFILES/NOR_196 3_Decree.pdf.

Norwegian Standing Committee on Energy and the Environment. 1996. "Recommendation #251, 1995–1996. Olje og gassvirksomheten [Oil and Gas Activities]." https://www.stortinget.no/no/Saker-og- publikasjoner/Publikasjoner/Innstillinger/Stortinget/1995–1996/inns-199596-251/?lvl=0.

Sandlos, John and Arn Keeling. 2013. "Zombie Mines and the (Over) Burden of History." *The Solutions Journal* 4 (3): 1–4.

Sommerfeldt, Amund, trans. 1908. *Pietro Querini: Beretningen om et Skibbrudd på Røst i 1432*. http://www.mobileread.com/forums/attachment.php?attachmentid=95292.

Steinberg, Philip, Jeremy Tasch, and Hannes Gerhardt. 2015. *Contesting the Arctic: Politics and Imaginaries in the Circumpolar North*. London: I.B. Tauris.

Szolucha, Anna. 2018. "Anticipating Fracking: Shale Gas Developments and the Politics of Time in Lancashire, UK." *The Extractive Industries and Society* 5 (3): 348–355.

Weszkalnys, Gisa. 2011. "Cursed Resources, or Articulations of Economic Theory in the Gulf of Guinea." *Economy and Society* 40 (3): 345–372.

_______. 2014. "Anticipating Oil: The Temporal Politics of a Disaster Yet to Come." *The Sociological Review* 62: 211–235.

_______. 2015. "Geology, Potentiality, Speculation: On the Indeterminacy of First Oil." *Cultural Anthropology* 30 (4): 611–639.

_______. 2016. "A Doubtful Hope: Resource Affect in a Future Oil Economy." *Journal of the Royal Anthropological Institute* 22 (S1): 127–146.

10

FROM LAND TO POLAR SEA

Russia's ill-starred quest for a new petroleum province in the Arctic offshore

Nina Poussenkova

In September 2014, Rosneft and ExxonMobil announced with great fanfare the discovery of the Pobeda (Victory) Field in the Kara Sea (Rosneft 27 Sept. 2014). It was a real victory both for Russia, who wanted to launch a petroleum province in the Arctic offshore and thus to confirm its status of an energy power capable of conquering a new Polar frontier, and for Rosneft who wished to demonstrate that it was more than a match for the global majors and capable of operating in the Arctic Seas. However, it was too early to celebrate Victory. ExxonMobil soon quit the joint venture (JV) with Rosneft because of the anti-Russian sanctions,[1] and the Pobeda project was shelved.

This event epitomized the challenges that Russia faced when it tried to commence hydrocarbon production in the Arctic seas. It seemed that fates conspired in creating insurmountable obstacles: anti-Russian sanctions, Russian domestic power games, low oil prices, shale revolution in the US, anticipated peak of oil demand and its decline because of COVID-19.

In general, Russia's desire to progress from the Arctic onshore (where oil is produced, e.g., in the Nenetsk Autonomous District, Yamal-Nenetsk Autonomous District, Krasnoyarsk Krai) to the offshore had strong political undertones. The country wanted to remind the world that it owned immense hydrocarbon treasures, hidden in the Polar seas; and that it was ready to extract them – a feat on par with the conquest of West Siberia in the 1960s–1970s where giant oil fields were discovered and commissioned under extremely tough climatic and geological conditions. West Siberia quickly began to provide a lion's share of the USSR's crude production and made the Soviet Union one of the key global energy players.

Also important was the consideration that Arctic oil, produced both onshore and offshore, would ensure full loading of the Northern Sea Route (NSR), a

DOI: 10.4324/9781003024125-10

"global transportation corridor that is used to transport national and international cargos, [whose] importance will grow with climate change" (Garant 26 Oct. 2020). Commercially, it would provide a shorter and cheaper route from Europe to Asia; politically, it would make Russia a strategic bridge between the two continents and thus raise its status in global affairs. Given the economic, trade, and political significance of NSR, Vladimir Putin stipulated that its cargo turnover should grow by a factor of 8 from 2017 to 2024 (Kryuchkova and Vedeneeva 2019), mainly due to the LNG of NOVATEK and oil of Gazpromneft and Rosneft.

Thus, in 2007, when the US Geological Survey published data on the enormous hydrocarbon riches of the Arctic continental shelf,[2] cold waters became the star of the global and Russian energy sector, particularly during the period of high oil prices and wide-spread fears of "peak oil." Tellingly, President Dmitry Medvedev predicted in 2008 that "The Arctic would be Russia's resource base of the 21st century" (Ilyin 2008). A decade later, Vladimir Putin stated in 2020: "During the next decade, Russia will 'grow' through the Arctic and Northern territories; the future of the country is based on this direction, including in terms of minerals extraction" (RIA Novosti 2017). Ironically, despite these predictions by top-level politicians, in addition to adverse macroeconomic factors, Russia's Arctic offshore quest, which, if successful, could have brought impressive political kudos for Russia, was largely undermined by Russian domestic political struggles and geopolitical aspirations.

Polar *Aqua Incognita*

Though crude production on Russia's continental shelf grew in 2008–2019 from 12.7 mln tons to 29.1 mln tons (Garant 9 June 2020), it still represents only some 5% of the total Russian output of 568 mln tons in 2019. Moreover, offshore oil is mainly extracted in the Okhotsk Sea and in the Caspian Sea, that is, in relatively warm waters. The situation with cold water oil remains quite dismal.

Oil and gas have been produced in the Russian Arctic onshore for a long time, and in 2017, 96.2 mln tons of oil were extracted in the Arctic zone, both onshore and offshore, and the share of Arctic oil grew from 11.8% in 2007 to 17.6% of the total Russian oil output in 2017 (CDU TEK 2019). However, there is only one Russian oil company that actually extracts crude in the Polar waters: Gazpromneft, the oil subsidiary of Gazprom. In contrast to its parent company that played a waiting game far too long and ultimately gave up its ambitious plans to commission the giant Shtokmanovskoye Field in the Barents Sea, Gazpromneft proved that it was a doer rather than a dreamer and, in 2013, launched the Prirazlomnoye oil field located in the Pechora Sea with some 70 mln tons of recoverable crude reserves. In 2013–2019, Prirazlomnoye produced 13 mln tons of oil, and the field is expected to yield some 5 mln tons/yr at the peak (Neftegaz 2020). Nevertheless, it is a negligible amount compared to Russia's total oil output. Meanwhile, other offshore licenses of Gazpromneft are still at the exploration stage: Gazpromneft-Sakhalin holds licenses for four plots on the Arctic shelf: Severo-Vrangelevskiy (East Siberian and Chukchi Seas), Kheisovskiy (Barents Sea), as well as Dolginskiy and Severo-Zapadniy plots (pechora Sea) (Gazpromneft, "About," 2018).

Polar seas of Russia hold enormous hydrocarbon potential; but even the official estimates differ a lot. Thus, Dmitry Kobylkin, former Minister of Natural Resources and Environment, wrote that oil reserves in the Arctic zone of Russia amounted to 7.3 billion tons, that of natural gas to 55 trillion m^3; and the Arctic offshore accounted for 41% of the total resources of the High North (Neft I Kapital 13 Nov. 2019). Russia's Arctic Strategy up to 2035 gives a much higher estimate of the Arctic offshore resources: 17.3 billion tons of oil and gas condensate and over 85.1 trillion m^3 of gas that are "the strategic reserve for the development of [the] mineral and resource base of Russia" (Garant 26 Oct. 2020). According to the Central Dispatching Unit of the Ministry of Energy, total initial recoverable resources of hydrocarbons in the Arctic seas amount to 120 billion t.o.e (tons of oil equivalent) (CDU TEK 2019). Presumably, this vast difference in estimates stems from their use of the terms "reserves" and "resources," respectively.

Still, the real petroleum potential of the Russian Polar offshore is an open question, since the Arctic continental shelf remains truly an uncharted sea. As prospecting and exploration in the Russian Arctic practically stopped after 1991, and the Arctic drilling fleet was largely working in foreign warmer waters after the USSR collapsed until the present day, the degree of exploration of the Polar offshore in Russia remains negligible: 20% in the Barents Sea, 15% in the Kara Sea and almost 0% in the East-Siberian, Laptev, and Chukchi seas (CDU TEK 2019). Tellingly, it was only in 2020 that Russia's first ever stratigraphic well was drilled in the Kara Sea by Rosgeologiya at Rosneft's request (Russian Shipping 2020); this will permit improving the quality of geological information in the region. To compare, Norway drilled over 150 such stratigraphic wells at its continental shelf (Russian Shipping 2020). Recognizing this strategic weakness, Yuri Trutnev, Vice-Premier, said "We should focus on exploration of new fields. We have dropped behind in this respect, particularly on the continental shelf" (Trutnev 2020). He admitted that the number of wells that Russia drilled on the continental shelf is smaller by two orders of magnitude than the number that Norway drilled (Trutnev 2020). So, this low level of exploration entails a high uncertainty concerning the real resource potential of the Polar seas.

Sometimes, these insufficient studies combined with complex geology played a nasty joke on Russian oil companies. Thus, Dolginskoye Field of Gazpromneft with 200 mln t.o.e. of reserves seemed to be ready for commissioning, but, in 2015, Gazpromneft lobbied for changes in the field license because of the unexpected results achieved while drilling an exploration well there. Therefore, crude production in the Dolginskoye Field will begin in 2031 rather than in 2019 as planned (RBC 15 Nov. 2015). As Dmitry Donskoy, then Minister of Natural

Resources, commented frankly in 2017: "Nature likes to surprise us even in well-known regions. Who could have expected that an exploration well at the Dolginskoye oil field would discover a gas layer that would result in the revision of the geological model not only of the field itself, but of the hydrocarbon systems of the whole Pechora Sea as well?" (Neft I Kapital 2017).

Moreover, climatic, geographical, economic, and financial challenges seem to make Arctic offshore oil production a "mission impossible." Russia's Polar seas are relatively shallow: their average depths range from 6m for the Pechora Sea (the shallowest sea in the High North) to 1600 m for the Bering Sea (the deepest) (Wikipedia 2021). However, this advantage is more than offset by extremely severe climate: for example, the Kara Sea is considered the coldest sea on Earth (National Travel 2021). In winter, the temperature drops to −46° C, while in summer, it does not rise above 16° C; the polar night lasts 4 months, the polar day 8 months, and the sea is covered by ice most of the year.

Moreover, the territories that border these Polar seas have a very low density of population, particularly to the east of the Urals, ranging from 1.2 persons/km^2 in Krasnoyarsk Krai ("Official Portal" 2020) to 0.07 persons/km^2 in Chukotka ("Official Website" 2021). Not surprisingly, production and transportation infrastructure is practically non-existent. Accordingly, there are no sea ports near the Khara-Tumus Peninsula in the Laptev Sea (where Rosneft drilled the first exploration well in 2017), and the navigation season lasts only two months a year at best. In 2016, during the summer navigation, 8000 tons of cargo were delivered there from Arkhangelsk (Vasilieva 2020). Two ice-class cargo vessels sailed 3600 km through the Pechora, Kara, and Laptev seas and transported a drilling rig, equipment, and materials as well as modular housing facilities for the oilmen (Rosneft 3 April 2017).

Global climate change, which has particularly serious implications for the Arctic,[3] creates new risks and opportunities for cold water oil. The anticipated melting of ice in the Arctic seas should improve access to offshore hydrocarbons and facilitate their transportation; but, greater weather and climate variability creates additional challenges for their development, largely because of the growing intensity and potential damage of natural disasters. And climate change (combined with human negligence) recently caused a grave technogenic disaster in Russia's Extreme North: in Spring 2020, there was a massive leakage of diesel from a power station operated by Norilsk Nickel on the Taimyr Peninsula, which brought into sharp focus environmental issues in the Arctic, particularly on the continental shelf, and stricter safety rules will probably be adopted for the region (Vasilieva 2020). It will be another obstacle for petroleum operations in the High North.

Also, in December 2011, there was a terrible maritime disaster when the *Kolskaya* drilling platform sank in the Okhotsk Sea with over 50 crew members drowning and only 14 saved (RIA Novosti 2017). It emphasized the dangers connected with operations in the cold waters and showed that Russia was not fully prepared to deal with such emergencies, as Lloyds stated in its 2012 report.

Moreover, Russia faces an acute shortage of qualified personnel capable of working in the Arctic offshore with due regard for environmental sustainability. Thus, Gazpromneft's Deputy General Director for the Development of Offshore Projects Andrei Patrushev noted that "implementation of technically challenging offshore projects requires unique competencies and expertise beyond the scope of regular training programs" (Gazpromneft 13 December 2018).[4] Gazpromneft tries to overcome this shortage by partnering with leading Russian and international education establishments. For example, it supports the joint master-of-science program at Gubkin Oil and Gas Academy in Moscow and the University of Stavanger (Gazpromneft 13 December 2018).

Subsequently, cold water oil is the most expensive and hard to produce crude in the world. It has the highest average break-even price among all other types of new petroleum projects: $90 in 2019 versus $100 in 2016, higher than Venezuelan extra-heavy oil ($70 and $85, respectively) (LUKOIL 2019). Meanwhile, drilling of wells is the most costly and labour-intensive operation on the continental shelf. Experts at the Gubkin Academy estimate costs of drilling a well in the Arctic offshore at $200–500 mln, while over $700 mln were spent on the Universitetskaya-1 (RBC 2016).

Evidently, oil production in the Polar seas is only commercially feasible with high oil prices. According to Raiffeisenbank, the development of the Arctic offshore needs either massive fiscal benefits or oil prices above $100/bar (TASS 2019). Similarly, the vice-president of LUKOIL, Leonid Fedun, said that oil extraction on the Arctic shelf would become profitable with oil prices over $100/bar (Finance Rambler 2019). It is becoming increasingly clear that it is useless to hope for such prices in the near future. The US Energy Information Administration (EIA) expects that Brent prices will average $49/bar in 2021 (EIA 2020). Still, Russian state-owned oil companies, the only domestic players that can operate on the continental shelf (see below), have successfully lobbied for significant fiscal benefits for their offshore activities, thus levelling out somewhat the negative impact of low oil prices.

In addition, the development of the Arctic offshore is strongly hindered by the dismal situation in Russian ship-building, which has been in decline since the early 1990s when, amidst the general economic collapse in the country, many qualified workers quit the sector to search for better-paid jobs ("Interview of Alexei Rakhmanov" 2018). Consequently, Russia experiences a serious shortage of ship-building technology and professionals, and for a long time Russian energy companies had to place orders for vessels abroad. By now, Russia's ship-building sector, with its high costs, inferior quality, and missed deadlines, lags far behind such global leaders as Japan, China, and South Korea (Bessel, V. et al 2019). As Dmitry Donskoy noted:

> Continental shelf is the next stage. It is a formidable task, and not only from the point of view of developing the fields. Our objective is to create a domestic fleet of seismic and Arctic-class drilling ships since Russia faces

> an acute shortage of such vessels (Gazprom flot [Gazprom fleet, subsidiary of Gazprom] has only two drilling rigs capable of operating in the Arctic). (Neft I Kapital 2017)

Recently, Rosneft joined forces with Gazprombank and Rosneftegas to establish a shipbuilding subsidiary named Zvezda (the Star), Russia's first super-wharf for large-tonnage ship-building, which is supposed to solve the problem. It will manufacture Arctic-class tankers, sea platforms, drilling vessels, and other equipment necessary for the development of the continental shelf, including the Arctic. Rosneft has invited leading international companies to participate in the project, such as General Electric, Hyundai Heavy Industries, Fincantieri, Damen and Keppel (Rosneft 2019).

However, these are challenges common to all Arctic (and non-Arctic) states that attempt to produce cold water oil, though other countries are better positioned in terms of the available state-of-the-art oilfield technology, equipment, and vessels. In Russia, these universal challenges are aggravated by a whole spectrum of specific problems – domestic power struggles and geopolitical games – that turn out to be no less detrimental for the development of the Arctic offshore. The saga of Rosneft's attempts to conquer the Polar seas shows how high politics hinder the extraction of petroleum riches of Russia's High North.

Per Aspera at Polar Astra

In the Soviet Union, it was clear to both the oilmen and the relevant officials that it was necessary to go beyond West Siberia. Two oil crises in the USSR when West Siberia failed to meet its oil production targets, in 1977–1980 and 1982–1986, revealed that grave problems were accumulating in the sector (Alekperov 2011). Tellingly, in 1974, Minister of the Oil Industry, Valentin Shashin wrote:

> We face as never before the issue of the need to discover enormous new oil provinces equal to the Volga-Urals and West Siberia. East Siberia, the Caspian Depression, and *the shelves of seas and oceans, which have enormous potential for the growth of reserves,* may be such regions. But at present we are alarmed by the fact that the USSR Ministry of Geology is not drilling enough exploratory wells in the new prospective provinces. (Gustafson 1989, 83)

The Soviet top leadership firmly believed that West Siberian potential was inexhaustible and insisted on producing as much oil there as quickly as possible (Tchourilov, Gorst, Poussenkova 1996). Despite Shashin's justified alarm, no real progress in moving oil production offshore was achieved under Socialism.

Even so, active prospecting and exploration efforts on the Polar shelf were underway in the 1980s. Arktikmorneftegasrazvedka (AMNGR, Arctic

Offshore Petroleum Exploration Company)[5] discovered the Prirazlomnoye oil field in the Pechora Sea in 1989 and the giant Shtokmanovskoye gas and condensate field in the Barents Sea in 1988. AMNGR was financed from the Soviet state budget and received several domestically manufactured vessels: ice-class drilling ships *Valentin Shashin* (now *Deep Venture*) and *Victor Muravlenko* as well as three semisubmersible and two self-elevating drilling rigs (*Murmanskaya* and *Kolskaya*). After the USSR collapsed and the state funding stopped, AMNGR was hired by subsurface users, such as Gazprom, and it made several discoveries in the Barents and Kara Seas.[6] However, after 2006, it gave up drilling in the Arctic offshore because Russian customers did not provide enough orders, and AMNGR relocated its drilling rigs to Vietnam.

Nevertheless, though AMNGR proved in the 1980s that the continental shelf of the Western Arctic contains vast hydrocarbon resources (Gornaya 2004, 25), their development never started because of the almost non-existent oil production infrastructure and, most importantly, because the USSR did not possess state-of-the-art offshore technologies (Milchakova 2018). As Sergey Donskoy noted, "the USSR was the first country in the world to develop petroleum reserves in the Arctic, albeit only onshore. As for offshore production, the West beat us to it. This technological gap became particularly pronounced in the 1990s, when Russia had to cope with other problems, besides the shelf" ("Interview" 2017).

After the USSR collapsed, Russian oil companies did all they could just to stay afloat during the turbulent 1990s. They were unable to think about conquering new frontiers amidst low oil prices and the profound crisis that hit Russia in general and the oil sector in particular, with crude production dropping from 570 mln tons in 1987 to the all-time low of 301 mln tons in 1996 ("BP Statistical Review 1965–2019"). Despite the vicissitudes of the 1990s, the Russian petroleum industry did make some tentative steps from onshore to offshore. Sakhalin-1 and Sakhalin-2 Production-Sharing Agreements (PSAs) were launched in the Okhotsk Sea. Natural conditions there are very challenging, though much milder than in the Arctic waters, and Sakhalin Island already had petroleum infrastructure available, since oil production there began onshore in 1928 (Rosneft, "About" 2021). Moreover, experienced foreign companies were operators of both PSAs: ExxonMobil at Sakhalin-1 (where Rosneft was initially involved with a 40% stake,[7] and Shell at Sakhalin-2 (Gazprom joined as operator only in 2006, when the project was already well underway). Thus, though Rosneft participated in Sakhalin-1, it was only learning the ropes of offshore activities.

In the 2000s, the situation in Russia was improving: economic and political stability was achieved, oil production began to grow, and oil prices were rising, at least up to the 2008–2009 crisis ("BP Statistical Review 2020" 26). Oilmen's ambitions to go offshore were gaining momentum. But they were hindered by domestic political games and turf battles of the powerful state-controlled companies.

Dogs in the manger

A typical case of such political battles was Rosneft's attempt to join the Arctic projects of Gazprom that it implemented through Rosshelf.[8] In 2001, a plan was hatched to develop five northern fields[9] through a 50:50 Sevmorneftegas JV, with Gazprom using Rosshelf, and Rosneft using its key subsidiary Purneftegas. Interestingly, Gazprom did not initially want to let other Russian companies participate in the development of the Prirazlomnoye and Shtokmanovskoye Fields, which were the main assets in the JV, but could not cope on its own (Neft I Kapital 2002). Ultimately, it had to invite Rosneft, since the latter accumulated substantial experience in shelf projects through its Sakhalinmorneftegas subsidiary and enjoyed the strong support of the state.

Rosneft's involvement initially gave the stalled Prirazlomnoye project a new lease on life. Rosneft created a transportation scheme for Prirazlomnoye crude through the tanker *Belokamenka* that it freighted and anchored near Murmansk (Neftegaz 2004). Then, in 2002, it purchased an old platform from the Hutton Field in the UK sector of the North Sea, whose upper part was to be used for the construction of the platform for Prirazlomnoye (Polivanov 2013). Rosneft expected that this would save it some $300 mln, but soon discovered that the platform was no longer useable (Gottemoeller and Tamnes 2008, 77). This event showed that though Rosneft gained some offshore experience on Sakhalin, it still urgently needed sophisticated international partners. While the amounts that Gazprom and Rosneft had invested in Sevmorneftegas became equal by 2002, and following an asset swap with LUKOIL, Rosneft consolidated a blocking interest in Rosshelf in 2005. The company had to sell its entire share in the JV to Gazprom to get funds to pay for Yuganskneftegas' acquisition.[10] As a result, while it seemed then that Rosneft's leadership at Prirazlomnoye might evolve into its leadership on the whole Arctic shelf, this did not happen until a decade later.

Another twist of fate that hindered Arctic offshore oil development was the desire of Russian state companies to secure monopoly rights to the Polar seas. During the early 2000s, the era of resource nationalism (which usually flourishes during high oil prices), Rosneft and Gazprom took an important initiative that had very controversial implications for the Arctic shelf. In April 2005, a deputy head at Gazprom, Boris Nikitin, put forward the idea of transferring the management of the continental shelf to Gazprom and Rosneft. In January 2007, during a government meeting chaired by Vladimir Putin, this proposal was supported by the authorities, and control over offshore developments was handed to the state companies (Vedomosti 2007) giving them exclusive access to the potential petroleum frontier. This trend was reinforced in July 2008, when President Dmitry Medvedev signed the federal law on "Introducing Amendments to the Federal Law on the Continental Shelf" (originally adopted in 1997). Now, companies could receive offshore licenses without contests or auctions, although the law was only applicable to Russian state-controlled entities. Additionally, the subsurface users had to have at least five years' experience of shelf operations and, unsurprisingly, only Gazprom and Rosneft met these criteria.

When oil prices were high, Russian companies were not short of funds to implement offshore projects, although the debts of Gazprom and – particularly – of Rosneft (resulting from its aggressive acquisitions of Russian private oil companies)[11] did limit their investment programs, but they did not have enough understanding of shelf operations. Consequently, the government officials and the companies realized, with regret, that they needed international partners. Furthermore, the financial crisis of 2008–2009, combined with the drop in oil prices, forced the Russian establishment to rethink its attitude towards Western involvement in shelf development. Rosneft had to admit, albeit indirectly, that it would be unable to cope on its own with offshore projects and it started hunting for strategic allies (see below).

This decision to limit development of the shelf to state companies was detrimental to future oil production in Russia also because it prevented LUKOIL from joining in Arctic offshore activities. Privately owned LUKOIL, the second-biggest oil producer in Russia, in contrast to the two state companies, had considerable experience operating on its own in the Caspian Sea, where it actually launched a new petroleum province, in the Baltic Sea, as well as abroad, where it participated in consortia with international oil companies (IOCs) and national oil companies (NOCs). As Vagit Alekperov, CEO of LUKOIL, complained, the state selects only those "who have the right" rather than the experience and technologies (Gavshina and Dzyadko 2011).

Vagit Alekperov tried to lobby for the liberalization of access to offshore resources, but Rosneft and Gazprom torpedoed his attempts. In 2012, they sent letters to the government stating their concern about the plans of the Ministry of Natural Resources to liberalize the continental shelf legislation. They claimed that the amendments proposed by the Ministry were aimed "not at the development of the shelf, but at raising capitalization of certain private companies with foreign capital" (Solodovnikova and Mordyushenko 2012). Alekperov's lobbying had very limited results, and he admitted in a recent interview that "the most promising [offshore] plots are already licensed to Rosneft and Gazprom" ("Interview" 2019). In 2020, Vice-Premier Yuri Trutnev proposed to include measures aimed at allowing private investors access to offshore projects in the all-Russian plan of economic recovery (Kozlov 2020). However, it seems that now the urgency of liberalizing access to the continental shelf diminished because of low oil prices and anti-Russian sanctions that made offshore petroleum projects commercially unattractive.

It takes two to tango

After the 2008–2009 crisis, Rosneft admitted to the urgent need to find solvent and experienced foreign partners for its Arctic ventures. By that time, Rosneft was quickly becoming the queen of the Polar seas by acquiring numerous offshore licenses (in tough competition with Gazprom). For example, in 2010 it bought five licenses: for a Yuzhno-Russkiy plot in the Barents Sea, for three Vostochno-Prinovozemelskiy plots in the Kara Sea, and for a plot in the Black Sea. After this shopping spree, Rosneft overtook Gazprom in their offshore race: it had rights for 16 plots, while Gazprom only had 12 (Nepomnyaschih et al. 2010, n.p.).

As of 1 January 2019, Rosneft accumulated 55 licenses for subsurface plots in the Arctic, Far East, and Southern seas of Russia with the total hydrocarbon resources estimated at 41 billion t.o.e (Rosneft, "Offshore Projects" 2021) (Table 10.1). Rosneft regards development of the Arctic offshore as its strategic direction for the continental shelf projects and optimistically believes that, by 2050, the Polar seas will account for 20–30% of total Russian oil production (Rosneft, "Offshore Projects" 2021).

TABLE 10.1 Rosneft's licences in the Arctic Seas

Area of the Arctic	*Arctic Sea*	*Name of plots*	*Fields discovered in Arctic Seas*	*Total recoverable resources as of 01.01.2019*
Western Arctic (19 licenses)	*Barents Sea*	Fedynskiy, Tsentralno-Barentsevskiy, Perseevskiy, Albanovskiy, Varnekskiy, Zapadno-Prinovozemelskiy, Gusinozemelskiy		16 billion t.o.e.
	Pechora Sea	Russkiy, Yuzhno-Russkiy,Yuzhno-Prinovozemelskiy, Zapadno-Matveevskiy, Severo-Pomorskiy 1 and 2, Pomorskiy, Medynsko-Varandeiskiy	Severo-Gulyaevskoye, Medynskoye-More, Varandei-More, Pomorskoye	
	Kara Sea	Vostochno-Prinovozemelskiy 1, 2 and 3, Severo-Karskiy	Pobeda	
Eastern Arctic (9 licenses)	*Laptev Sea*	Ust-Olenenskiy, Ust-Lenskiy, Anisinsko-Novosibirskiy, Khatangskiy, Pritaimyrskiy	Tsentralno-Olginskoye	18.3 billion t.o.e.
	East Siberian Sea	Vostochno-Sibirskiy-1		
	Chukchi Sea	Severo-Vrangelevskiy 1 and 2, Yuzhno-Chukotskiy		

Source: https://www.rosneft.ru/business/Upstream/offshore/

While Gazprom preferred the "wait-and-see" attitude towards operating in the Polar seas, Rosneft wanted to move forward and further strengthen its positions in the Russian oil sector through its Arctic offshore ventures.

Searching for foreign partners, it first made an obvious choice by trying to join forces with BP in 2011. Though Rosneft was then headed by Eduard Khudainatov, presumably it was Igor Sechin, Vice-Premier in charge of the energy sector (and BoD Chairman of Rosneft), who masterminded the deal. Rosneft already had considerable experience working with BP in the Sea of Okhotsk, where they participated in Sakhalin-4 and Sakhalin-5 – though both projects failed for geological reasons (Yagova 2010). It looked like a bad omen because the fiasco that happened later with this partnership remains one of the greatest mysteries in the Russian oil sector.

The partners' main goal was to jointly develop oil fields in the Arctic offshore. Their JV (Rosneft – 66.7%, BP – 33.3%) was to explore and produce hydrocarbons at three areas in the Kara Sea: Vostochno-Prinovozemelskiye-1-3. Furthermore, BP was to bear the cost burden ($1.4–2 bln) at the exploration stage (RIA Novosti 2011), making it initially a risk-free option for Rosneft. The deal was underpinned by a share swap between the two companies (Melnikov 17 Jan. 2011). BP, probably, wanted to get access to huge new resources in Russia and to establish a strategic relationship with the powerful NOC. Rosneft, presumably, chose BP because it had impressive experience working in Alaska and offshore. The 2010 Macondo disaster in the Gulf of Mexico also made BP more eager to do a deal and therefore more flexible in negotiations with Rosneft. The deal received a blessing from the very top in Moscow: Premier Vladimir Putin publicly stated that "the government intends to create the most favourable fiscal and administrative regime for the project of developing the Arctic shelf" (Derbilova 2011).

On January 26, 2011, in Davos, the respective heads of Rosneft and BP, Eduard Khudainatov and Bob Dudley, signed a cooperation agreement (Gavshina and Derbilova 2011). On the same day, the Russian shareholders of TNK-BP, the Alfa-Access-Renova (AAR) consortium, appealed to the London Supreme Court to halt the deal within the framework of their own agreement with BP. AAR revealed that it had "the right of first refusal" for partnership with BP in Russia (Gavshina and Derbilova 2011). The London courts transferred the lawsuit to the Stockholm arbitration court, which ultimately recognized that BP had no right to establish a strategic alliance with Rosneft without the approval of TNK-BP's board of directors (Melnikov 25 Mar. 2011). Consequently, the share swap and partnership in the Arctic were forbidden.

It is unclear why the Russian shareholders of TNK-BP acted in such a disruptive manner. However, AAR would never have risked a public conflict with Igor Sechin (and attempted to undermine a deal officially supported by Vladimir Putin) without having the support of somebody at the very top. Probably, Putin was reminding Sechin that, for all his undoubted influence, he was not omnipotent. In any case, this failure was a serious stumbling block (caused by the domestic intrigues) in the development of the Kara Sea petroleum reserves.

After the BP fiasco, Rosneft did not waste any time in searching for new partners. Rosneft quickly chose ExxonMobil from several contenders, probably because the major's offer was attractive but also because the partnership with the world's biggest public oil corporation was a politically appealing proposition as it would significantly raise Rosneft's status. Furthermore, ExxonMobil had clearly proved itself as a Sakhalin-1 operator and had a reputation for "getting the job done." Even Vladimir Putin was impressed by the unique technological potential of ExxonMobil, noting that "I was surprised that your platform in Canada can withstand the blow of an iceberg with a weight of 1 mln tons" (Kolesnikov 2011).

As a result, in the summer of 2011, Eduard Khudainatov and the Head of Development at ExxonMobil, Neil Duffin, signed a strategic cooperation agreement (Melnikov 31 Aug. 2011). The partners established a JV (Rosneft – 66.7% and ExxonMobil – 33.3%), and, in addition to the Kara Sea licences, they were to work in the Black Sea. Total investments were estimated at $3.2 bln, with ExxonMobil bearing the burden of expenses for exploration, while assisting Rosneft in some oil production spheres (Table 10.2). Vladimir Putin said: "It is fearsome to name such figures[...] Direct investments in the project might amount from $200 bln to $300 bln[...] And if we take infrastructure into account, they might reach $500 bln" (Kolesnikov 2011).

Significantly, Igor Sechin stated during the presentation of Rosneft's strategic alliance with ExxonMobil (presumably, mainly speaking about the Arctic

TABLE 10.2 Rosneft's offshore strategic partnerships

Foreign companies	*Main assets*	*Other partnership elements in Russia*	*International projects offered to Rosneft*
ExxonMobil	Kara Sea (Vostochno-Prinovozemelskiye 1–3), Black Sea	ExxonMobil to assist Rosneft in developing hard-to-recover reserves in West Siberia	Exploration projects in the Gulf of Mexico, the US and Canada
Eni	Barents Sea (Fedynskiy and Tsentralno-Barentsevskiy) Black Sea		International projects primarily in North Africa
Equinor	Barents Sea (Perseevskiy block) Okhotsk Sea	Equinor to assist in the development of shale oil in Samara region and the highly viscous oil of the Severo-Komsomolskoye field in Yamal-Nenetsk Autonomous District	Norwegian offshore and international projects of Equinor

Source: Overland and Poussenkova (2020): 243.

offshore) that it exceeded in terms of scale such projects as the first space walk, or the flight to the Moon, and in terms of investments, the development of the Brazilian shelf, or the North Sea (Melnikov 19 Apr. 2012). The political undertones of the Arctic venture were thus emphasized.

During a meeting with Vladimir Putin, Rex Tillerson, CEO of ExxonMobil and later the US Secretary of State, received assurances that the offshore projects would be implemented under a special regime. Indeed, in April 2012, due to the strong lobbying efforts of Rosneft and Igor Sechin personally, the government introduced generous fiscal and tariff incentives for the development of the Russian shelf. Offshore ventures were to be exempt from export duty, and a lower rate of mineral production tax would be applied (5–15% depending on the complexity of the project). Moreover, guarantees of the stability of these terms were provided for 15 years, a particularly important caveat for such expensive endeavours with a long pay-back. Experts believe that the Russian fiscal regime for the continental shelf became one of the most attractive regimes in the world (Melnikov 16 Jan 2013). However, analysts with Skolkovo Business School calculated that Russian state petroleum companies, primarily Rosneft, would be the main beneficiaries of the offshore oil production projects rather than the Russian state (Barsukov 2012).

In 2013, Rosneft and ExxonMobil expanded the scope of their strategic cooperation by including seven additional licensing plots with a total area of 600,000 km^2 in the Chukchi, Laptev, and Kara Seas. Igor Sechin (who was appointed CEO of Rosneft in May 2012) commented, "With the signing of today's agreements, the partnership between Rosneft and ExxonMobil, which is unprecedented in terms of its scale, reaches a totally new level" (Rosneft 13 Feb. 2013). In addition to ExxonMobil, on 25 April 2012, Eni signed an agreement with Rosneft on the joint development of two blocks in the Barents Sea and an area in the Black Sea (Gavshina and Derbilova 2011) (Table 10.2). The foreign company was obliged to finance the exploration efforts to the tune of $2 bln. Rosneft candidly admitted that the government had helped it to reach an agreement with Eni by providing generous tax benefits. Blessing the deal, Putin added: "I am sure that the projects will be successful, and I want to assure you that the government will do everything to support such projects" (Gavshina and Derbilova 2011). Soon after that, on 5 May 2012, Rosneft and Statoil (now Equinor) signed a Cooperation Agreement. Vladimir Putin again promised that "the project would enjoy the unstinting support of the government" (Melnikov 10 May 2012).

It is noteworthy that Rosneft took an initiative by proposing a joint declaration about the careful development of the Arctic shelf. On 23 November 2012, Igor Sechin and Helge Lund (Statoil's CEO at the time) signed a Declaration on Environmental Protection and Preserving Biodiversity in Exploration and Development of Mineral Resources of the Arctic Continental Shelf of Russia, and Helge Lund confirmed that "we fully support this initiative of Rosneft." They emphasized their commitment to the sustainable development of the Arctic, including minimizing the impact on the life of Indigenous peoples and on climate change (Rosneft 23 Nov. 2012). Emulating this example, Rosneft and ExxonMobil signed a similar declaration on 7 December 2012 (Rosneft 7 Dec. 2012), and Rosneft and Eni followed suit in June 2013 (Rosneft 21 June 2013). This initiative signifies that Russian oil companies are becoming increasingly aware of environmental hazards inherent in the development of the Polar seas.

In addition to its Western partners, within the framework of Russia's "Pivot to Asia,"[12] Rosneft invited companies from China, India, and Vietnam to the Arctic shelf with rather disappointing results. Thus, in 2013, before the sanctions, Rosneft signed agreements with the China National Petroleum Company (CNPC) on joint activities in the Barents and the Pechora Seas (Kezik and Dzyadko 2013). However, CNPC was not eager to start implementing these challenging polar projects because of disagreements with Rosneft on prices, high costs, the unclear economics of Arctic exploration, and the rigid terms of partnership that limit opportunities for the Chinese company to participate in the management of these projects (Kozlov 2016). CNPC, who is normally interested in getting access to foreign equity oil,[13] probably preferred to work in warmer waters and with less intransigent partners.

In 2017, after the second round of technological and financial sanctions against Russian oil companies was introduced by the US (see below), Gazpromneft also began negotiations with India's ONGC and China's CNOOC about potential joint operations in the northern seas. These ventures are still at the discussion stage (Petlevoi 2017).

Oilmen propose, but sanctions dispose

In August 2014, Rosneft and ExxonMobil began drilling the northernmost well in Russia, the $700 mln Universitetskaya-1, some 250 km from the shore at depths of 80 metres, using the *West Alpha* drilling platform (Rosneft 27 Sept. 2014). The area of the Universitetskaya structure covered 1200 km^2, and resources of three Vostochno-Prinovozemelskiy plots were estimated at 13 bln t.o.e (Rosneft 27 Sept. 2014). Anticipation was high, as the reserves of the Kara province could have exceeded the Gulf of Mexico, the Brazilian shelf, and the Arctic shelf of Alaska and Canada, and were comparable to the current resource base of Saudi Arabia. Igor Sechin said: "The beginning of exploratory drilling in the Kara Sea is one of the key events in the global oil and gas industry. As a result, we hope to discover a new Kara offshore oil province. The development of the Arctic shelf has a tremendous multiplicative effect for the whole Russian economy" (Rosneft 9 Aug. 2014).

In September, Rosneft announced the discovery of a new field. Igor Sechin stated:

> it is excellent light oil that is comparable to Siberian Light. This is a unique result from the first prospective well at an absolutely new field. This is our joint victory: our friends and partners from ExxonMobil, Nord Atlantic

> Drilling, Schlumberger, Halliburton, Weatherford, Baker, Trendsetter, and FMC participated, and we want to name this field Pobeda (Victory). (Rosneft 27 Sept. 2014)

Pobeda had initial estimated reserves of 130 mln tons of oil and 499 bcm of gas. It proved that, geologically, the shelf of the Kara Sea was a continuation of the West Siberian petroleum province (Rosneft 27 Sept. 2014). But the celebration of "Victory" was soon cut short: the US introduced anti-Russian sanctions, and ExxonMobil had to halt operations on the Arctic shelf in Russia. Meanwhile, Rosneft was unable (or unwilling) to develop Pobeda on its own.

The US and the EU sanctions introduced following the accession of Crimea and the Ukrainian crisis targeted Russian individuals and companies that were close to the Kremlin as well as key segments of the economy. In the oil sector, restrictions were mainly aimed at shale, deepwater, and Arctic offshore projects, and among the Russian energy companies, Rosneft was hit the hardest, as it was included in every "black list" (Table 10.3).

TABLE 10.3 The US and EU sanctions of 2014 and 2017 that affect the Russian oil sector

	US-2014	**EU-2014**	**US-2017**
Financial sanctions	Debt and equity with maturity over 90 days	Debt and equity with maturity over 30 days	Debt and equity with maturity over 60 days
Sanctions applicable to	***Rosneft*** NOVATEK Transneft ***Gazpromneft***	***Rosneft*** ***Gazpromneft*** Transneft Their subsidiaries (over 50% controlling interest)	***Rosneft*** NOVATEK Transneft ***Gazpromneft***
Technological sanctions	Deliveries of oil-production equipment for deepwater shelf (over 150 m), for ***Arctic shelf*** and for shale oil Drilling rigs, equipment for horizontal drilling, ***offshore platforms for operations in the Arctic***, software for hydrofracking operations, ***remote-control underwater equipment,*** high-pressure pumps, drilling pipes and other state-of-the-art equipment.	Deliveries of equipment for oil exploration and production on deepwater shelf (over 150 m), ***Arctic shelf,*** and shale oil production with the use of hydrofracking.	Deliveries of products or technologies for supporting exploration or production at new deepwater, ***Arctic shelf*** or shale projects. Opportunity for introducing sanctions for sales of equipment, technologies and services, as well as investments in export pipelines.

Sanctions applicable to	***Rosneft*** LUKOIL Gazprom Surgutneftegas Their subsidiaries (over 50% controlling interest) in Russia	***Rosneft*** ***Gazpromneft*** Transneft Individuals or companies with over 50% stake in companies named in the sanctions list	***Rosneft*** LUKOIL Gazprom Surgutneftegas Their subsidiaries with controlling interest over 33% worldwide Any person selling equipment, technologies or services for pipeline projects

Source: Mitrova, T. et al (2018), 11.

Interestingly, the Western sanctions did not affect the current oil production in Russia because the projects to which they were applicable (i.e. deepwater, Arctic offshore, and shale) accounted for an insignificant proportion of the total Russian crude output. There was a slight slowdown of oil production after the onset of the first wave of sanctions in early 2014.[14] However, this may just as well have been due to the concurrent oil price collapse and later also due to the OPEC+ deal. Whatever caused Russian oil production growth to stagnate, it is clear that sanctions did not cause it to collapse (Overland and Poussenkova 2020, 186).

Still, sanctions did undermine Rosneft's offshore ventures with its international partners because relevant Russian technologies and equipment simply did not exist. Vice-Premier Yuri Trutnev said: "Regrettably, Russian industry does not have now a whole series of technologies[...] We should focus on developing technologies, learn how to manufacture subsea complexes, how to drill horizontal wells, build ice-resistant platforms; all these are foreign technologies for us, and we are not allowed access to them because of sanctions" (2020).

Indeed, all Russian Arctic offshore projects involved either foreign partners or international service companies and providers of equipment and technology. Tellingly, Gazpromneft was lucky (or wise) in not wasting time in political intrigues and launching Prirazlomnoye before sanctions hit the oil sector. Though Gazpromneft developed the field on its own, many international contractors and suppliers participated in the project. Baker Hughes, Halliburton, and Schlumberger were involved in drilling wells; Aker Solutions, National Oilwell Varco, Cameron, and FMC Technologies helped with the design. In total, foreign contractors' share in *Prirazlomnaya* platform's operations and servicing was some 50% (Mitrova et al. 2018).

Thus, Arctic offshore projects that were implemented by Russian oil companies on their own proceeded, while those that involved foreign partners collapsed. ExxonMobil withdrew from its exploration JVs with Rosneft altogether in 2018 (Rogtec 2018). By that time, Equinor also gave up its plans to work on Russia's continental shelf (Kozlov 15 Dec. 2020); only Eni remained in business, but it was then expected that it would have to halt its joint activities with Rosneft (Kozlov 2018). True, Russian oil companies were not the only victims of the sanctions; their foreign colleagues also suffered because they had to give up lucrative deals.

In July 2014, Rosneft and North Atlantic Drilling Ltd. (NADL), subsidiary of the Norwegian Seadrill, signed long-term agreements on drilling on Russia's shelf by six NADL's drilling rigs up to 2022 (RBC 14 March 2015). Moreover, Rosneft intended to buy a 30% stake in NADL. However, because of sanctions, in November 2014, Seadrill stated that the deal was postponed till May 2015; in March 2015, Rosneft cancelled its order for a West Navigator drilling ship from NADL (RBC 14 March 2015). The financial position of Seadrill deteriorated, and its shareholders filed a lawsuit against it because of its deals with Rosneft (Neftegaz 2017).

Ultimately, sanctions and low oil prices forced Rosneft to downgrade its offshore plans. High costs of drilling, problems with raising credit, and the shortage of sea drilling rigs and ice-breakers resulted in missed deadlines established by licenses. Therefore, in 2016, Rosneft asked the Ministry of Natural Resources for permission to delay exploration at the offshore fields, and Rosnedra allowed it to postpone exploration at 19 plots in the Arctic, Far East, and Southern seas for 2–5 years (and Gazprom and Gazpromneft to postpone activities at 12 plots) (RBC 10 June 2016). Given these missed deadlines, in 2016, the Ministry of Natural Resources imposed a temporary moratorium on issuing new offshore licenses until the obligations under existing licenses are met (TASS 2016). It was expected then that this time-out would last 3–4 years (Tass 2016). It probably meant that Russia no longer considered the development of the Arctic offshore petroleum reserves its top priority.

Still, Rosneft intended to continue its Arctic adventure on its own and, in April 2017, began drilling the Tsentralno-Olginskaya-1 well in the Laptev Sea. It announced, in June 2017, that it discovered a field with geological reserves of 298 mln tons of light and sweet oil that might mean the launch of a new petroleum province in the Eastern Arctic (Lenta.ru 2017), but its ability to manage and afford field development is in doubt without an international partner.

Other Russian actors also decided to try their luck in exploring the uncharted Polar seas. In Autumn 2019, Gazpromneft and NOVATEK established a JV to explore and develop Arctic offshore fields (Neft I Kapital 6 Sept. 2019). The JV chose the Severo-Vrangelevskiy plot in the Chukchi Sea as a pilot project. Thus, NOVATEK will be the first Russian privately owned company to work in the Arctic offshore. Now, with low oil prices and falling demand for oil because of COVID-19, the timing is not optimal for offshore petroleum operations, particularly in the wilderness of the Chukchi Sea. However, both companies conclusively proved that they are capable of operating in the Arctic: Gazpromneft on the Prirazlomnoye Field and NOVATEK with its onshore LNG projects. This

Chukchi Sea acid test will demonstrate their ability to get the job done without foreign partners in a very challenging geographical and macroeconomic environment (Neft I Kapital 6 Sept. 2019).

Soon afterwards, on 22 July 2020, Rosneft began working in the Kara Sea on its own: it is drilling prospect wells Vikulovskaya and Ragozinskaya within the Vostochno-Prinovozemelskiy-1 and -2 plots. According to Rosneft's website, "two high-tech drilling platforms are operating there that correspond to the strictest requirements of industrial and environmental safety. Importantly, they are equipped with systems of ice monitoring that permit following movements of icebergs, thus helping to avoid accidents in the challenging climatic conditions" (Rosneft 18 Aug. 2020). However, these two high-tech drilling platforms are foreign: the Chinese platform *Nan Hai Jiu Hao*, owned by China Oilfield Services Limited (that is controlled by CNOOC) is drilling at Vikulovskaya, and *Oriental Discovery* at Ragozinskaya (Kozlov 31 Aug. 2020). Ironically, Western sanctions against the Russian oil sector forced IOCs and international service companies to cede the capacious Russian market to their Chinese competitors.

Thus, although Rosneft succeeded in attracting foreign partners to explore the Arctic offshore, it failed in implementing production projects (at least in the short-to-medium term) because of an unfortunate mixture of domestic political games and geopolitics, as well as low oil prices and declining demand for oil. Lured by potential petroleum wealth, the IOCs were prepared to share their expertise and invest money, but ultimately were not prepared to ignore political realities. Thus, although impressive crude reserves were discovered in the Polar seas, their rapid development is unlikely until relations between Russia and the West have improved. Rosneft searched for solvent partners in Asia, but Chinese and Indian companies do not have enough offshore expertise, particularly in the Arctic. They will hardly be able to help Rosneft fulfil its (and the government's) ambitions on the Russian continental shelf. Consequently, anti-Russian sanctions undermined the development of a new petroleum province, even if they will affect Russian crude production only in the long term given the time horizons for cold water oil.

Back to earth

Sanctions and low oil prices forced Russia to shelve the shelf projects. Its ambitious plans to produce 35–36 mln tons of oil in the Arctic seas by 2035, as formulated by the Draft Energy Strategy up to 2035 (RBC 10 June 2016), were seriously downgraded. Tellingly, Deputy Minister of Energy Pavel Sorokin said in December 2019: "by 2035 oil production on the Arctic shelf could grow to 30 mln tons provided legislation is changed" (Neft I Kapital 13 Nov. 2019). Meaning if tax benefits are granted to companies – presumably, the super-liberal fiscal regime for shelf operations introduced in 2012 is not enough in the era of sanctions and low prices.

It is probably not surprising that, with all these adverse macroeconomic and political factors, the Arctic offshore seems less commercially attractive than the Russian onshore where there are at least three viable alternatives.[15] The first alternative is enhancing oil recovery ratio, which is less than 25% in Russia versus 40–50% in Norway and the US. According to the Ministry of Economic Development, a 1% increase in the oil recovery ratio in Russia would let it annually produce an additional 20 mln tons of crude (Ministry of Economic Development 2018, (Khimiya et al. 2017) 21), a little less than what the Arctic offshore is expected to yield by 2035 (with new fiscal benefits). Besides, raising oil recovery ratio will permit extending life of the massive mature fields in West Siberia in the regions with well-developed social and production infrastructure, as well as helping to resolve the problem of old oil towns built near the aging giants, thus alleviating potential social tension.

The second alternative is the development of hard-to recover reserves including heavy oil, highly viscous oil, oil from low porosity sediments, and shale oil (the latter of which is found in the Bazhen formation in West Siberia as well as Khadum, Domanic, and Abalac formations). They can yield up to 80 mln tons by 2035 (versus 38 mln tons in 2017), according to the Ministry of Energy (Bashkatova 2018), considerably more than the Arctic shelf.

Small and mid-size non-integrated oil companies (some 250 firms operating in Russia) that work on small or depleted fields that are not interesting for big players are the third alternative to the Arctic offshore. Currently, they produce roughly 14 mln tons/yr; the Energy Center of Skolkovo estimates that if they are given certain benefits, their production can increase to 42 mln tons/yr by 2030 (Vygon et al. 2014), again more than the Arctic continental shelf can provide. These three options are simpler, cheaper, and environmentally safer (because they do not entail operations in the fragile Polar marine environment) than the development of the Arctic offshore. They are less "sexy" for the image of Russia as an energy power, and not as politically appealing, but they are socially oriented and commercially viable.

The Energy Strategy of Russia up to 2035 that was finally adopted in 2020 states that Russia needs to maintain oil production up to 2024 in the range of 555–560 mln tons and up to 2035 in the range of 490–555 mln tons (Garant, "Russia's Energy Strategy" 2020). It places a strong focus on incentives for production from mature fields, commissioning of small fields, operation of low-yield wells, hard-to-recover reserves (including shale), and promoting small and mid-size enterprises, i.e. aims at the intensive mode of oil industry development. It does mention as key measures the creation of long-term investment conditions for accelerated development of the fields in the Arctic zone and on the continental shelf, the need to liberalize access to offshore reserves, and promotion of the domestic market of high-tech services for exploration and development of the continental shelf (Garant, "Russia's Energy Strategy" 2020), but, in general, the document does not convey a sense of urgency to produce cold water oil.

Indeed, given other, more attractive investment opportunities available to the Russian companies, the new Arctic offshore production projects will probably not be launched before 2030 (LUKOIL 2019). Now many Russian officials and oil executives have a very practical attitude towards Polar offshore oil. Dmitry Kobylkin, former Minister of Natural Resources, recently said: "I am convinced that the development of the Arctic region is the project for decades, if not centuries, ahead. [...] I continue to believe that the Arctic shelf should be regarded as a strategic reserve for the development of the resource base and hydrocarbons production in Russia" (Neft I Kapital 13 Nov. 2019). Similarly, Leonid Fedun frankly stated "Arctic [offshore], in my opinion, is a reserve for future generations. Realistically, we could return to the Arctic after 2035, and maybe even after 2050, if there is a need in such volumes of hydrocarbons by that time" (Finance Rambler 2019). But will this strategic reserve be needed in the foreseeable future, with the peak of oil demand anticipated by mid-2020s, and with the plans of the EU and China (the key customers for Russian oil) to become carbon-neutral by 2050 and 2060, respectively? So, the Polar seas have a chance to keep their cold waters clean from petroleum activities.

Notes

1 The technological sanctions against the Russian oil sector introduced by the EU in 2014 and the US in 2014 and 2017 targeted deep water, shale, and offshore Arctic projects.

2 See US Geological Survey, "Circum-Arctic Resource Appraisal: Estimates of Undiscovered Oil and Gas North of the Arctic Circle," (2008). https://pubs.usgs.gov/fs/2008/3049/fs2008-3049.pdf

3 See GreenFacts "Arctic Climate Change," https://www.greenfacts.org/en/arctic-climate-change/index.htm

4 Gazpromneft is particularly aware of the importance of environmental safety after much-publicized incidents with Greenpeace activists who boarded the *Prirazlomnaya* platform in 2012 and 2013 protesting against petroleum operations in the Arctic waters. See https://www.greenpeace.org/russia/ru/news/2013/18-09-action-on-Prirazlomnaya/.

5 Arktikmorneftegasrazvedka was established in the Ministry of the Gas Industry in 1979 in order to prospect and explore hydrocarbons in the Arctic seas. The state-owned enterprise was corporatized in 2011 and transferred to Zarubezhneft, a 100% state-owned company mainly dealing with overseas oil projects.

6 Including three oil fields (Varandeiskoye-More, Medynskoye-Mor, and Dolginskoye) in the Barents Sea.

7 In 2001, it sold half of its stake to Indian ONGC; so now Sakhalin-1 shareholders are ExxonMobil (30%), SODECO (30%), Rosneft (20%) and ONGC (20%). https://sahalin1.rosneft.ru/about/Glance/OperationalStructure/Dobicha_i_razrabotka/Dalnij_Vostok/sahalin1/

8 Initially, Rosshelf (its shareholders were major defence enterprises located in the North of Russia, R&D institutes, and Severstal, a big metallurgical company) created in 1992 was to develop Prirazlomnoye Field. In 1994, Gazprom joined the project by acquiring a 51% stake in Rosshelf; Arkahngelskgeoldobycha had 13.6%, and Sevmashpredpriyatiye and Severstal – 12.5% each (Poussenkova 2007).

9 Prirazlomnoye, Shtokmanovskoye, Kharampurskoye, Vyngayakhinskoye and Etypurovskoye.

10 See Poussenkova 2007.
11 See Poussenkova 2016.
12 Because of Russia's worsening relations with the West as a result of the Ukraine crisis, Moscow began making a stronger focus on the Eastern dimension of its foreign policy and economic relations by building closer partnerships mainly with China and India, including in the energy sphere.
13 See analysis of overseas acquisitions of Chinese NOCs in EIA, "China Country Analysis" (2015).
14 See dynamics of oil production in 2010–2019 in BP, "BP Statistical Review of World Energy" (2020).
15 See Poussenkova (2019) 86–108.

References

Alekperov, Vagit. 2011. *Oil of Russia: Past, Present and Future*. Moscow: Kreativnaya Ekonomika Publishing House.

Barsukov, Yu. 2012. "Northern Liberal Ocean." *Kommersant*. 21 Sept. 2012. https://www.kommersant.ru/doc/2026740.

Bashkatova, A. 2018. "Russia Drills a Well into Nowhere." *Nezavisimaya Gazeta*. 24 Oct. 2018. https://www.ng.ru/economics/2018-10-24/2_7339_oil.html.

Bessel, V., et al. 2019. "Shipbuilding Priorities for the Russian Oil and Gas Industry." *Bureniye I Neft*. Sept. 2019. https://burneft.ru/archive/issues/2019-09/3.

BP. 2020. "BP Statistical Review of World Energy, All Data, 1965–2019." https://www.bp.com/en/global/corporate/energy-economics/statistical-review-of-world-energy.html.

_______. 2020. "BP Statistical Review of World Energy – 2020." https://www.bp.com/content/dam/bp/business-sites/en/global/corporate/pdfs/energy-economics/statistical-review/bp-stats-review-2020-full-report.pdf.

CDU TEK. 2019. "Petroleum Klondike of the Arctic." 14 Feb, 2019. https://www.cdu.ru/tek_russia/issue/2018/12/545.

Derbilova, E. 2011. "Bolshoi Petroleum." *Vedomosti*. 17 Jan. 2011. https://www.vedomosti.ru/politics/video/2011/01/15/bolshoi-petroleum.

Energy Information Administration. 2020. "Short-Term Energy Outlook." 8 Dec. 2020. https://www.eia.gov/outlooks/steo/report/prices.php.

_______. 2015. "Country Analysis China." Last updated 14 May 2015, http://www.eia.gov/beta/international/analysis.cfm?iso=CHN.

Finance Rambler. 2019. "LUKOIL Named the Breakeven Price for Oil Production in the Arctic." 17 Dec. 2019. https://finance.rambler.ru/markets/43362319-v-lukoyle-nazvali-rentabelnuyu-tsenu-nefti-dlya-dobychi-v-arktike/.

Garant. 2020. "Russia's Energy Strategy up to 2035." 9 June 2020. https://www.garant.ru/products/ipo/prime/doc/74148810/.

_______. 2020. "Strategy for the Development of Russia's Arctic Zone and Ensuring National Security up to 2035." 26 Oct. 2020. https://www.garant.ru/products/ipo/prime/doc/74710556/.

Gavshina, O. and E. Derbilova. 2011. "London's Shelf." *Vedomosti*. 28 Jan. 2011. https://www.vedomosti.ru/business/articles/2011/01/28/londonskij_shelf.

Gazpromneft. 2018. "About." https://sakhalin.gazprom-neft.ru/business/about/.

_______. 2018. "We Accumulated a Unique Experience in Development of Offshore Reserves." 13 Dec. 2018. https://www.gazprom-neft.ru/press-center/lib/2112700/.

Gornaya Promyshlennost. 2004. "25th Anniversary of Arktikmorneftegazrazvedka." https://mining-media.ru/ru/article/yubilyary/1380-25-let-arktikmorneftegazrazvedka.

Gottemoeller, Rose and Rolf Tamnes. 2008. *High North – High Stakes. Security, Energy, Transport, Environment*. Fagbokforlaget.

Greenpeace. 2013. "Greenpeace Activists Boarded Prirazlomnaya Platform." 18 Sept. 2013. https://www.greenpeace.org/russia/ru/news/2013/18-09-action-on-Prirazlomnaya/.

GreenFacts. 2006. "Arctic Climate Change." https://www.greenfacts.org/en/arctic-climate-change/index.htm.

Gustafson, Thane. 1989. *Crisis Amid Plenty*. Princeton Princeton University Press.

Ilyin, A. 2008. "Borders will be Drawn for the Arctic." *Rossiiskaya Gazeta*, 18 Sept. 2008, https://rg.ru/gazeta/2008/09/18.html.

"Interview of Alexei Rakhmanov: Ship-building Business Is Very Difficult and Dangerous." 2018. *Kommersant*, 20 Dec. 2018. https://www.kommersant.ru/doc/3835729,

"Interview of Sergei Donskoy: Keys to the North." 2017. *Kommersant*. 27 Mar. 2017. https://www.kommersant.ru/doc/3247645.

"Interview of Vagit Alekperov: We do not Need Benefits, We do not Have Benefits." 2019. *Kommersant*. 25 Dec. 2019. https://www.kommersant.ru/doc/4206663.

Kezik I. and T. Dzyadko. 2013. "Rosneft Invites the Chinese to the Shelf." *Vedomosti*. 19 Feb. 2013. http://www.vedomosti.ru/companies/news/9243391/shelf_kitajcam#ixzz2LKACtuze.

Khimiya I., et al. 2017. "Chemical Industry" Kommersant 15 June 2017. https://www.komersant.ru/doc/3325258.

Kolesnikov, A. 2011. "It is Frightening to Name Such Figures." Kommersant. 31 Aug. 2011. https://www.kommersant.ru/doc/1761397.

Kozlov, D. 2018. "Drilling Freeze." *Kommersant*. 14 May 2018. https://www.kommersant.ru/doc/3628125.

_______. 2020. "Oilmen Reached an Agreement Onshore." *Kommersant*. 15 Dec. 2020. https://www.kommersant.ru/doc/4614745.

_______. 2020 "Rich Is the Drilling Bit." *Kommersant*. 31 Aug. 2020. https://www.kommersant.ru/doc/4474262.

_______. 2013. "The Firm Political Fate." *Kommersant*. 31 May 2016. https://www.kommersant.ru/doc/3000838.

_______. 2020. "The White House will Work on the Shelf." *Kommersant*. 16 June 2020. https://www.kommersant.ru/doc/4379221.

Kryuchkova, E., and A. Vedeneeva. 2019. "The Arctic Was Sent to the Far East," *Kommersant*. 19 Jan. 2019. https://www.kommersant.ru/doc/3859135.

Lenta.ru. 2017. "Rosneft Discovered a New Petroleum Province." 18 June 2017. https://lenta.ru/news/2017/06/18/olginskaya/.

Lloyds' Report. 2012. "Arctic Opening – Opportunity and Risk in the High North." 12 April 2012. https://www.lloyds.com/news-and-risk-insight/risk-reports/library/natural-environment/arctic-report-2012.

LUKOIL. 2019. "Main Development Trends of the Liquid Hydrocarbons Market up to 2035." https://lukoil.ru/Business/Futuremarkettrends.

Melnikov, K. 2012. "Igor Sechin Threw Skeletons out of Cupboards." *Kommersant*. 19 Apr. 2012. http://www.kommersant.ru/doc/1918809.

_______. 2011. "Rosneft had a Glimpse of the New World." *Kommersant*. 31 Aug. 2011. https://www.kommersant.ru/doc/1761897.

_______. 2011. "Russian Petroleum." *Kommersant*. 17 Jan. 2011. https://www.kommersant.ru/doc/1568681.

_______. 2012. "Statoil Occupied a Niche at the Shelf." *Kommersant*. 10 May 2012. https://www.kommersant.ru/doc/1930535.

_______. 2011. "The Deal of the Century to be Closed in Courts." *Kommersant*. 25 Mar. 2011. https://www.kommersant.ru/doc/1607492.

_______. 2013. "The Shelf Was Handed to State Companies." *Kommersant*. 16 Jan. 2013. http://www.kommersant.ru/doc/2105645

Milchakova, N. 2018. "Shale Is Ours!" *Neft I Kapital*. 21 Sept. 2018. https://oilcapital.ru/article/general/21-09-2018/slanets-nash?ind=450.

Ministry of Economic Development. 2018. "Forecast of Social and Economic Development of the Russian Federation up to 2036." November 2018, 21. https://www.economy.gov.ru/material/directions/makroec/prognozy_socialno_ekonomicheskogo_razvitiya/prognoz_socialno_ekonomicheskogo_razvitiya_rossiyskoy_federacii_na_period_do_2036_goda.html.

Mitrova T. et al. 2018. "Prospects of Russian Oil Production: Life under Sanctions." *Skolkovo Business School*. March 2018. https://energy.skolkovo.ru/downloads/documents/SEneC/research04-ru.pdf.

National Travel. 2021. "The Kara Sea." https://national-travel.ru/karskoe-more.

Neft I Kapital. 2019. "Gazpromneft and NOVATEK will Jointly Develop Severo-Vrangelevskiy Offshore Plot." 6 Sept. 2019. https://oilcapital.ru/news/upstream/06-09-2019/severo-vrangelevskiy-uchastok-na-shelfe-sovmestno-budut-razrabatyvat-gazprom-neft-i-novatek

_______. 2017. "Interview with Sergei Donskoy: We Do Not Exclude the Possibility of Major Discoveries in West Siberia." 1 Nov. 2017. https://oilcapital.ru/interview/01-11-2017/sergey-donskoy-my-ne-isklyuchaem-vozmozhnosti-krupnyh-otkrytiy-v-zapadnoy-sibiri-bd2b3be2-b288-493d-940e-4c49f3e00de0.

_______. 2002. "Shareholders' Meeting of Rosshelf Decided not to Transfer to Sevmorneftegas the License for the Development of Prirazlomnoye Field." 2 Feb. 2002. https://oilcapital.ru/news/markets/02-07-2002/sobranie-aktsionerov-rosshelfa-reshilo-ne-peredavat-sevmorneftegazu-litsenziyu-na-osvoenie-prirazlomnogo-mestorozhdeniya.

_______. 2019. "The Ministry of Natural Resources Estimated Oil Reserves of the Arctic at 7.3 bln tons." 13 Nov. 2019. https://oilcapital.ru/news/markets/13-11-2019/v-minprirody-otsenili-neftyanye-zapasy-arktiki-v-7-3-mlrd-tonn.

Neftegaz. 2004. "Belokamenka: A Miracle of Miracles." 23 Mar. 2004. https://neftegaz.ru/news/companies/305906-belokamenka-chudo-chudnoe/.

_______. 2020. "Over 6 Years of Prirazlomnaya Platform Operations, Gazpromneft Produced 13 Mln Tons of ARCO Oil." 17 April 2020. https://neftegaz.ru/news/dobycha/543022-za-6-let-raboty-platformy-prirazlomnaya-gazprom-neft-dobyla-13-mln-t-nefti-sorta-arco/.

_______. 2017. "Two More Years for Thinking. Rosneft, Seadrill and NADL Reached an Agreement on the Extension of the Framework Contract." 2017. https://neftegaz.ru/news/Acquisitions/209908-eshche-2-goda-na-podumat-rosneft-seadrill-i-nadl-dogovorilis-o-prodlenii-ramochnogo-soglasheniya/.

Nepomnyaschih, A. et al. 2010. "Rosneft Overtook Gazprom." *Vedomosti*. 15 Oct. 2010. http://www.vedomosti.ru/newspaper/article/247730/gazprom_obstavili#ixzz12VDGMlhg.

"Official Portal of the Krasnoyarsk Krai." 2020. http://www.krskstate.ru/about.

"Official Website of the Chukchi Autonomous District." 2021. www.чукотка.рф.

Overland, Indra and Nina Poussenkova. 2020. Russian Oil Companies in an Evolving World: The Challenge of Change. Cheltenham: Edward Elgar.

Petlevoi, V. 2017. "Gazpromneft Looks for Partners for the Arctic." *Vedomosti*. 29 Mar. 2017. https://www.vedomosti.ru/business/articles/2017/03/29/683288-gazprom-neft.

Polivanov, A. 2013. "Thirty Six Holes in the Pechora Sea." *Lenta.ru*. 30 Sept. 2013. https://lenta.ru/articles/2013/09/30/pri/.

Poussenkova, Nina. 2007. *Lord of the Rigs*. J.Baker Institute for Public Policy.

_______. 2016. *The Rosneftization of the Russian Oil Sector.* Oxford Institute of Energy Studies publication, November 2016. https://www.oxfordenergy.org/wpcms/wp-content/uploads/2017/01/OEF-107.pdf.
RBC. 2015. "Gazpromneft Negotiated a Record Extension for Commencement of Oil Production on the Shelf." 15 Nov. 2015. https://www.rbc.ru/business/15/11/2015/5645a9429a7947c868dcadf9.
_______. 2015. "Rosneft Cancelled its Order for the Drilling Vessel of North Atlantic." 14 Mar. 2015. https://www.rbc.ru/business/14/03/2015/5503702e9a79472a9bc852b5.
_______. 2016. "The Arctic without Wells: Russia Gave up its Plans for Intensive Production of Hydrocarbons on the Shelf." 10 June 2016. https://www.rbc.ru/newspaper/2016/06/10/57593ed59a79476c142e7256
_______. 2020. "Trutnev Admitted that Russia Lagged behind its 'Neighbours' on the Continental Shelf." 18 Feb. 2020. https://www.rbc.ru/politics/18/02/2020/5e4b92d69a79478d29addfc0.
RIA Novosti. 2017. "Accident on the Kolskaya Floating Drilling Rig in the Sea of Okhotsk in 2011." 3 May 2017. https://ria.ru/20170503/1493612941.html.
_______. 2011. "Rosneft and BP Agreed to Cooperate on the Continental Shelf in Russia." 15 Jan. 2011. https://ria.ru/20110115/321772704.html.
Rogtec. 2018. "ExxonMobil: Exits JVs with Rosneft over Offshore Sanctions." 1 Mar. 2018. https://rogtecmagazine.com/exxonmobil-exits-jvs-with-rosneft-over-offshore-sanctions/.
Rosneft. "Offshore Projects." 2021. https://www.rosneft.ru/business/Upstream/offshore/.
_______. 2020 "Igor Sechin Reported to the RF President on the Company's Activities in the Current Market Conditions." 18 Aug. 2020. https://www.rosneft.ru/press/releases/item/202375.
_______. 2013. "Rosneft and Eni Signed a Declaration about Careful Development of the Russian Arctic Shelf." 21 June 2013. https://www.rosneft.ru/press/releases/item/1228/
_______. 2014. "Rosneft and ExxonMobil Began Drilling Operations in the Kara Sea." 9 Aug. 2014. https://www.rosneft.ru/press/today/item/153568/.
_______. 2013. "Rosneft and ExxonMobil Expand their Strategic Cooperation." 13 Feb. 2013. https://www.rosneft.ru/press/releases/item/177813/.
_______. 2012. "Rosneft and ExxonMobil Signed a Declaration about Careful Development of the Russian Arctic Shelf," 7 Dec. 2012. https://www.rosneft.ru/press/releases/item/177751/.
_______. 2012. "Rosneft and Statoil Signed a Declaration about Careful Development of the Russian Arctic Shelf." 23 Nov. 2012. https://www.rosneft.ru/press/releases/item/18661/.
_______. 2017. "Rosneft Began to Drill the Northern-Most Well on the Russian Continental Shelf." 3 Apr. 2017. https://www.rosneft.ru/press/releases/item/186075/.
_______. 2014. "Rosneft Discovered a New Field in the Kara Sea." 27 Sept. 2014. https://www.rosneft.ru/press/releases/item/153712/.
_______. 2019. "The Construction and Assembly Works were Completed in the Sapphire Plant." 28 Jan. 2019. https://www.rosneft.ru/press/news/item/193625/;
_______. 2017. "Vladimir Putin Took Part in the Ceremony of Commissioning of the Heavy Staple of SSK Zvezda." 8 Sept. 2017. https://www.rosneft.ru/press/news/item/187665/.
Rosneft Website. 2021. "About: RN-Sakhalinmorneftegas." https://rnsmng.rosneft.ru/about/Glance/OperationalStructure/Dobicha_i_razrabotka/Dalnij_Vostok/rnsmng/.

Russian Shipping. 2020. “The First-ever Stratigraphic Well in Russia was Drilled in the North of the Kara Sea.” 26 Oct. 2020. https://rus-shipping.ru/ru/prof/news/?id=42718&utm_source=yxnews&utm_medium=desktop&utm_referrer=https%3A%2F%2Fyandex.ru%2Fnews%2Fstory%2FV_Rossii_vpervye_proizvedeno_stratigraficheskoe_burenie_skvazhin_na_severe_Karskogo_morya--70946fcd1c06e685e21d2672358197a3.

Solodovnikova, A. and O. Mordyushenko. 2012. “Rosneft and Gazprom Joined Forces.” *Kommersant*. 19 Oct. 2012. https://www.kommersant.ru/doc/2047685.

TASS. 2019. “Experts: The Development of the Arctic Shelf is Unprofitable with Oil Prices Below $ 100/bar.” 25 Sept. 2019. https://tass.ru/ekonomika/6927654,

_______. 2016. “The Cabinet of Ministers Introduced a Temporary Moratorium for Issuing Offshore Development Licenses.” 7 Sept. 2016. https://tass.ru/ekonomika/3601709.

Tchourilov, L., I. Gorst, and N. Poussenkova. 1996. *Lifeblood of the Empire: a Personal History of the Rise and Decline of the Soviet Oil Industry*. PIW Publications.

Trutnev, Yu. 2020. “Russia is Lagging Behind in Exploration of Offshore Fields.” *Kommersant*. 18 Feb. 2020. https://www.kommersant.ru/doc/4259583.

Vasilieva A. 2020. “The Arctic Train: What will Follow the Norilsk Accident.” *Kommersant*. 4 June 2020. https://www.kommersant.ru/doc/4366214.

Vedomosti. 2007. “The Kremlin will Allocate all Shelf Fields to Gazprom and Rosneft.” 22 Jan. 2007. https://www.newsru.com/finance/22jan2007/national.html.

Vygon, G. et al. 2014. “Do Russian Independent Oil Companies have a Future?” *Skolkovo Energy Center*. http://www.assoneft.ru/activities/developments/459/.

Website of Arktikmorneftegazrazvedka. 2021. “About the Company.” http://www.amngr.ru/o-predpriyatii/.

Wikipedia. 2021. “The Bering Sea.” Last modified 2021. https://ru.wikipedia.org/wiki/%D0%91%D0%B5%D1%80%D0%B8%D0%BD%D0%B3%D0%BE%D0%B2%D0%BE_%D0%BC%D0%BE%D1%80%D0%B5.

Wikipedia. 2021. “The Pechora Sea.” Last modified 2021. https://ru.wikipedia.org/wiki/%D0%9F%D0%B5%D1%87%D0%BE%D1%80%D1%81%D0%BA%D0%BE%D0%B5_%D0%BC%D0%BE%D1%80%D0%B5.

Yagova, O. 2010. “BP and Rosneft Made a Geological Decision.” *Kommersant*. 19 Feb. 2010. https://www.kommersant.ru/doc/1324535.

11

SVALBARD AND OIL

Dangerous games in the Arctic?

Helge Ryggvik

In his memoir, the former Norwegian Foreign Minister Knut Frydenlund points out how central, but also burdensome, the Svalbard issue has been at Norway's Ministry of Foreign Affairs (Frydenlund 1982). He writes about when he, as a younger parliamentary representative, rode an elevator with a Labour Party colleague and the Conservative Party's then Foreign Minister, Jon Lyng. Lyng commented: "I do not know which of you will take over after me, but I just want to tell you that if the phone rings with us at night, then my first thought is that now it applies to Svalbard" (Frydenlund 1982, 63). Lyng, who had been Prime Minister for a few days in 1963, was Norway's Foreign Minister between 1965 and 1970. Frydenlund, having held a dominant position in Norwegian foreign policy in the 1970s and 1980s, clearly believed that Lyng described the same reality he experienced. Because Svalbard's Russian population was twice as big as Norway's in the years after World War II, a fractious relationship with the Soviet Union (and later Russia) was Norway's main concern. From the mid-1970s, the economic rights to the ocean surrounding Svalbard became a particular potential flash point. Here, Norway's opponent was not only Russia but also all other countries that initially signed the Spitsbergen treaty. On the surface, the main issue was fishing rights. However, in 1977, when Norway claimed the right to fish in a large 200 nautical mile zone around Svalbard, this also had implications for its right to explore for oil and gas in the same waters.

Although Frydenlund made no secret of the importance of future oil rights in 1977, he knew that oil exploration in the ocean around Svalbard would be an issue in the distant future. In the 1970s, the main credo for Norwegian oil policy was to develop its potential resources slowly (Ryggvik 2010). Drilling activities were restricted to areas in the North Sea south of 62° latitude. This caution was one of the reasons the US government opposed Norway's claim for sovereign

DOI: 10.4324/9781003024125-11

rights to the continental shelf surrounding Svalbard. A newly released secret 1976 CIA document notes:

> If sovereign rights on the shelf reside in Norway without qualification, then, in addition to exclusive control over access, Norway might seek to delay exploration indefinitely for environmental or security reasons; Norway's economic interest could be satisfied by oil from the North Sea and Norwegian Sea for some time to come. (CIA 1976, 7)

The CIA was well informed. However, what was considered the distant future in 1977 was not so distant in the early 2020s. With the rest of the world preoccupied by the COVID-19 crisis, the Norwegian government introduced breathtaking measures that would radically expand Norwegian oil activities in the Arctic, right to the cusp of the disputed Svalbard Zone.

In many histories of oil, the main agents for change, often also the villains in the narrative, are the international oil companies (Sampson 1975; Yergin 1991; Coll 2013; Tarbell 2015). However, during the 2010s, large international oil companies like Shell, Exxon, and Chevron withdrew from new drilling activities in the Barents Sea. This was partly because geological expectations no longer outweighed cost concerns: building production installations that can safely withstand the forces of nature in such unfriendly, remote waters is challenging. In the late 2010s, Arctic oil exploration also became a potential political liability. Even large banking institutions stopped financing Arctic oil. Goldman Sachs' decision in December 2019 to stop investing in oil sent a strong signal to the oil industry (Kirschgaessner 2019) with JP Morgan and Barclays following suit in 2020 (Barclays 2020, 3; J.P. Morgan Chase & Co. 2020, 6). These policy shifts demonstrated that environmental and climate movement opposition towards Arctic oil activities had got a wide hearing.

So, why did the Norwegian state ignore considerations from the world's largest and supposedly most competent, international oil companies? What made a further oil expansion to the North so important that the state was ready to destroy Norway's reputation for being an environmentally friendly nation? And was Norway really ready to risk a dangerous international confrontation by, in practice, occupying the Svalbard Zone and handing out drilling rights in the area? In 2017, following Russia, the US, Britain, Iceland, and Spain who vocally opposed Norway's position, the EU Parliament called for a ban on oil drilling in "icy waters," with clear aim at potential Norwegian activities in the northern Barents Sea near Svalbard (European Parliament 2017, 14).

The Svalbard Treaty

Most issues in relation to Svalbard and the ocean around the archipelago come down to interpretations of the Spitsbergen Treaty of 1920.[1] In article 1 of the treaty, Norway was given "full and absolute sovereignty over the archipelago of

Spitsbergen," including Bear Island and Hopen. Most of the following articles strongly limit Norway's sovereignty, however, giving it the role of environmental caretaker. Article 9 states that no military activities and bases are allowed. Article 3 states that Norway must administer the archipelago based on a non-discriminating principle. Companies from all signatory countries possess hunting rights and could start business activities on Svalbard. The Norwegian state could not tax these activities beyond what was necessary to finance its administration of Svalbard, and Norway must not require visitor visas.

The treaty's non-discrimination articles and limits on taxation seriously concerned Norway. Shifts in international maritime law in the late 1970s prompted Norway to establish 200 nautical mile "economic zones" around its long mainland coast. An economic zone gave a state control over the seabed, the ground beneath (oil and gas), and the water column above (fish). If Svalbard had an economic zone, and if the Svalbard Treaty became effective, Norway would have to struggle with restrictions on who was allowed to explore and produce potential resources as well as the right to tax these activities. Without the right to tax, oil production would be of little interest for the state.

Instead of establishing an economic zone around Svalbard, Norway declared a 200 nautical mile fisheries protection zone around the archipelago in April 1977. Norway's formal justification for this step was its "full and absolute sovereignty of Svalbard," according to the Svalbard Treaty (Ministry of Foreign Affairs 1977). Norway also stated that it was still within its rights to establish an economic zone and that the Svalbard Treaty would not be effective within it if or when it did so (Ministry of Foreign Affairs 1977).[2]

Taking control of fishing rights in an ocean, that before had been free for all, was a challenge. In addition to the original members of the Svalbard Treaty, countries with growing international fishing fleets, like Spain, Portugal, and Iceland, protested. Getting to control the rich waters around Svalbard was a large economic asset. Norway argued that, without strict regulations, a growing fleet of trawlers would destroy the stock.

None of the original signatory powers has since accepted Norway's position,[3] making Svalbard an area of constant tension and prompting several dangerous confrontations (Christensen 2014). The most serious encounters have involved Russian trawlers, though the Norwegian Coast Guard has also confronted Icelandic, Spanish, and Baltic fishing vessels. However, the conflict has been subdued to an extent by the fact Norwegian regulations seem to have had a positive effect on the fish stocks. Norway has also administered fisheries as if the Svalbard Treaty is in effect, handing out significant fishing rights to Russian and EU-based fishing vessels. This raises the question of why Norway needed to "occupy" the waters of the Svalbard Zone in the first place. An alternative would have been to administer a similar regime in accordance with the Svalbard Treaty.

Fishing rights were of immediate interest, though the right to control petroleum resources was considered to be of even more value in the future. According to Pedersen, signals from the US influenced Norway's decision to refrain from declaring a full economic zone (Pedersen 2008, 187; Pedersen 2011). The US made it clear that it would oppose a Norwegian claim of sovereign rights to oil resources in the Svalbard Zone. Norway's delicate diplomatic position became even more complicated when it partly shifted its strategy by claiming that the seabed and the ground beneath it around Svalbard was an extension of the Norwegian continental shelf.

Rather than taking the issue to the International Tribunal for the Law of the Sea (ITLOS), Norway has, on several occasions, signalled that its main tool when it comes to oil and gas rights will be to carry out an "occupation" (Støre 2005; Fife 2014). An occupation means dividing the area into blocks and handing out drilling concessions. The idea is that rights will be created gradually by facts on the ground. If other states do not act in protest, the occupation can eventually be claimed a success.

Compared to assigning fishing rights, administering oil activities under such conditions is difficult and dangerous. For companies like British BP, Dutch Shell, and American ExxonMobil, it is difficult to be involved in the Svalbard play if their governments oppose Norway's position. EU opposition to oil activities in "icy waters" adds to such difficulties (European Parliament 2017, 14). The main opponent to Norwegian expansion in the Svalbard Zone has been Russia. Before the Svalbard Treaty, in the Soviet period, and even more so in the post-Soviet 2000s, Russian nationalists have claimed that "Spitsbergen" historically belongs to Russia. Russia's formal position is to insist that the Svalbard Treaty's conditions apply to both the continental shelf and the water column in oceans around Svalbard.

In this sense, Russia opposes Norway's view on the same formal grounds as other countries. However, from Russia's perspective, full Norwegian control of the ocean around Svalbard is of great strategic military and geopolitical importance; as a countermeasure to a potential NATO first strike, it would be essential for Russia to get nuclear submarines into the Atlantic quickly, without being discovered. When the Norwegian Petroleum Directorate (NPD) started extensive seismic surveys in the Svalbard Zone in 2012, Russia initiated similar surveys on its own without seeking permission from the Norwegian government. If Russia were to react similarly to Norwegian regulated drilling in the zone by starting drilling from a Russian rig, Norway would have either to respond with force or accept lawlessness in the zone. In such an event, it would be a serious concern that Russia built a modern military base at Franz Josef Land in the 2010s, northwest of Svalbard, much closer to the disputed area than any NATO base on Norway's mainland.

Svalbard and oil

There are many ways to deescalate potential conflict around the Svalbard Zone. For Norway, one obvious alternative would be to bury all ambitions to exploit oil and instead work for a ban against any oil activities. The best way to achieve this would actually be to accept the position that all clauses of the Svalbard Treaty must apply in the zone. Norway then could do what it has already done on Svalbard's mainland: block all new oil activities based on environmental concerns.[4]

The delicate historical balance between Norway's treaty-bestowed role as a caretaker and the Soviet state's interest was first challenged when the American oil company Caltex signalled in 1960 that it wanted to drill for oil on Svalbard (Bjørklund 2008, 31). Given that Caltex's sudden interest in Svalbard coincided with an intensification of the Cold War, one might suspect that US strategic interests were involved. The thought of adding American oilmen to the delicate power balance on Svalbard was disturbing for the Russians and also for some in the Norwegian Ministry of Foreign Affairs. Nevertheless, historical records seem to suggest that Caltex's interest in Svalbard was real (Bjørklund 2008, 31). It was well known that the archipelago consisted of geological layers dominated by sedimentary rocks from periods with large organic deposits. Major discoveries in Alaska had awakened the oil industry's interest in the Arctic. Svalbard was unique as it made it possible to drill far north in the Arctic and, of course, the prospect of finding oil without having to pay large taxes was tempting for any oil company.

Drilling on Svalbard started two years ahead of the first drilling in the North Sea (Johannesen and Stenløkk 2004, 78; Senger et al. 2020, 10). Negotiations with Caltex were led by the Norwegian Foreign Ministry's Chief Justice, Jens Evensen, who also led work to develop a concessionary system in the North Sea. Evensen's first strategy was to look for ways to by-pass the Svalbard Treaty and acquire as much revenue for the state as possible if oil was found. However, major oil discoveries in the North Sea from the early 1970s prompted Norway's policy towards Svalbard to shift radically. In the North Sea, the Norwegian state was free to tax as much as it liked. Since none of the first wells on Svalbard showed very promising results, incentives to push for further developments were weakened. At the same time, with a newly awakened environmental awareness in Norway, potential oil finds on Svalbard could create more problems than they solved (Ryggvik 2021, 404). There were reasons for concern: because of permafrost and other issues, Caltex's first well led to serious pollution (Johannesen and Stenløkk 2004). In a fjord, otherwise covered with ice during winter, leaked gas and drilling mud caused a visible contaminated hole.

Article 2 of the Svalbard Treaty gave Norway all it needed to deal with the challenge. In 1973, much of Svalbard's western coast was converted into national parks, and northeast Svalbard and southeast Svalbard were made nature reservations, with even stronger protections. The strict regulations also covered Edgeøya, where both Fina and Caltex had just completed dry wells. When the Svalbard Environment Act came into force in 2002, new environmental regulations for the rest of the archipelago were introduced and a series of new national parks and protected areas was established. This brought 65% of Svalbard's land area under protection from oil activities in the foreseeable future (Arlov 2011, 380).

With the new regulations, not a single new drilling permit has been granted on land (Direktoratet 2001, 13). Indeed, the Norwegian Government's April

2009 Svalbard Report, and its May 2016 successor, both state that oil drilling on Svalbard is banned, and no drilling permit has been granted in Svalbard's territorial waters where the same regulations apply (Meld. St. 2008–2009, 92; Meld. St. 2015–2016, 94). In 2004, Svalbard's territorial waters were enlarged from 4 to 12 nautical miles (Ot.prp. 2002–2003). By accepting that Svalbard had an economic zone beyond the territorial waters, Norway could easily ban oil activities in the much larger 200 nautical mile zone around Svalbard.

While Norway's policy towards Svalbard's mainland was based on an understanding that there was oil and gas elsewhere, it never lost sight of the potentially large resources in Svalbard's offshore. The 200 nautical mile zone around Svalbard is three times as big as that opened for oil exploration in the Norwegian part of the North Sea. When Norway introduced its slow pace policy in the early 1970s, it assumed that developing resources too fast could leave the country too dependent on rent from the oil sector and compromise safety and the environment (Meld. St. 1973–1974). A slow pace would also permit the Norwegian industry the necessary time to adapt to new challenges and develop skills to compete with the established international offshore supply industry. In 1974, while discussing an important white paper on future implications of the petroleum sector for Norwegian society, the Parliament set 90 million tonnes o.e. per year as Norway's yearly oil and gas production cap (Meld. St. 1973–1974). In early 1988, when production levels were about to exceed this amount, Parliament instead established a 25 billion NOK cap on yearly investment (Meld. St. 1987–1988, 60).

The Norwegian oil industry's success undermined the initial goals of the slow pace policy. Petroleum companies needed a continuous stream of new projects to maintain and enlarge their capacity, and supply firms became dependent on ever more and larger investment projects, making it impossible to stay under the caps on both production and investments. The same dynamics ended up overriding Norway's early climate policy. These contradictions were clearly expressed through Gro Harlem Brundtland's parallel role as head of the UN report "Our Common Future," basis for the UN's first climate summit, and Norway's Prime Minister in the critical second half of the 1980s and first half of the 1990s (UN 1987). Inspired by Brundtland's leadership, the Norwegian Parliament decided in 1989 that Norway's CO_2 emissions should be stabilized at 1989 levels until the year 2000. However, from 1988 to 2000, Norway's oil production tripled. Production of gas increased fivefold until 2005 (NPD 2021). Early strict regulation on flaring gas in the oil fields had some effect, but producing oil on large installations out in cold seas demanded a large amount of power. With gas and water pumped down into reservoirs to keep up pressure, older fields gradually needed more power. Norwegian authorities refused to take any responsibility for emissions from Norwegian oil exported and then burned elsewhere. Large oil fields were powered by gas turbines that sometimes could burn as much CO_2 as Norway's largest towns. Drilling platforms and supply vessels also burned a lot of diesel. This meant that increased oil activities led to a major increase in Norway's total emissions.[5]

When Norway's total production of oil and gas peaked in the early 2000s, industry pressure to access new acreage increased substantially. For a period, the industry's main priority was to open the Lofoten, Vesterålen, and Senja areas (LoVeSe) for drilling. The area had been temporarily protected because it had long been Norway's most important fishing ground and was the Atlantic cod's main spawning place. It was also particularly vulnerable because the continental shelf there was narrow. Drilling rigs would come very close to shore. Wind and currents would very likely lead potential oil spills to the famously beautiful coast in hours.[6]

Even though the industry struggled hard to open the area, it was very likely aware that the LoVeSe region did not possess enough resources to keep the Norwegian offshore industry alive in the coming decades. This led the industry and state supporters to make the Barents Sea the main focus for future developments.

The rich circle of life

While the Lofoten area is well known to all Norwegians, who start learning about the importance of cod from common children's rhymes, the Barents Sea appears somewhat more distant. Compared to the narrow continental shelf outside Lofoten, it is much bigger – stretching from around the 70th degree of latitude off the northern Norwegian coast, up to close to the 80th degree of latitude outside the northern parts of Svalbard. To the east, the Barents Sea stretches out to the Russian island of Novaja Semlja. When the famous climatologist James Hansen visited Norway in 2017, the leader of the oil lobby institution Norway Oil and Gas, Karl Eirik Schjøtt-Pedersen, criticized Hansen and other "foreigners" for having false assumptions about the Barents Sea, arguing that because of the Gulf Stream, the Barents Sea was not really in the Arctic (High North News 2015; NRK.no 2015; Stavanger Aftenblad 2015).

Hansen, of course, was well aware of the Gulf Stream's effect on climate. Nevertheless, Schjøtt-Pedersen had a point in the sense that the Gulf Stream makes both Norway's northern coast and the Barents Sea unique. It is this unusual combination of relatively warm water streaming far to the north and meeting a cold stream from the north that makes the Barents Sea so fantastically rich and vulnerable. When cod and other species spawn in the Lofoten area, the eggs and fry follow the water current up to the Barents Sea. It is here the fish find the nutritious diet that makes them grow. In fact, there are very few ocean areas with a similarly intense growth of marine biomass.[7]

The Barents Sea, like other areas close to the Polar Front in the Arctic and Antarctic, has a significant population of whales. It also supports one of the largest populations of sea birds worldwide (Miljødirektoratet 2017). Since birds need a land base, the Barents Sea's size is a challenge. Hence, they are concentrated in a radius around Bear Island, on and around Svalbard's Islands, and on the Bird Cliffs in northern Norway. Access is also the main challenge for mammals like polar bears, walrus and seals. The Barents Sea's ice zone appears an extension of the Arctic icecap. However, it is different in the sense that it freezes and melts on a yearly basis. The ice is an important part of the rich ecosystem. Beneath it, both plankton and algae survive the winter. When the ice melts around April, both the fresh water and the "seeds" of algae and plankton are essential in the new growth period. For seals, the ice is a platform from which to fish, while polar bears hunt the seals. If oil platforms are to be installed in the Svalbard Zone, they will also need to withstand encounters with icebergs, adding to their cost. However, the main challenge for potential oil installations in the Barents Sea is not huge icebergs of the kind that threaten petroleum infrastructure on Canada's Grand Banks. The chief challenge from an environmental point of view is their proximity to the unique and vulnerable marine life found within the Polar Front. The northern part of the Norwegian sector in the Barents Sea that has been opened for drilling since the 2010s is within this circle. The Svalbard Zone is even more environmentally sensitive than the currently exploited Norwegian sector.

Schjøtt-Pedersen was also in some sense right when he claimed that the Gulf Stream makes large parts of the Barents Sea accessible for the oil industry. With an average depth of 300 metres, oil companies do not have to struggle with challenges related to deep sea wells, like those in the Gulf of Mexico, or off the coasts of Brazil and West Africa. However, there are also significant limits to this accessibility, including ones with serious implications for safety. Twenty-four hours of daylight in some weeks during the summer do not actually help visibility when tight sea fog covers the area for long periods. Substantial distances from bases on land would make large rescue operations difficult, as would the fact that GPS signals are unreliable from around 71 degrees of latitude, and from 73 degrees signals are often non-existent.[8] This compromises safety equipment, including dynamic positioning, onboard drilling platforms and support vessels. GPS limitations would also create problems should a company need to drill a relief well after a blow-out.

Most importantly, in the winter, Arctic conditions return. Long periods with no natural light and extremely low temperatures create working conditions that are very different from what the industry is used to handling. While the Barents Sea is accessible for drilling platforms in the summer and early fall, production installations have to operate year round and need to withstand all operational challenges for years. This is why drilling and production in the Barents Sea have been two vastly different scenarios.

Drilling in the Barents Sea

The first drilling in the undisputed Norwegian part of the Barents Sea started as early as 1980 after the ban against drilling north of 62 degrees was lifted. This happened without too much controversy, partly because the environmental movement in Norway had not yet gained momentum. At the same time, the spectacular Bravo blow-out on the Ekofiskfield in the middle of the North Sea in 1977 actually dampened public fears in relation to large oil spills. Contrary to other large oil spills,[9] the *Bravo* blow-out did not lead to any visible environmental damage (Ryggvik 2021, 58). After a mild storm, the oil spill evaporated. Because the spill happened far away from the coast and in an area with no concentrations of seabirds or other marine life, the accident influenced the general conception of an oil spill as solvable. While oil spills in the US and many other countries are typically considered as a poisoning of nature, in Norway, geologists and other oil industry-related experts succeeded in creating a perception of oil as "a product of nature" (NOFO 2011). Norway invested in oil spill contingencies, with lenses that were supposed to collect oil in waves up to 2 metres.[10] The general idea was that bad weather and higher waves would solve the problem by breaking down the oil spill.

One other reason that drilling in the Barents Sea got relatively little attention until the 1990s was the fact that very little oil was found. In some of the first wells drilled, Statoil discovered several medium-sized neighbouring gas fields that later were connected to become the Snøhvit Field. International companies like Shell, BP, and ConocoPhillips took part in some of the drilling prospects as minority participants in license groups. However, none of these companies lobbied the Norwegian state for a more aggressive opening of new areas in the north; their focus was still the North Sea and the Norwegian Sea.

From the early 1980s, industrialist politicians were the driving force behind the oil industry's pushing further into the Arctic. The policy of expansion to the north was at the same time deeply embedded in state apparatus, particularly in the Ministry of Oil and Energy, and the NPD. The dominant Labour party politician Bjartmar Gjerde, worked for opening new areas north of 62 degrees latitude, first as Minister of Industry and then as Norway's first Minister of Oil and Energy from 1977. His main motive was to make sure that regions in the north would get the same opportunity to develop oil-related industries as those in southern and western Norway.

Nevertheless, after almost 20 years of drilling, oil companies had still not discovered large enough finds that could defend investment in production facilities. Part of the problem was the costs related to northern infrastructure. When Statoil finally decided to develop the Snøhvit Field, it was influenced by political pressure and manoeuvring from forces in the Labour Party and from local politicians and business communities in northern Norway. Adjustments to the tax system by Schjøtt-Pedersen, in his short period as Finance Minister, made Statoil make the final decision. Since the Snøhvit Field was developed as a tie-in solution, with remote-controlled subsea wells connected by pipelines to process facilities onshore near the city of Hammerfest, the project avoided having to operate as a yearlong installation in open waters. But without access to pipelines like those in the North Sea and the Norwegian Sea to transport the gas to the main markets, Statoil had to build a LNG-plant, creating a lot of problems. To acquire the energy that was needed to freeze the gas, Statoil had to use large gas turbines that burned so much gas that the plant ended up emitting as much CO_2 as the total emissions from Norway's capital, Oslo (Equinor 2019).

Plans and science

The first drilling campaign in the Barents Sea started without much knowledge of what kind of environmental impact a potentially large oil find could have. Initially, the oil sector got an exemption when Norway passed several new laws that demanded a formal system of impact assessments for large investments beginning in 1985 (Andersen 2017). In the 1990s, however, impact reports were often superficial and came at a point when all major decisions already had been made. Their main aim was to prevent or anticipate conflict rather than influence decisions (Andersen 2017). However, with the 2002 decision to develop the Snøhvit Field and the new discussion around opening LoVeSe, a short moratorium for new licenses in the Barents Sea was introduced. Subsequently, the framework for all new concessions was to be based on a thorough "forvaltingsplan" (administration plan). This plan would assess all fundamental environmental concerns in relation to factors such as the fisheries and risk of large accidents (Meld. St. 2005–2006). The research involved in this plan was instrumental in strengthening knowledge about marine life in the Barents Sea as well as in the wider Arctic ecosystem. The Norwegian state also increased financing of relevant research institutions.[11]

Although the new "forvaltingsplans" highlighted both the richness and the vulnerability of marine life in the Barents Sea, they did not block the state from laying the foundations for an intense new drilling campaign in the area. This was possible because of the use of risk assessments in which the probability for accidents and spills was set to nearly zero (Meld. St. 2005–2006). When the 19th concession round on the Norwegian continental shelf was announced in 2006, a new epoch of drilling in the Barents Sea could start. In the concession rounds that followed every second year, the government opened up a record number of blocks for applications in the Barents Sea. When Norway and Russia finally agreed in 2011 on the dividing line between the two countries' offshore economic zones, a large new area east of the former drilling sites was opened up.

It took some years before the actual drilling in the Barents Sea took off in earnest. The temporary fall in oil prices following the financial crisis in 2008 contributed to the postponement of some projects. In every round, the number of production licenses was much lower than the number of blocks open for applications. With the high infrastructure cost and the shale gas revolution in the US that almost destroyed the market for LNG gas, it was clear that the companies now mainly sought oil, not gas.

There was a period of some optimism between 2011 and 2015 when several small and medium-sized oil fields were discovered north of the Snøhvit Field. However, in 2020, 40 years after a large part of the undisputed Norwegian section of the Barents Sea was opened for the oil industry, and after more than 140 wells were drilled, none of the findings compared to the large oil fields in the North Sea, which still constituted the "bulk" of Norway's oil and gas production. In the 22nd and 23rd concession rounds, there were high expectations for particularly promising blocks based on seismic surveys (the Korpfjell and the Gjøkåsen license) (Ryggvik 2021, 358). Both turned out to be disappointments, with no oil and only little gas. It was not until 2016 that the first production of oil in the Barents Sea began in the Goliat Field, though its development was tainted by scandals. The cost of building and the production installation was around 5 billion dollars, more than 40% over budget (Acona 2019, 9). A report criticized the ministry and the Petroleum Safety Authority for giving the Italian operator (Eni), permission to start production when there were many obvious breaches of regulations (Riksrevisjonen 2018–2019). Then a report from Knut E. Rosendahl showed that even if the project might give the operating oil company some profit, a very advantageous tax system meant that the project ended in the red for the Norwegian state (Bellona 2017; Upstream 2017). This was without including risks related to possible accidents, including oil spills, and potential CO_2 emissions. Snøhvit and Goliat accounted for only 6% of Norway's total production of gas and less than 2% of the country's production of oil, respectively (Oljedirektoratet 2021). If the Johan Castberg Field becomes active in 2022–2023 as planned, oil production in the Barents Sea might reach around 4% of Norway's total.

The drilling campaign in the undisputed Norwegian part of the Barents Sea in the 2010s could be seen as a fulfillment of the "Scramble for the Arctic" atmosphere that got significant international coverage after Russia planted a flag beneath the North Pole in 2007. This atmosphere was enforced when the 2008 US Geological Survey claimed that 10% of the world's remaining oil resources and 30% of its remaining gas could be found in the Arctic (US Department of the Interior 2008, 4). However, neither the undisputed part of the Norwegian Barents Sea nor the disputed Svalbard Zone were well represented in the report. The report concluded that the area north of Alaska and areas northwest and northeast of Greenland had the most oil, while areas north of Russia contained the most natural gas. Given the potential risk and cost related to developing fields in the Svalbard Zone, if the assessment was right, it would be difficult to argue that any extension of the Norwegian oil industry into this zone was worth the effort.

Partly as a response to the disappointing US numbers, the NPD started its own assessment of the eastern part of the Svalbard Zone in 2013. When a report was presented in the spring of 2017, it was surely what enthusiasts of northern expansion of the Norwegian industry were hoping for (Oljedirektoratet 2017). Before 2017, the NPD's official assessment was that there were still much larger undiscovered resources in the North Sea and the Norwegian Sea than in the Barents Sea. With the new calculations, 60% of undiscovered oil and 66% of undiscovered gas were expected to be found in the Barents Sea. Most of this was in the disputed Svalbard Zone. This clearly signalled that Norway was making preparations to move oil installations into the Svalbard Zone, raising the stakes for other players with interests in the area.

A road to conflict?

The first time the Svalbard issue was raised in a serious way in relation to further expansion of offshore oil activities was during a grandiose speech by the newly appointed Norwegian Foreign Minister, Jonas Gahr Støre, just after the 2005 election (Støre 2005). Gahr Støre announced the Labour Party led government's "northern area policy," which would become a trademark of his 7 years as Foreign Minister. In the speech, Gahr Støre announced that:

> Seeking prior approval from other countries would not harmonize with the Svalbard Treaty, which enshrines full and unrestricted Norwegian sovereignty over the archipelago [...]. We often read that few states have expressed open support for the Norwegian view [...]. In isolation, that is correct. However, at the same time, it turns the issue upside down. It is one of the rarities that states issue statements of support in relation to other states' establishment of zones [...]. Moreover, let me emphasize: Activity builds credible exercise of sovereignty. The fact that we are present as a responsible fishing nation strengthens our position as a responsible manager of fisheries. The same will be the case for the energy chapter that is now opening. (Støre 2005)

Gahr Støre offered the old Norwegian position in relation to the rights in the ocean around Svalbard. However, never before had the argument been presented in so loud and activist a form. In response to the speech, the British Ministry of Foreign Affairs arranged a London meeting with key participants in the Svalbard Treaty, including Russia, but without Norway (*Aftenposten* 2007). This reaction made Gahr Støre more careful. In the following years, he did not raise the argument with the same aggression. Norway went back to the policy of gradually asserting its right to the area by claiming space in the zone. The conservative Erna Solberg government took the same approach from 2013.

While the 2020 version of the "forvaltingsplan" discussed which areas would be opened up for drilling, the Svalbard issue was not mentioned at all. Until the 22nd concession round in 2013, the question was not of acute relevance since the industry was not ready to drill that far north. In 2013, further expanding the Norwegian oil sector to the north was linked to the definition of the "ice edge," when the liberal party Venstre made identifying this limit a condition for supporting what was then a new minority government (Bjørndal 2016). With global warming causing the ice to fluctuate during the year and from year to year, it was a highly controversial way to regulate expansion to the north. Environmental and science communities wanted to define the ice edge so the activity zone was placed as far to the south as possible, while the oil lobby and their political supports tried the opposite. The conflict escalated in 2015 when the government "moved" the ice edge far to the north, without any scientific support, just before the 23rd concession round was announced (Meld. St. 2014–2015). The parliamentary majority reacted and sent a hasty version of the "forvaltningsplan" back to the government. However, this was done without changing the terms for the 23rd round. Hence, the government and the oil lobby got what they wanted: the opportunity to drill in the promising Korpfjell blocks.

The "small-steps-and-not-too-loud-strategy" meant that it was never openly discussed that the government defined the ice edge, which also constituted the overall limit for where drilling was allowed, far into the Svalbard Zone. Again, as long as the blocks were well inside the undisputed Norwegian economic zone, it was not an immediate issue. Nevertheless, Norway was now closing in. The fact that Norway started a significant survey inside the zone was in itself a small step forward, even if this, to some degree, was nullified by the fact that Russia started a similar (but much less thorough) survey. In the 22nd and the 23rd concession rounds, there were blocks in an overlapping zone between Norway's 200 nautical mile economic zone and a 200 mile zone around Svalbard. Russia reacted with a sharp diplomatic protest.[12] However, these blocks did not prompt a strong outcry from other nations, most likely because Norway had a much stronger case in the area. Given that Norway had "full sovereignty" at Svalbard, the Norwegian Zone had been active since 1977, and that "Svalbard" had not yet claimed an economic zone, neutral outsiders might agree that Norway was in a strong position to define where the border between the two zones would be.

Norway, COVID-19, and *The Shock Doctrine*

In *The Shock Doctrine*, Naomi Klein writes about how reforms that have led the world in a more market liberal direction have been introduced in connection with crises or shocks, when opposition to such measures has often been put out of play (Klein 2007). The shocks in question could be anything from wars, to terrorist attacks, to natural disasters. The COVID-19 crisis in the spring of 2020 was such a shock. While the environmental and climate movement in Norway was quarantined with limited opportunities to regroup and mark its opposition through actions and demonstrations, Erna Solberg's government implemented the most comprehensive series of measures to date to facilitate the conditions for further oil expansion to the north.

In April 2020, the government published a new version of a "forvaltningsplan" which presented a new definition of the ice edge (Meld. St. 2019–2020). In the lead up to this, Norsk Polarinstitutt and Havforskningsinstituttet, the two institutions responsible for most of the research determining environmental status in the Barents Sea plan, recommended a safety zone much further to the south than had previously been the case (Ryggvik 2021, 46). However, the government still ended up with a definition further north and well inside the Svalbard Zone. In early June, a clear majority of the Norwegian Parliament voted on an economic stimulus plan for the oil sector that was strongly related to furthering expansion to the north. A week later, the Ministry of Oil and Energy announced a large number of blocks under the so-called TFO system, where blocks could be handed out without any political involvement (Regjeringen.no

19 June 2020). Most of them were in the northern part of the undisputed Norwegian part of the Barents Sea. In late June, when most Norwegians were on summer vacation after COVID restrictions eased, the most extensive and aggressive northward-pointing concession round to date was announced (Regjeringen.no 24 June). In the 25th round, 20 blocks were inside the grey overlapping area between the Svalbard Zone and Norway's economic zone. As many as 14 blocks had a direct border in the fully disputed Svalbard Zone. Norway could not come closer without actually stepping into the Zone. The most northern blocks were within the area that all relevant science institutions wanted to protect.

The oil industrial complex in action

The discussions that led up to tax relief for the oil industry make clear the contradictory situation Norway was in. With the decline in demand for oil in the wake of the pandemic and an ever stronger global understanding that radical measures needed to be introduced to curb CO_2 emissions, a rational response would have been to use the crisis as an opportunity to divest from Norway's dependency on oil and gas. As large oil companies like Shell officially revised their expectations for future demand, oil prices would decline (Randall and Warren 2020).

The Norwegian oil sector's tax system has served the country well in the sense that most of the revenues from large oilfields end up in state coffers. A concrete expression of this is the internationally well-known Norwegian oil fund (in April 2021 worth 1,300 billion US dollars). However, even though the state has received most of the resource rent from Norwegian oil and gas, it has also taken on an increasing amount of the risk involved in new projects because of a number of lucrative tax relief rules. In the 2010s, economists in the Ministry of Finance complained that the system subsidized investments in new oil field development projects. In fact, with an increasing number of marginal fields with potentially costly installations, like in the Barents Sea, it was a clear possibility that the number of projects that would be profitable for operating oil companies would run at a loss for Norwegian society.

Even though Norway's oil production was halved from around 2000 until 2014, the Norwegian offshore supply industry experienced its most intensive growth period to date. This was somewhat related to the industry's international success. Still, the majority of expansion was related to activities on the Norwegian continental shelf. Part of the reason for increased investment while production actually was falling was the fact that new fields were smaller and, as in the Barents Sea, in more remote locations. Norway had an ever-larger offshore supply industry, dependent on ever-smaller oil resources.

Due to overheating earlier in the 2010s, the Norwegian offshore industry was hit quite hard when oil prices fell in the autumn of 2014. Nevertheless, large field development projects that were already underway helped to dampen the effect of the crisis. This applied to the large, 2.5 billion barrel, Johan Sverdrup Field in the North Sea, which was profitable in most price scenarios. Two of the projects that kept the level of investment in the oil sector almost as big as before were the Goliat and Johan Castberg Fields in the Barents Sea. Both were marginal fields in the sense that they were vulnerable to falling oil prices and cost overruns in the construction phase, but with investments up to 7 billion US dollars for both, they were truly mega projects (St-prp. 2008–2009; Prop. 80 S 2017–2018). The industry's major problem, however, was that after the completion of the Castberg installations, there were no correspondingly large projects that could maintain a similar level of investment in the long term.

When the COVID-19 crisis hit hard from March 2020 onwards, parts of the offshore supply industry were immediately affected. At the same time, large offshore yards were working on field development contracts that lasted for years in some cases. This meant that the immediate effect of the crisis was less severe than it was for hotels, restaurants, and the aviation industry. But knowing that the industry would be without large new projects in the medium term, key players in the industry took advantage of the crisis. Camouflaged as a pandemic measure, a group of lobbyists led by the owner of the large offshore companies Kvaerner and Aker Solutions, supported by unions with members of the supplier industry, won approval for a tax proposal that aimed to reduce the risk of new investments dramatically.

Previously, when the state took significant risks in the exploration and development phase, it was justified because it would eventually get the money back through income from large future discoveries. But with the COVID-19 crisis, challenges related to global warming, the large development costs, and the uncertainty associated with the geology of the Barents Sea, this was no longer a "selling" argument on the part of the Norwegian oil lobby. This time, the measures were justified on the grounds that they were necessary to sustain production environments that would be central when Norway necessarily had to go through a "green shift." But the crisis measures were not just about making sure that the competence and capacity of the offshore industry was kept alive. It was directly linked to the strategy of forcing Norwegian oil operations north into the Barents Sea.

The establishment of actual production facilities fell significantly short of the areas up to 74 degrees latitude and even further north into the Svalbard Zone. The only operating oil field, Goliat, was at 71 degrees north, at the height of the North Cape. The Johan Castberg Field would be installed at 72 degrees north, which already represented major challenges both in terms of temperatures and distance from land. By 2020, the industry had not yet demonstrated that operations so far to the north could take place painlessly. Nevertheless, during the negotiations in Parliament, Jonas Gahr Støre succeeded with the support of offshore industry lobbyists and trade unions affiliated with the Labour Party (of which he had now become leader) to gain support for the development of the Wisting Field at 73.3 degrees north (Ryggvik 2021, 510).

The probability that the Wisting Field would ever be profitable was small. For this reason, Equinor, which was the operator of the field in January 2020, concluded that the company would postpone a decision on its development

(Helgesen 2020). Since the Wisting Field was well inside the Polar Front, it would face extreme temperatures in winter. With distances of 360 kilometres to land, it would be very costly to link the field to electricity cables. Since the actual production installation and the oil in the reservoir (which kept a particularly low temperature) required a lot of heating, the production itself would lead to extensive emissions of CO_2. In the North Sea and in the Norwegian Sea, where there is existing infrastructure, there are projects with a relatively short timeline, where production can be stopped before the world's carbon emissions must go to zero before 2050. A project like the Wisting Field, and certainly possible developments within the Svalbard Zone, will have timelines with production at best lasting several decades after 2050. In other words, it was a project that would undermine Norwegian as well as the world community's climate ambitions.

However, while an investment decision for Wisting was risky from a purely financial point of view, it made sense for the Norwegian oil lobby for two reasons. First, investment in the project would ensure a high level of activity in the offshore supply industry during the 2020s. Second, it would create infrastructure crucial to making whole year activity in the Svalbard Zone feasible. With more and more people reliant on such activities, Norway had a large social group with immediate self interest in pressing or supporting a dangerous expansionist policy, whereby Norway might trigger an international crisis around the Svalbard Zone with an unpredictable, potentially dangerous outcome.

Notes

1 The name Svalbard was introduced by Norwegian nationalists and became the formal Norwegian name for the whole archipelago when the treaty was ratified in 1925.
2 Norway's position was at the time defended by Carl August Fleischer, "The Exclusive Economic Zone under the convention regime and in state practice" (The 1982 Convention on the Law of the Sea. Proceedings 1983).
3 See Churchill and Ulfstein (1992); Petersen (2011) and Rossi (2015).
4 It may seem surprising that a treaty from 1920 places so much emphasis on environmental issues. The concern for the environment on Svalbard was real. Many years of intense foreign use, combined with no real authority and regulation, had left Svalbard an industrial junkyard. The treaty's second article states that: "Norway shall be free to maintain, take or decree suitable measures to ensure the preservation, and, if necessary, the re-constitution of the fauna and flora of the said regions, and their territorial waters."
5 Norway's CO_2 emissions increased from 8.2 million tonnes in 1990 to 15.5 million tonnes in 2007. Most of the increase was related to the oil sector. In the 2010s, emissions stabilized around 14 million tonnes per year.
6 The struggle around oil in Lofoten is described in Kristoffersen et al and Dale and Farquharson in this volume.
7 A similar effect, although further to the south and in a more limited area, accrues at the Grand Banks off Newfoundland and Labrador, Canada.
8 Global Positioning Systems (GPS) need access to satellites to calculate exact positions. Most GPS satellites orbit the earth from west to east at lower latitudes than the Arctic circle.
9 For example: the 1969 St. Barbara spill in California, spills from tankers off the coast of Spain and elsewhere throughout the 1970s, and later the 1989 *Exxon Valdez* disaster.

10 Waves on the Norwegian continental shelf are often much higher than that.
11 Partly related to that research, Tromsø University and other knowledge institutions in the north were strengthened.
12 The Russian protest note is commented on by the Vice Secretary of State in the Norwegian Ministry of Oil and Energy (Pedersen 2015).

References

Acona, Hovedrapport. June 2019. *Utredning av feltutbyggingsprosjekter på norsk sokkel.*

Aftenposten. 12 April 2007. "Gahr Støre gikk i taket. Storbritannia har drøftet Svalbards fremtid med EU-land, USA og Russland, men ikke med Norge."

Andersen, Gisle. 2017. "Parlamentets natur, Produksjon av en legitim miljø- og petroleumspolitikk (1945–2013)." Universitetsforlaget. Phd, Universitetet i Bergen.

Arlov, Thor B. 2011. *Den rette mann. Historien om sysselmannen på Svalbard.* Trondheim. https://www.miljostatus.no/vernet-natur/

Barclays. April 2020. "Position on Climate Change," 3.

Bellona, Nyheter. 10 November 2017. Historien om Goliat: En varslet milliardsprekk. Teknisk Ukeblad. Berge, Unni. 2005. "Petroleumsaktivitet i Barentshavet: Konflikt eller sameksistens? Ei organisasjonsteoretisk analyse av to avgjerdsprosessar." Masteroppgåve, Institutt for statsvitenskap. Universitetet i Oslo.

Bjørklund, Kristoffer. 2008. *Caltex-saken og norsk oljepolitikk på Svalbard fra 1960 til 1973.* Masteroppgave i historie Institutt for arkeologi, konservering og historie Universitetet i Oslo.

Bjørndal, Gjesdal. 2016. "Iskanten: På grensen mellom vitenskap og politikk." Masteroppgave, Universitetet i Oslo.

Christensen, Pål. 2014. *Havet, Fisken og oljen: Norges fiskeri og kysthistorie 1970–2014.* Bind IV. Bergen.

Churchill, Robin and Geir Ulfstein. 1992. *Marine Management in Disputed Areas. The Case of the Barents Sea.* London.

CIA. 1976. *US Policy toward Svalbard (Spitsbergen).* February 32, NSC-U/SM-162A. 7.

Coll, Steve. 2013. *Private Empire: ExxonMobilø and American Power.* New York.

Direktoratet for Mineralforvaltning. 2001. "Bergmesteren for Svalbard, Bergverksvirksomhet på Svalbard." *Veileder.* LOV-2001-06-15-79.

Equinor. 27 May 2019. "Årsrapport 2018 til Miljødirektoratet for Hammerfest LNG." Klimautslippvch fra Oslo.

European Parliament. 2017. "An Integrated EU Policy for the Arctic European Parliament Resolution of 16 March 2017 on an Integrated European Union Policy for the Arctic." 2228(INI). § 14.

Fife, Rolf Einar. 12 Dec. 2014. "Folkerettslige spørsmål i tilknytning til Svalbard." *Regjeringen.* https://www.regjeringen.no/no/tema/utenrikssaker/folkerett/svalbard/id2350955/.

Frydenlund, Knut. 1982. *Lille land – Hva nå?.* Universitetsforlaget Oslo.

Helgesen, Ole Ketil. 20 Feb. 2020. "Equinor Rips up the Script at Wisting." *Upstream Online.* https://www.upstreamonline.com/field-development/equinor-rips-up-the-script-at-wisting/2-1-758023.

High North News. 29 October 2015. "Klimaforskning med politisk synsing."

J.P. Morgan Chase & Co. 2020. "Environmental and Social Policy Framework as of February 2020."

Johannesen, Geir Mossige and Jan Stenløkk. 2004. "Oljedrømmer og rå geologi på Norges ytterkant." Norsk Oljemuseum Årbok.

Kirschgaessner, Stephanie. 2019. "Goldman Sachs to Stop Financing New Oil Drilling in the Arctic." *The Guardian*. 16 Dec.

Klein, Naomi. 2007. *Sjokkdoktrinen, Katastrofekapitalismens fremarsj*. Oslo.

Meld. St. 20. 2019–2020. "Helhetlig forvaltningsplaner for de norske havområdene. Barentshavet og havområdene utenfor Lofoten, Norskehavet, og Nordsjøen og Skagerrak."

______. 32. 2015–2016. Svalbard.

______. 20. 2014–2015. "Oppdatering av forvaltningsplanen for Barentshavet og havomr. dene utenfor Lofoten med oppdatert beregning av iskanten."

______. 2008–2009. Nr. 22. Svalbard.

______. 2005–2006. "Helhetlig forvaltning av det marine miljø I Barentshavet og havområdene utenfor Lofoten (forvaltningsplan)." Nr. 8.

______. 1987–1988. "Nasjonalbudsjettet 1988." Nr. 1.

______. 1973–1974. "Petroleumsvirksomhetens plass i det norske samfunnet." Nr. 25.

Miljødirektoratet, Forslag til blokkutlysning i den 24. konsesjonsrunde. Miljødirektoratets vurdering av de foreslåtte blokkene. Oslo 26.04.2017.

Ministry of Foreign Affairs. 15 June 1977. "Regulations on fish Protection Zones off Svalbard." Lovdata. LOV-1976-12-17-91-§1.

NOFO. 2011. Effektiv og robust oljevernberedskap. Brosjyre.

NPD. 2021. "Fact Pages." Last modified 4 May 2021. https://www.npd.no/en/about-us/information-services/open-data/

NRK. October 2015. "Norsk olje og gass i strupen på klimaguru." No. 28.

Oljedirektoratet. 2017. "Barentshavet nord, Geologisk undersøkelse av petroleumsressursene i østlige deler av Barentshavet nord."

______. 2021. "fakta app."

Ot.prp. 2002– 2003. "Om lov om Norges territorialfarvann og tilstøtende sone." Nr. 35

Pedersen, Bård Glad. 3 March 2015. "Tendensiøst fra Klassekampen."

Pedersen, Torbjørn. April 2008. "Conflict and Order in Svalbard Waters." Ph.D. University of Tromsø.

______. 2011. "International Law and Politics in U.S. Policymaking: The United States and the Svalbard Dispute." *Ocean Development & International Law* 42, 120–135.

Prop. 80 S. 2017–2018. "Utbygging og drift av Johan Castbergfeltet."

Randall, Tom and Hayley Warren. 1 Dec. 2020. "Peak Oil Is Suddenly Upon Us." *Bloomberg*. https://www.bloomberg.com/graphics/2020-peak-oil-era-is-suddenly-upon-us/?fbclid=IwAR1OyYkBUe64NzbV2RMTR4CtPp8kmzDDv5OJcRPsMl7cwz0EI82FXjTBM10

Regjeringen.no. 19 June 2020. "Utlysing av TFO 2020." Pressemelding.

______. 24. juni 2020. "Høring om 25. konsesjonsrunde."

Riksrevisjonen. 2018–2019. "Riksrevisjonens undersøkelse av Petroleumstilsynets oppfølging av helse, miljø og sikkerhet i petroleumsvirksomheten." Document 3.6, 2018–2019. https://www.riksrevisjonen.no/globalassets/rapporter/no-2018-2019/petroleumstilsynet.pdf

Rossi, Christopher R. 2015. "'A Unique International Problem': The Svalbard Treaty, Equal Enjoyment, and Terra Nullius: Lessons of Territorial Temptation from History." *Washington University Global Study Law Review* 15 (17).

Ryggvik, Helge. 2010. *The Norwegian Oil Experience: A Toolbox for Managing Resources*. Oslo.

______. 2021. *På kanten. Norsk oljevirksomhet i Barentshavet og hvorfor det må settes en strek*. Oslo.

Sampson, Anthony. 1975. *The Seven Sisters: The Great Oil Companies & the World they Shaped.* London.
Senger, Kim et al. 4 May, 2020. "Petroleum, Coal and Research Drilling Onshore on Svalbard: a Historical Perspective." *Norwegian Journal of Geology* 99 nr. 3.
Stavanger Aftenblad. 16 Nov. 2015. "Norsk Arktis er forskjellig."
St-prp. 2008–2009. "Utbygging og drift av Goliatfeltet." Nr. 24.
Støre, Jonas Gahr. 11 Nov. 2005. *Nordlys.* https://www.nordlys.no/nyheter/gahr-stores-tale/s/1-79-1820864
Støre, Jonas Gahr. 10 Nov. 2005. "Et hav av muligheter – en ansvarlig politikk for Nordområdene." Speech at the Univeristy of Tromsø.
Tarbell, Ida. 2015. *The History of the Standard Oil Company: All Volumes.* New York.
UN. 1987. "Our Common Future." un-documents.net/wced-ocf.htm
Upstream. 6. Nov. 2017. "Goliat set to be a 'loss maker'"
US Department of the Interior. 2008. "U.S. Geological Survey, Circom-Arctic Resource Appraisal: Estimates of Undiscovered Oil and Gas North of the Arctic Circle."
Yergin, Daniel. 1991. *The Prize, the Epic Quest for Oil, Money, and Power.* New York.

12

RAW, DENSE, AND LOUD

A whale's perspective on cold water energy

Amy Donovan

"Eliminating the Ghost"[1]

When she is four years old and they all are moving north, a bloom of long, thin fronds attacks her mother. There is no warning, or none that she can yet identify. Dangling from the fronds there is something heavy like dense and hardened tentacles all massed up. It wrenches and bobs; her mother seems to fight it. But the fronds are coarse and they grip. What does a whale know of grip? She does not grasp, but she learns that other beings do.

She hears her mother's movements strain as the fronds tighten. Over days, her mother becomes thin, weary, less buoyant. Her mother's calls are quieter and they hold pain, and the daughter's body holds this pain too, in her bones, in her blubber, in the increasing effort it takes to keep moving, to keep eating under the weight of all this sadness. She slows to match her mother's pace. She drifts alongside her mother as half of her brain sleeps. The other half retains a reductive sense of her body in the world. One eye remains open. It reminds her to breathe. It reminds her that her dying mother is here. Often a loud roar takes over, sound without voice, filling her body, deafening her ears and blearing her eyes. When she was younger, her mother would raise her own volume to remind her daughter she was still there, that the roar was not everything even if it felt that way. Now her mother hardly calls, she never sings.

Perhaps, the four-year old whale even now bears a memory of ropes, a fear that is generations deep. Perhaps, she watches when her mother takes her final breath. Perhaps, her mother tells her she will not have the strength to surface again.

Somehow, the young whale mourns.

Maybe it is now that she encounters whatever blunt trauma will sever half of her right tail fluke. Maybe the hacked-off piece will dangle for weeks, months.

DOI: 10.4324/9781003024125-12

She may by now be used to dragging heavy things with her body, but is it different when the thing weighing her down is in fact a part of her body?

But maybe she has already experienced this. Maybe the fluke healed long ago, and her mother was there to buoy her as she learned the delicate balance of swimming with an unevenly weighted tail. Maybe now when she tilts her caudal peduncle to propel herself through the water – toward luscious humming krill or toward her sister or north or south with the sun – she tilts the left side a little less.

The tilt does not kill her. She grows to the age when she can have her own calves. She learns time and again what killed her mother because the traplines entangle her too. Time and again she frees herself and she is so tired. She frees herself and perhaps she is finished grieving but she is so tired, too tired to eat much, or move much. She is tired and the ocean is so loud, louder than she remembers, constant interference now, the sea floor throbbing. Her body throbbing with it.[2]

She'll have an obituary in the *Globe and Mail* that explains her migration path and refers to her as Starboard and then as 3603 even though 3603 was what the humans called her first and neither of these, surely, are the name her mother used when she called to her over the roar of the ships.[3]

When she becomes a specimen she'll be EG#4. The information provided with her will be from June 22, 2017 and will state that "She is anchored in fishing gear (most likely snow crab gear) with rope through her mouth and around her right flipper and leading to depth."[4]

For a whale to be anchored is to drown and this will be her final diagnosis. In the several pages of necropsy analysis that follow this, the whale becomes a carcass, referred to not as *she*, but as *it*.[5]

***

Of fewer than 400 living North Atlantic right whales, 17 deaths were recorded along with EG#4 in the same year. Scientists now estimate that number to be 42, bringing the total amount of known deaths between 2017 and 2020 to roughly 56. In November 2020, the population count is 356 (Davie 2020). They die from entanglement and ship strikes (NOAA 2020). Some people point to other factors: speaking out against a proposal to do seismic testing off the coast of Nova Scotia, long time activist Mary Gorman told the *Nova Scotia Advocate* that

> Fishermen will tell you that whales and tuna used to be able to navigate in between nets when the fishermen were out. The whales would come right in, they were so smart. They could snatch those herring before those nets did.
>
> Everyone thinks it's ship strikes and fishing gear, and they are an issue, but they're much less an issue if whales have their senses, if they haven't been blasted into madness, basically. (Devet 2019)

What is madness, for a whale?

***

Their whistles trace worlds.

They move through the salt water like water itself. The voices go high and low. The slopes between these poles are gentle; the whistles oscillate from one note to the next with grace. The sounds travel through air sacs and soft fatty tissues.

For toothed whales, like the pilot whales, dolphins, and sperm whales who live in the North Atlantic, underneath the whistles is a reassuring pulse of clicks. The tissues vibrate. Fat conducts. Each whale shapes the sounds she emits by changing the shape of her melon, a mound of wax and special fats in the forehead that their bodies never burn for energy. The shapes she makes tell the sounds where to go, what to describe. In this way, the whales inscribe the geographies of their worlds.

These geographies have also shaped the ways that toothed whales hear them. Their ears are not connected by bone to their skulls. When their calls return to them, slipping through fats in their jaws, through the sinuses that isolate the ear, the directional information is not skewed by attachment to a vibrating skull. Echoes locate.

Baleen whales like the right whale likely do not echolocate. Nonetheless, they call to one another over tens to hundreds of kilometres. These calls enable them to socialize, to find mates, to keep track of offspring, to express joy or sorrow. Whales' attunement to sound is powerful and intentional, like our human attunement to the visual dimensions of the worlds we encounter. It is through sound that whales focus their attention. It is through sound that they meet what is around them.[6]

When robotic, foreign, repeating sounds, much louder than whales' own calls, an eerie and unfamiliar frequency, bludgeon their sonic environment, the sensory response is manifold and intense. Auditory ganglion cells – nerve cells, information carriers, far more of them than we humans have – in each inner ear send waves of panic through the fluids there. Hair cells light up from all the stimuli. Stereocilia, the organelles we call "hairs," respond by bending. Under a microscope, it might look like a reckless, sinister dance. There is so much sound, too much. It is not stopping.

The bludgeon is anthropogenic sonar. The whales have no way to understand it, and in the "senses-wilting [miasma]" it creates, they are also bereft of their ability to understand anything else (Siebert 2009, n.p). They likely can't hear well enough to communicate their fear to others anyway.

During the panicked cellular dance, sometimes, the whales raise the volume of their voices.[7] Sometimes, they go silent. Sometimes, they flee.[8] It is the latter response that humans most often witness, when panicked and disoriented whales turn up in foreign bodies of water and are unable to find their way out, or end up piled in abstract, reeking heaps on beaches, rock-beaten, and fatally warm. In

2008, for example, roughly 100 melon-headed whales lurched into Madagascar's Loza Lagoon system at least 40 kilometres from the shallowest depth of water – 100 metres – where that species would normally be comfortable (Southall et al. 2013, 11). Some of them travelled 65 kilometres inland and ended up in mangrove estuaries, a place they are equally as ill-equipped to navigate as would be the multi-beam echosounder whose unceasing calls drove them there, or the ExxonMobil contractor operating it.

Whales use sound as a basic function of navigating daily life. Petroleum-dependent human economies have come to "use" it too, as per the quotation below. Oil and gas companies find underwater stores of extractable resources by sending huge blasts of noise into the ocean and listening to how the seafloor responds. ExxonMobil's current website details:

> Sound has been an integral part of oil and gas exploration for the last five decades[...] For years, only a fraction of sound waves were considered usable as computer processing techniques could not yet handle the complexity. Today, our proprietary and patented algorithms, coupled with investments in supercomputers, accomplish the massive computational task of analyzing all seismic data to create a picture of the subsurface in richer detail. (ExxonMobil, "Exploration Technology" n.d.)

The language on the Petroleum Geo-Services (PGS) website becomes almost romantic in its explanation:

> The surface of the ocean acts like a perfect acoustic mirror. Every signal that is reflected back from the subsurface bounces off the sea-surface, creating echoes, or ghosts, that are visible on the seismic data.
>
> GeoStreamer® was designed to mitigate the effect of the ghost reflections coming from the perfectly reflecting sea surface. The ability to remove the ghost accurately allows us to tow the recording sensors deeper, further away from the noise generated by the ocean's surface. (PGS, "Eliminating the Ghost" 2020)

What of the ghosts of the right whales whose home waters, in recent years, PGS' technology has filled with echoes so loud they render the communications even of living whales spectral, too?

★★★

What kinds of energies are depleted when humans and our computers search the layers of sediments under the ocean for stores of petroleum? "The need for energy is universal," declares Exxon's current website. I click through to the "Outlook for Energy" page and learn that "Almost half of the world's energy is dedicated to industrial activity" like building homes and roads. Half: a quantity that, flippantly deployed, denotes a world sapped of vitality beyond the human, in which the universal need for energy is reduced to industrialized society's need for petroleum. Yet, the world is much more than human, and energy is much more than the kind of which half can be used industrially. In this chapter, I problematize the assumption that such publicity statements rest on: that resource exploration adds to the world's stocks of usable energy. In the ocean, resource exploration is also depletion of energy.

I make this argument through attunement to the sensory experiences of the whales who live and die surrounded by anthropogenic sound. This draws attention to the realities whales face, as well as demonstrating that multi-species attunement makes possible new ways of looking back at ourselves and, in this way, may open other avenues for "[living] with each other in the flesh" (Haraway 2003, 17). Such thinking is particularly important for scholars of energy. While the traces I seek here differ from the western, Anglo, human world that is Cara New Daggett's focus in *The Birth of Energy*, I hold with her argument that finding and making visible the "traces" of energy, thermodynamic and otherwise, may present "other ways of knowing and living energy" (Daggett 2019, 7, 11) that the world badly needs.

★★★

June 2017, when EG#4 fought her way through the Gulf of St. Lawrence, struggling, scarred, fell within the May–November "temporal boundaries" – to use the industry's language – for a 3D seismic survey off the Grand Banks of Newfoundland's underwater plateau (LGL Limited 2017). The *MV Ramform Titan* survey vessel was part of PGS' fleet and equipped with GeoStreamer technology to, in the words of a PGS web header, "eliminate ghosts." *Ramform Titan* can carry 24 streamer reels, with a "back deck layout [...] augmented by six independent source array handling booms." PGS' website claims that, together, these features "make it possible to fully utilize the operational weather window." The source array is an array of airguns. The streamers are buoyant cables that drag hydrophones which, after the blasts reach the sea floor, record the airguns' echoes (PGS, "Ramform Titan" 2020).

The North Atlantic was once a haven for myriad marine species because of the nutrients from river systems that fed it and because of the density that bears these nutrients along: slow-moving cold water holds more dissolved gases than fast-moving warm water molecules. Off north-eastern North America, shallow waters from the continental shelf and banks nourish photosynthesizing plankton, and strong currents circulate that food; the murk of cold salty waters is, or at least was historically, a "rich soup [...] the base of the ocean food chain" (Gulf of Maine Aquarium 1995). Because of this, the Grand Banks have historically been known as "one of the richest fishing grounds in the world" (Clarke and Drinkwater 2006). Right whales are among the species who nourish themselves from this rich soup, sucking plankton and krill into waving baleen plates as they swim.

EG#4 and the other whales who died in 2017 were mostly in and around the Gulf of St. Lawrence: not that far offshore. But underwater blasting from vessels like the Ramform fleet can be heard from thousands of kilometres away. A study analysing a decade's worth of underwater recordings from the Mid-Atlantic Ridge found that, in some parts of the North Atlantic, those noises were "present 80–95% of the days/month for more than 12 consecutive months in some locations. When several surveys were recorded simultaneously, whale sounds were masked (drowned out), and the airgun noise became the dominant part of background noise levels" (Weilgart 2013, 1, citing Nieukirk et al. 2012). Even without airgun noise, right whales have lost to ships over 60% of the communication space they are estimated to have had in the middle of the 20th century – a time some of them may be old enough to remember (Hatch et al. 2012, 991). Scientists know from measuring stress hormones that these whales' stress increases when there is shipping noise, which is almost all the time (Rolland et al. 2012), and much of it related to extractive industries: Even without calculating the fossil fuel cost of sundry material goods that arrive in Canada via ship, "crude oil and petroleum products are the largest commodities handled by Canadian ports, representing over 20% of total tonnage" and 30% of global maritime trade. In Canada, most of that traffic – 85%, amounting to 283 million tonnes per year – is along the Atlantic coast, with at least a third of that moving through the Gulf of St. Lawrence (ClearSeas 2020).

EG#4 grew up in the acoustic cloud of this traffic. For roughly half her life, she also may have periodically heard blasting from the seismic program initiated in 2011 by Newfoundland and Labrador's energy corporation, Nalcor, and funded partially by the provincial Department of Natural Resources (Nalcor 2011). Airguns are louder than ships, and their sounds are rhythmic and repeating. Biologist Lindy Weilgart told the *NS Advocate* seismic blasts "raise ocean background noise levels 1,000-fold over areas the size of New Brunswick" (quoted in Devet, 2019). They also kill zooplankton, reducing populations by a quarter as far away as 1,100 metres from the guns and by half between 500 and 650 metres away (McCauley et al. 2017, 3). Almost everyone who lives in the ocean depends indirectly on these tiny organisms for food, but baleen whales do so directly. 2017 was a record-setting year for seismic exploration off Newfoundland and Labrador. Early that June, the province's Minister of Natural Resources told CBC that NL's 3D program was one of the largest in the world, and its 2D program was "unrivalled in the modern exploration era" (Roberts 2017).

In 2018, Nalcor's board of directors voted not to invest in new seismic data collection that year. The provincial government directed them otherwise and gave them $20 million to follow through (Fitzpatrick 2018). I go to the homepage of Nalcor Energy Oil & Gas and click on "Capturing the Insights of Newfoundland and Labrador's Offshore." The introductory sentence is "The North Atlantic is a big place, we're making it a lot smaller".

When EG#4 died, she was thin and weak. What insights died with her?

EG#4's body was found floating roughly 1,000 km away from what the Environmental Assessment Update of the MKI Southern Grand Banks Seismic Program called the Potential 2017 Area of Interest within the Project Area where the *Ramform Titan* was allowed to blast.

Can the human imagination, can our empathy, travel over such distances, as whales do?

***

In singling out the depletion of whales' energy, my most basic meaning is that resource exploration reduces the abundance and diversity of life in the ocean. The energy humanities have room for such thinking. When, for example, Graeme Macdonald argues that writers and scholars must be more attentive to flows of energy, he includes not just petroleum but also food and "the most primary and fundamental energy form: organic muscular exertion" (Macdonald 2013, 21). However, the energy depletion I am thinking of is also of a more complex order: how whales can become so exhausted by what Rob Nixon calls "slow violence" – violence that is not spectacular in any given moment, but is accretive and chronic; violence that is, perhaps, obscured by the biases of human exceptionalism or the western understanding of evidence (Nixon 2011).

This slant approach to thinking the petroculture of the ocean responds to the recent multi-species turn in critical theory, the perspective that we cannot understand human social worlds without taking into account our nonhuman "messmates" (Haraway 2008, 4).[9] It also takes seriously calls like those of Bruno Latour, who argues that it is time to "stop taking nonhumans as objects, [that] as soon as we allow them to enter the collective in the form of new entities with uncertain boundaries, entities that hesitate, quake, and induce perplexity, it is not hard to see that we can grant them the designation of actors" (Latour 2004, 76). This works toward one of Latour's investments in actor network theory: making visible whose interests are really at stake in a given event and thus making heard the political voices of nonhumans.

As Kim TallBear points out, such thinking is not new. She wrote in 2011 that western science studies scholars may be "finally getting on board with something that is closer to [citing Vine Deloria, Jr.] an American Indian metaphysic" (TallBear 2011). This metaphysic can "extend the range of nonhuman beings with which we can be in relation" beyond the conventions of science studies (TallBear 2011). One example of this is Zoe Todd's work: her position that we must decentre the human also resonates with the multi-species approaches I take inspiration from here. But it extends those approaches in several crucial ways, one of which is especially important for my approach: Todd writes that fish "carry stories in their bones" – stories that differ across cultures, but all of which reflect how, in differing Indigenous legal orders, "fish exist simultaneously as many different things and beings and agents" (Todd 2018, 66). So, I would argue, do whales.

Todd is guided in her approach by the words of Leroy Little Bear when he says: "I wonder what scientific formula the fish has discovered. We should ask the fish. They've survived" (cited in Todd 2018, 60). Todd is writing specifically about Indigenous scholarship, but her words and those of Little Bear underline the importance of seeking nonhumans' stories and taking them seriously. If western thinking is ever to shift in a way that enables the planetary care necessary to address the environmental crisis we all share (albeit with very different levels of vulnerability and culpability), western scholars must find ways of hearing those stories too. In Todd's words, "we need to consider that a land without fish is one that will not only leave us physically bereft, but also intellectually and spiritually bereft as well" (Todd 2018, 73).

What of an ocean without whales, or an ocean which, like the contamination in the rivers where Todd's fish run, strips whales of their ability to be intellectual and spiritual beings as well as physical ones? How can I theorize this, and what is at stake when I do? Vanessa Watts' guidance is helpful: she argues that the hierarchical way in which many multi-species scholars frame agency is also a failure of western thought and its commitment to a split between epistemology and ontology. For her, this results in abstractions that displace thought from land and render Indigenous scholars less capable of hearing when the land speaks (Watts 2013). I take such arguments as a call for western scholars like myself to try to undo that divide.

Western bioscience has made great insights in recent years about whales' intra-species cultures.[10] Their brains, much more complex than ours, have "the cell that makes us human," spindle neurons that are the locus of empathy (Coghlan 2006).[11] What might these insights teach me if I could think with them in a way that begins to move beyond the bounds of the western abstract thinking Watts so roundly critiques? To this end, I turn to fleshly "speculative fabulation" and to story (Haraway 2012, 302). If, as Macdonald argues, flows of energy must be taken into account for the cultural study of humans,[12] the same is true of whales, who are intricately and intimately impacted by petroleum energy's extraction. Our human exhaustions are not the only exhaustions. What could whales tell us about energy, if we knew how to listen?

***

The thick, reaching mangrove roots must have been alien to the Madagascar whales, curving and tangling in the water. See the roots on one side, blurry black lines, inestimable expanses of them, a lens of murky water making them eerier still. See, on the other side – perhaps more roots; perhaps also more whales, sluggish now in their confusion, fatigue, hunger. Probably, those other whales are family members or community members. Probably now, at least, it's quiet enough to hear them if they call.

I wonder which was more difficult for them to interpret: these webs of roots grasping at mammals who do not grasp, or the fan of sound waves sweeping their usual habitat. Dead voices, hardly voices at all, "acquiring" the seafloor to "render" it mineable (Rutledge and Leonard 2001). Humans are used to having a single, unified field of vision. Whales, because of the placement and structure of their eyes, have four (Madrigal 2013). What else might such a being notice, stranded in the swamp? There would be some familiar species; there would be some salt, swelling on the tide, lingering as mangrove roots shun it. There would be little plausible food for melon-headed whales, but likely they are too stressed to think about that anyway.

They bump up against the roots. They collapse against rocks. They are limp on the sand. They reel into the round concrete supports for an algae-crusted wharf. These are the images currently available online. In the one I am thinking of,[13] the whale is draped over some kind of rusty pipe, the oxidation broken by patches of dirty, peeling white paint. Or it is the trunk of a sapling shorn of its branches. Anyway, it is low tide, and the ground the body lies on is mud or muddy concrete and the mud encases things like pebbles and broken glass. There are some human onlookers on the wharf above, which is harsh, bringing to mind brutalism or dystopia. It is a poor place for a funeral. Likely, even when the tide comes in, there will be no other whales to keep the body company.

★★★

Before whaling, there would have been 10,000 to 20,000 right whales living in the region of the North Atlantic EG#4 and her mother would have called home (Monsarrat et al. 2016, 1). Of the 350-odd who remain, some have begun to call more loudly to make themselves heard over ship noise (Parks et al. 2011, 33) and likely airgun noise too. They are constantly stressed, something we know because their stress hormones dropped dramatically with the slowdown in ship traffic after the 9/11 terrorist attacks (Rolland et al. 2012). Probably, they are lonely too: So few of them remain. Those who do face diminishing nutritional energy, coupled with the high energy costs both of shouting to be heard and of dragging fishing gear around for days to years – something virtually all of them will do. The ropes break baleen plates, cut off blowholes, chafe to the bone (Kershaw 2019). Even if the whales were not "blasted into madness" (Gorman, cited in Devet 2019), how would they find any energy to think, socialize, or care for one another?

I began to think about this chapter around the beginning of the COVID-19 pandemic that also briefly quieted the oceans. At the time, five companies held the first ever permits for airgun-based geophysical exploration along the Eastern Seaboard. The permits, issued by the National Oceanic and Atmospheric Administration under the direction of the Trump administration, were "incidental harassment authorizations (IHA) [...] to incidentally harass marine mammals during geophysical survey activities in the Atlantic Ocean" (NOAA Fisheries 2018). Now, it is November 2020 and, because of dragged-out court challenges and likely also the spring 2020 pandemic lockdown, the permits have expired, and Trump's presidency is coming to an end.

In environmentalist circles, the mood around this news is celebratory. But around the Grand Banks, seismic exploration continues. Newfoundland's offshore seismic program has been approved until 2023 (LGL 2020, 2). In summer 2020, PGS surveyed about 8,500 square kilometres in the waters around Newfoundland and Labrador (Europétrole, "PGS Completes 2020"). The location of one of those surveys, South Bank 3D, has no landmass separating it from the entrance to the Gulf of St. Lawrence (LGL 2020, 1). Even if the right whales stayed in the Gulf, even as they may have enjoyed slightly less danger from ship strikes and slightly less noise from ship sound during this strange pandemic summer, nothing sheltered them from the sounds of those airgun blasts.

Airgun arrays, according to bioacoustician Christopher Clark, work like lawnmowers. Thirty or forty guns blast simultaneously every few seconds, moving back and forth across large areas of the ocean for months at a time (Clark 2016). In this metaphor, the spatial and visual dimensions – indeed, it is through the *visual* that these machines are mapping, making images with their own echoes – overshadow the acoustic one: lawnmowers are loud, and it is on this level that "mowing the lawn" with airguns does the most damage. In either mode, the phrase begets its own powerful ecological metaphor, lawn mowing displacing ecosystems, bumblebees, grazing herbivores, and vegetable gardens in a bid for visual uniformity and the kind of rationalization that is also the goal when oil companies map the depths of the ocean.

Strandings like the one in Madagascar, like the many other documented instances when whales, having gone silent or become disoriented, often fatally, in response to the disabling of their primary sense,[14] are a warning for as yet less-explored northern waters like the area around the Grand Banks. Two Canadian government reports point out that seismic noise must be distressing to species who live in and move through the North Atlantic, including right whales and Sowerby's beaked whales.[15] Yet, in North Atlantic waters under the jurisdiction of Newfoundland and Labrador regulations and data are sparse. Seismic companies are not required to do acoustic monitoring to check for marine mammals near their surveys, and most do not. In 2017, the year EG#4 and 41 other right whales died, none did (Butler 2019). Testing must stop if a threatened or endangered animal is observed nearby (Butler 2019), but "nearby" means within 500 meters (LGL 2020, 62), and the Grand Banks are the foggiest place on earth (Migiro 2018).

"Seeing" through this fog, on the level of policy or of critique, will require us to reorient our sensory priorities, to take foggy connections seriously and to listen and feel across distances and differences that are sure to disorient us too. If we cannot know with certainty that right whales can hear surveys outside the Gulf, that is precisely why we need critical and artistic tools to make collective decisions when uncertainty is the very condition of possibility for the kind of expansive imagination we need to give the right whale a chance at survival.[16]

Animating the ghost

Harry Brower, Sr. was an Iñupiaq whaler and community leader. Late in his life, he was in a hospital in Anchorage, Alaska when a baby whale came to him and took him to Utqiagvik (formerly Barrow), Alaska. "I wasn't dreaming," he told oral historian Karen Brewster, "I was dead!" (Brower 2004, 156). "I swam with that whale, but I never had no trouble breathing, like I do now. He took me anyplace[...] He told me stories about where his mom and he were shot, how they were wounded, suffered" (Brower 2004, 157). With the calf, Brower watched as his sons killed the calf's mother, in springtime when, according to Brower, his sons should not have been whaling out of concern for precisely this scenario. The calf also showed him an accident that happened during this whaling journey: a rope broke and "hit three guys" (Brower 2004, 159). Brower woke with this knowledge, and it was confirmed via telephone. He tells of how he realized later, with the help of a cleric, that the reason he did not die in the hospital bed was so that he could bring the whale's stories to the whalers so they could correct their harmful practices (Brower 2004, 158).

I am not Indigenous. My writing practice cannot approach such proximity to the whale's psyche. But Brower's story demonstrates both the power and the reality of listening to nonhumans and encourages my belief that it is essential to seek their perspectives. To borrow Thom van Dooren's words:

> The work of holding open the future and responsibly inheriting the past requires new forms of attentiveness to biocultural diversities and their many ghosts. But beyond simply listening, it also requires that we take on the fraught work—never finished, never innocent—of weaving new stories out of this multiplicity: stories within stories that bring together the diversity of voices necessary to inhabit responsibly the rich patterns of interwoven inheritance that constitute our world. (van Dooren 2013, 20)

How might I swim with whales, hear their stories and voices? A starting point may be Watts' insight that "flesh becomes action not because it is material but because it must do so for ears to remain open and low to the ground" (Watts 2013, 33). As a terrestrial animal, working to "get to know the inhabitants of the landscape [...] [using] all the learning practices I can think of" (Tsing 2015, 159), I turn to the digital archive to attempt to access whales' ground.

***

There are YouTube videos of underwater seismic airgun blasts. I put on my AirPods, thinking of the sonic and etymological resonances between the words, thinking of *pods*. The blasts are not quite rhythmic, but they have a rhythm to them, like breath. Their intensity rises and ebbs. When it is louder, even without very good headphones, it seems to fill my consciousness. I stop typing until it quiets. I think I stop thinking.

On the recording, the loudest of the noises shakes. I can tell from its quavering quality the sound is traveling through water. It is like wind, but I have lived next to the ocean for most of my life and have never heard such menace on the wind. The sounds do not relent. Their uniformity becomes more and more monstrous as I listen. I begin to feel them in my stomach. They are not innocent. I am not innocent.

Conventionally, legal thresholds for marine noise levels are those that will not damage whales' hearing permanently. But as Clark points out, there is so much noise now it should be considered chronically debilitating (Clark 2016). For Nixon, part of the challenge of engaging slow violence is that of apprehension and apprehensibility. "Imperceptible threats" pose a sensory, empathetic challenge, and classic environmental thought – like most of the western tradition – privileges "sight above all" (Nixon 2011, 14), thus obscuring the oppression as well as the knowledge that rests on other sensory planes.

I try another exercise: if sound is primary for whales the way vision is for me, what would be the visual equivalent of airgun blasts? Flashes of blinding light come to mind, but I think that it might be more like claustrophobic darkness. Like finding yourself inside a giant throat, muscles contracting around you. There would be a moment in every swallow when you don't know if you can take it, when you wonder if this time you yourself will stop breathing. Maybe the unnatural darkness would stop its rhythmic folding and unfolding over you. Maybe you could wait it out, or live differently somehow, within it. Maybe you would hide, or run away – as I said, sometimes whales do that. They can't always find their way back.[17]

The airgun recording is labelled by the YouTuber who uploaded it as ASMR – "autonomous sensory meridian response" – for humans. It doesn't have this effect for me. But its allusion to what happens during an ASMR experience is poignant: sound engenders, in the human body, the effect of touch, often tingling on the back or neck.[18] I wonder if this odd synaesthesia may offer a small window onto the world of a whale. Sound is always vibration, but in the water, it is more explicitly so. Remember the delicate floating ears, the soft sound lens of fat and wax. The chorus of whales' voices, their echoes, remind me how sound makes geography. In recordings, I can feel space folded into their songs. I can feel how the songs glance off underwater mountains, how they follow the current. I cannot precisely translate the kind of knowledge I think I am receiving when I listen to whales. Yet, "the sphere of the voice is constitutively broader than that of speech: it exceeds it" (Cavarero 2005, 13).

Sensory attunement to the bodily experiences of nonhuman animals is essential for any robust conceptualization of what it means to seek and reap energy from the deep ocean. Tsing writes that "allowing only human protagonists into our stories is [...] a cultural agenda tied to dreams of progress through modernization." Attending to other kinds of stories demonstrates "other ways of making worlds;" it "enlarge[s] what is possible." All this seems to me especially important in petrocultures-oriented criticism, given the close historical ties between petroleum and "progress through modernization" (Tsing 2015, 155). Yet, other stories do not only enlarge what is possible. They expand our understanding of what *is*, and therefore how, and perhaps what, we can think about it.

In this spirit, I ask: What about the whales' listening? For me, the airgun blasts are a powerful, sensory-affective example of how sound can colonize, and how the colonization of sound can reduce space. Listening to them is exhausting, even for me. For the whales, these blasts are perhaps an example of what Lisa Stevenson and Eduardo Kohn call "voice decoupled from speech" (Stevenson and Kohn 2015, 50). Such decoupling, as they argue and as I have attempted to demonstrate in the attunements of this chapter, can be emancipatory, revelatory; it allows an expanded range of voices to resonate, outside logocentrism. The airgun disrupts this. Its voice is devoid of most meaning except the callous logic of extraction and the punch of abstruse fear. It helps to make the North Atlantic smaller.

***

The poet E. J. Pratt grew up on the shores of the North Atlantic in a series of Newfoundland outports (Peach 2006). His 1936 poem "Silences" begins thus: "There is no silence upon the earth or under the earth like the silence under the sea." In the 1930s, Pratt had no reason to know just how incorrect this statement was, but his assumption points to how what Tsing calls "ordinary human bias" is also skewed by the paucity of our range of attention (Tsing 2015, 155). "There are other worlds around us," Hugh Raffles writes. "Too often, we pass through them unknowing, seeing but blind, hearing but deaf, touching but not feeling, contained by the limits of our senses, the banality of our imaginations, our Ptolemaic certitudes" (Raffles 2010, 12). How can I write against the desire for certainty?

Despite some technical inaccuracies, Pratt's poem works against these limits, drawing on gestural, sensory observations and speculations to imagine culture beyond the human world. He lauds the "exquisite edge of the feeling" that is evident in a silent exchange the poet observes between rival cats. "A word would have dulled" that exquisiteness, he writes:

> For only such culture could grow in a climate of silence —
> Away back before emergence of fur or feather, back to the unvocal
> sea and down deep where the darkness spills its wash on the
> threshold of light, where the lids never close upon the eyes, where
> the inhabitants slay in silence and are as silently slain. (Pratt 1936)

In the early to mid-1930s, stress – after an all-time high around the turn of the decade – was low for whales. The Depression had slowed whaling, engendering a period of calm before the sound-storm of World War II (Trumble et al. 2018, 4). Exceeding the limits of his era's scientific awareness, and unconcerned with the gaps across which he equivocates, Pratt's words take on even more significance now, when humans not only fail so frequently to hear – metaphorically or otherwise – the calls of whales below the ocean's surface, we also literally silence their voices by the sheer blare of our presence in their lives. They strand in silence "and are as silently slain."

★★★

There's another poem that seems eerily prescient of contemporary whales' distress and, perhaps, loneliness: D. R. Wagner's 1968 "Occasional Poem #1: The Whales Dream." The poet narrates floating "down the ferby sea," with "mile old blubber," rum and liquor flowing from sea caves:

> [...] and myn self surfing
> tail fin under me sitting
> or standing as it were up
> right or wrong no difference
> made it right and I was spearm.
>
> down the ferby ski
> I drafted far out untill
> when looping back I could
> know longer see what was
> behind me other than
> myself.

The poem's off-kilter grammar, its phonetic or fantastic spellings, its weird two-columned form: all this disoriented me in addition to the drunken drift of the content. I came across the nearly forgotten half-page of verse – a poor-quality scan, the text seeming to almost speak itself, and crackle aloud – from the *Wormwood Review* when looking for a grey whale sleep study I didn't find. Sometimes, images accost us: It was as if, in seeking scientific data to inform a sensory narration of a whale's dream, I was instead thrown *into* the perspective of a disoriented whale. The experience was powerful in part because, like a whale accosted by sonar pings, I wasn't expecting it.

That is the power of poetry, but more broadly, it is the power of an image, its "affective hold" that can render us captive (Stevenson 2014, 45). Lisa Stevenson argues that "psychic life can *only* be captured on the page imagistically, in words or visual images" or, as she demonstrates, the narration of other sensory experiences like sound (Stevenson 2014, 45, emphasis in original; vii–viii). To begin to attune myself to the psyche of the whale, then – that psyche that lives parallel to our own, infused indelibly in our own cultural archive – I need to think through whales' images. How to do so is not obvious, and as much as this chapter is an attempt to think whales' sensory experiences in order to render my writing practice able to hear them, it is also a practical attempt at figuring out how such listening could even be possible. By "image" I mean less something visual than a sensory flash, accessible across taxa. If, as Stevenson (citing Freud) points out, humans think pictorially (Stevenson 2014, 44), the images whales think with are likely precipitated mostly in sound. To forge a conversation across orders of mammals, I navigate intersensorially, interspersing analytical work with my efforts to make narrative that renders "the opacity of an image that can match the density of our feelings [...] something to hold us" (Stevenson 2014, 12).

***

In its coverage of the Madagascar stranding, the International Union for Conservation of Nature (IUCN 2013) published a photo of one of the whale carcasses on a rough wooden cart. The whale's body is patched by sun. Jagged scars are visible across most of its skin. The tail, caudal peduncle – the part of the whale's body where the flukes meet the torso – and the posterior third or so of the dead whale flop off the cart without ceremony, except that the flukes are stiff, curved in a subtle S like inverted violin sound holes. The curve is elegant despite the dry, grassy debris coating the tip of the top fluke. The flipper on the whale's right side is outstretched too, like a drowning person reaching a hand above water.

Roughly 30 human onlookers are gathered around this whale. Many of the adults self-consciously pose – indifferent, concerned or with strained smiles. The children are different. Their faces show sorrow. A wide-eyed boy strains toward the whale; I see the strain in the curve of his eyebrows. The whale's eyes are not visible in the image, but they would be to him. They would still be open.

It's the girl standing just in front of his right arm whose expression seems to evince the extent of the scene's tragedy. Her eyebrows are creased with worry, the left one in a shape that almost perfectly mirrors that of the whale's fluke, even down to the place where the curve is widest: slightly inward from the widest part of her eye, slightly upward of centre on the fluke. Even in the pixelated photo, it's evident that tears are streaming down her cheek. Like the boy's, her lips hang limp, shapeless, seemingly soundless, like the whale whose dead eye I think she is staring into.

The girl's torso is angled away from the whale. Yet, its dead eye has arrested her gaze.

***

Pratt's beginning stanza – "No cries announcing birth,/No sounds declaring death/[...]/The drama is silent" (Pratt 1936) – may point to what philosopher Cora Diamond calls "the difficulty of reality [...] experiences in which we take something in reality to be resistant to our thinking" (Diamond 2008, 45). The surface of the ocean refracts sound; from above, human ears hear the sounds coming from below at greatly diminished volume if we can hear them at all. The reality that under the surface is an almost constant caterwaul is difficult, especially because we depend so heavily on the resources we move around and extract via the blaring machines.

Here, I follow Diamond's invitation to let such difficulties hold us, by taking seriously "the livingness and death of animals" not as facts but "presences that may unseat our reason" (Diamond 2008, 73). Writing nonhuman animals as protagonists[19] defies what Latour calls "one of the main puzzles of western history:" our insistence on *de*-animating the nonhuman world in the face of a liveliness that should be obvious (Latour 2014, 8). I think lively stories, stories which attempt to harness the "coming apart of thought and reality [that] belongs to flesh and blood" (Diamond 2008, 78), open enough space to hold difficult realities in abeyance long enough to be held by them, long enough for us, too, to feel tight plastic ropes sloughing off our skin as the ropes' weight pulls them tighter. This, in turn, makes room for empathy; to borrow the words of Donna Haraway, "If we know well, searching with fingery eyes, we care. That is how responsibility grows" (Haraway 2008, 287).

Some stories we can feel with our fingers. The sensory narration of a whale's panic may enable what John Berger describes when he writes:

> The events of a body's perceptions, when read, become images in the mind [...] The whole of nature is a telltale filter for the intelligence that has passed through it. Our bodies are part of the same filter, and from our bodies come our minds with which we read the tale. (2008, 165)

There's no shortage of critical work cautioning against the reductionism and inaccuracy of falsely equating nonhuman animals' ways of perceiving the world with those of human animals. Yet, the work of the energy humanities is a compelling indicator that, across cultures and time periods, the fact of having some kind of culture of energy is something humans share,[20] an attitude reminiscent of anthropologists' faith in the revelatory potential of studying kinship cross-culturally. In this light, the leap I propose – to extend the lens of energy to our studies of whale cultures – is not so radical. Whether or not they have explicit storytelling cultures – which it seems likely they do – they, like all animals, require energy to live. For them, energy is plankton and krill, is their relationships with their families, is their ability to engage with one another and with their surroundings. If whales do not have the specific energy cultures that arise from the entwinement of energy and economy, this may be a present absence (Stevenson 2014, 40) that makes their cultural attunements all the more interesting. What can be thought, what can be sung, in what ways can relationships happen, in what ways can individuals care for one another, outside of the pervasive fraught presence of petroleum-based fuel?

***

Neuroscientist John Lilly, who famously and infamously worked with dolphins, speculated that aside from the paths humans have taken, there may be "other paths for large brains to take, especially if they live in some other element than air"; for the Cetacea, "which are without benefit of hands or outside constructions of any sort, they may have taken the path of legends and verbal traditions rather than that of written records" (Lilly 1961, 110). These speculations should be inspiring. But if whales' cultural knowledge is in their songs and stories, that knowledge dies with them when they fling themselves onto beaches, starve quickly in muddy lagoons or starve slowly as the effort of carrying a snow crab trap depletes their energy for finding food.

Human science does not have the tools to measure this tragedy. But I make myself pause again because, perhaps, the biggest difficulty of reality of all is that the human experience of time is not nearly as linear, nor so often linear, as the western tradition's narration of it. Berger decries this phenomenon as "a uniquely modern form of egoism" that has broken the interdependence of living and dead, having "disastrous results for the living, who now think of the dead as eliminated" (quoted in Nixon 2011, 18). In Nixon's words, the slow violence in "imposed official landscapes" – an umbrella that comfortably contains the seascape of resource extraction – "[severs] webs of accumulated cultural meaning and treat[s] the landscape as if it were uninhabited by the living, the unborn, and the animate deceased" (Nixon 2011, 18).

A deceased whale is, to gesture toward Claude Lévi-Strauss' famous claim that the empirical realities of some animals are "good to think" (Lévi-Strauss 1962, 89), a good animacy to think with. A single whale fall is an ecosystem unto itself, providing as much nutrition for the ocean as thousands of years' worth of marine snow (Mascarelli 2009). I can conceive the giant body falling to the sea floor, then swathes of flesh detaching and lifting away from all the weight. I can picture the other animals who would consume this flesh and burrow into the body. Even gravity is cyclical. If culture is contained in song, could it persist, as in the throat singing in Stevenson's (2014) ethnography, beyond a potentially dying individual? Could the ocean hold their songs after they die? What do the airgun blasts add to the polyphony? Might stories seep out of the whales' bones[21] and live on in that way?

Sometimes, slow violence in the ocean takes away even this animacy from the deceased and from those left behind. Many whales' fat stores are isolated from the rest of their bodies, so a given whale can carry extremely high levels of toxic pollution without falling ill – that is, unless, like EG#4, the whale becomes hungry and is forced to consume some of her own fat, in which case she risks poisoning herself. Many whales who beach are no longer returned to the ocean by their human finders. For example, belugas in the St. Lawrence River have been categorized as toxic waste. When they die on shore, they are brought to landfills (Giggs 2015).

What does all that organic energy do in a landfill? What song might persist there?

***

In the bottom right corner of the group photo from the Madagascar stranding, only partially in the frame, is a dog caught mid ear-shake. Its ear is open to the whale. The dog is not looking at this scene of death, but perhaps it is listening.

Kohn, with reference to his Runa interlocutors, writes: "How jaguars see us matters to us." Cultural scholarship[22] "cannot limit itself to asking how people see the world [...] by returning the feline's gaze, we allow jaguars the possibility of treating us as selves. If, by contrast, we were to look away, they would treat us as, and we may actually become, objects" (Kohn 2013, 92). It is this risk, perhaps, that the listening dog in the photo is aware of as it cocks its ear. And if it is this that I am working against in this chapter, the potential I am working toward is the same which is activated when "jaguars and humans, in this act of looking back at each other [...] become persons to each other[...] We can become puma by returning a puma's gaze" (Kohn 2013, 93).

One way of returning that gaze, I suggest, is through imagistic engagement with the textures of whales' lives. When we draw images of those we encounter in field research, Michael Taussig argues, we acknowledge and attend to "the charge they put into the world" (Taussig 2011, 78). Placing himself in the position of the viewer, Taussig writes that, looking at a drawing, "I am likely to feel like stretching out my hand to the thing pictured" (Taussig 2011, 24). Images, perhaps, extend the worldly charge they figure so that others who were not there can feel it too. When Kohn returns the jaguar's gaze, so do I. Perhaps, it is this possibility that makes "style and voice [...] so important, for it is they that do the heavy lifting of analysis" (Taussig 2011, 148). Analysis without style and voice – perhaps, in Watts' formulation, epistemology decoupled from ontology (Watts 2013) – cannot hold the weight of what is happening to the whales. And the charge they put into the world is also a kind of energy. Believing, with Kathleen Stewart, that "the densely felt textures of sensory worlding [...] fuel generativity" (Stewart 2011, 451), I want to refocus that energy, against the petroleum-fuelled currents of contemporary capitalism in which both the whales and I are caught.

***

In Olga Tokarczuk's novel *Flights*, Charlotta, the assistant, caregiver, and daughter of a well-known embalmer, travels to an Amsterdam port to deliver specimens for transit to an exhibit. She becomes transfixed by the tattoos of a sailor working on the ship. I think of the little girl in the IUCN image, staring at the dead whale, when one tattoo in particular arrests Charlotta:

> She sees on his shoulder a colourful fish, a great whale, and because the sailor's muscles are working, she has the impression that this whale is alive and that it lives with this man in some sort of unprecedented symbiosis[...] This big, sturdy body makes an enormous impression on her. She feels her legs get slow and heavy, and her body opens up from below, that's how it feels—it opens up, to that shoulder, to that whale. (Tokarczuk 2017, 221)

At first blush, the "big, sturdy body" making the impression on Charlotta would seem to be that of the man. But the paragraph's grammar suggests that Charlotta may, in fact, be impressed by the body of the whale, its image so powerful it rises off of the shoulder on which it is inscribed, ceasing to be an illustration and becoming instead an agent of symbiosis, a "messmate" (Haraway 2008, 4) on whom the sailor may depend as much as the image depends on him. In the face of this, Charlotta

> clenches her teeth so tight her head roars. She starts to walk along the canal toward him, but in the end she slows and stops. She's overcome by a strange feeling that the water here is overflowing onto the banks. Gently, at first checking with the first waves the place of its expansion, then it grows bolder, pours out onto the paving stones[...] Charlotta feels distinctly the weight of the element—her skirts absorb the water, become leaden, she can't move. She feels this flood in every inch of her body and sees the surprised boats battering the trees; always lined up with their bows against the current, now they've lost their direction. (Tokarczuk 2017, 221)

In this scene, the image of the whale incites several transformations in Charlotta. Her "body opens up [...she] clenches her teeth [...] her head roars." She feels that water is filling the area around her. She recognizes water's self-awareness, as it evaluates what space it has then grows bold. The water's weight arrests her movement. She absorbs it into her skin and feels the flood in her body. The scene must only take place over a few minutes, but time during these moments seems to work differently.

For me, the power of this scene is in how, when Charlotta locks her gaze on the whale in first a witnessing (in the man) and then an experience (her own) of intense bodily empathy, all of a sudden the earth asserts itself, overtaking both Charlotta's presence on its surface and the human infrastructure that surrounds her. The agency of the nonhuman world becomes not only obvious but also clamorous and wildly powerful. Charlotta and the boats become disoriented. For scholars of the very extractivism that incites and exacerbates such earthly agency, attunement to that agency as it works around us *always*[23] – not only in what Walter Benjamin calls "moment[s] of danger" (Benjamin 2003, 391), although perhaps all moments are to an extent dangerous – might alert us to what is at work, and what we might build or become, beyond extractivism or petroculture. What, I wonder, is the ecological ethos of Tokarczuk's tattooed sailor?

The novel never returns to Charlotta; instead, in the section that follows, the narration stays with the embalmed specimens on the ship. Charlotta remains on-shore, separated from the sailor and the specimens of which she's taken so much care. The vibrations in her head, their sound, the sudden inundation, the water filling her body but without apparently harming her: Had Charlotta, briefly, become the whale?

Notes

1 Petroleum Geo-Services (PGS) website "Eliminating the Ghost," 2020.
2 The sources for the narrative in these paragraphs include Daoust et al.'s 2018 right whale necropsy report, and Tu Thanh Ha's 2017 *Globe & Mail* article.
3 Ha 2017.
4 Daoust et al. 2018, 161 (p. 1 of EG#4's diagnostic report).
5 Daoust et al. 2018, 161 (p. 1 of EG#4's diagnostic report).
6 The sources for the narrative in this paragraph and those which immediately precede and follow it are: Ketten 1994, 2000; Koopman et al. 2006; Lee 2015; Norris 1980; Wartzok and Ketten 1999.
7 See Hildebrand 2005.
8 On the effects of marine noise on whales, see also Cox et al. 2006; Rolland et al. 2012; Weilgart 2018 (offering an extensive bibliography), Weir 2008; Wright et al. 2007.
9 See also Tsing 2015; Latour 2004, 2005, 2014; Kohn 2013.
10 See Rendell and Whitehead 2001.
11 See also Hof and van der Gucht 2007.
12 Macdonald is referring to literary texts, but I believe his argument can be extended to broader cultural studies. The inter-species leap is my own.
13 International Fund for Animal Welfare 2013.
14 See Weilgart 2013; Siebert 2009; McKenna 2011; Parks et al. 2011; Helble et al. 2020; Hildebrand 2005.
15 See Fisheries and Oceans Canada 2018; Committee on the Status of Endangered Wildlife in Canada 2006.
16 See Latour 2014.
17 See Mazzariol et al. 2011.
18 The phenomenon is little studied, but sometimes also happens as a result of visual stimuli; see Barratt and Davis 2015.
19 See Tsing 2015.
20 See, for example, Jennifer Wenzel's introduction to *Fueling Culture* (Szeman, Wenzel and Yaeger 2017), as well as other works in this volume.
21 See Todd's account of fish bones (Todd 2018, 66)
22 The word Kohn uses is "anthropology," but his broader argument speaks to a much wider disciplinary range.
23 See Latour 2014; Ghosh 2016.

References

Barratt, Emma L., and Nick J. Davis. 2015. "Autonomous Sensory Meridian Response (ASMR): A Flow-like Mental State." *PeerJ* 3: e851. doi: 10.7717/peerj.851.

Benjamin, Walter. 2003. *Selected Writings Volume 4: 1938–1940*, edited by Howard Eiland and Michael W. Jennings. Cambridge, MA: Harvard University Press.

Berger, John. 2008. *From A to X: A Story in Letters*. New York: Verso.

Brower, Harry Sr. 2004. *The Whales, They Give Themselves*, edited by Karen Brewster. Chicago: University of Chicago Press.

Butler, Patrick. 2019. Offshore Oil Exploration in N.L. Often Occurs without Wildlife Monitoring. *CBC Newfoundland & Labrador*. https://www.cbc.ca/news/canada/newfoundland-labrador/offshore-oil-exploration-environmental-monitoring-1.5322388.

Cavarero, Adriana. 2005. *For More than One Voice: Toward a Philosophy of Vocal Expression*. Stanford, CA: Stanford University Press.

Clark, Christopher. 2016. "How Ocean Noise Pollution Wreaks Havoc on Marine Life." Interview by Richard Schiffman. *Yale Environment 360*, 31 March. https://e360.yale.edu/features/how_ocean_noise_pollution_wreaks_havoc_on_marine_life.

Clarke, R. Allyn and Kenneth F. Drinkwater. 2006. "Grand Banks." *The Canadian Encyclopedia*. https://www.thecanadianencyclopedia.ca/en/article/grand-banks

ClearSeas. 2020. Oil Tankers in Canadian Waters. https://clearseas.org/en/tankers/.

Coghlan, Andy. 2006. "Whales Boast the Brain Cells that 'Make us Human.'" *New Scientist*, 27 November. https://www.newscientist.com/article/dn10661-whales-boast-the-brain-cells-that-make-us-human/.

Committee on the Status of Endangered Wildlife in Canada. 2006. "COSEWIC Assessment and Update Status Report on the Sowerby's Beaked Whale *Mesoplodon bidens* in Canada." https://www.sararegistry.gc.ca/virtual_sara/files/cosewic/sr_Sowerbys_Beaked_Whale_0807_e.pdf

Cox, T. M., T. J. Ragen, A. J. Read, E. Vos, R. W. Baird, K. Balcomb, J. Barlow, J. Caldwell, T. Cranford, L. Crum, A. D'Amico, G. D'Spain, A. Fernández, J. Finneran, R. Gentry, W. Girth, F. Gulland, J. Hildebrand, D. Houser, T. Tullar, P. D. Jepson, D. Ketten, C. D. MacLeod, P, Miller, S. Moore, D. C. Mountain, D. Palka, P. Ponganis, S. Rommel, T. Rowles, B. Taylor, P. Tyack, D. Wartzok, R. Gisiner, J. Mead, L. Benner. 2006. "Understanding the Impacts of Anthropogenic Sound on Beaked Whales." *Journal of Cetacean Resource Management* 7 (3): 177–187.

Daggett, Cara New. 2019. *The Birth of Energy*. Durham: Duke University Press.

Daoust, P.-Y., Couture, E.L., Wimmer, T., and Bourque, L. 2018. Incident Report: North Atlantic Right Whale Mortality Event in the Gulf of St. Lawrence, 2018. Collaborative report produced by: Canadian Wildlife Health Cooperative, Marine Animal Response Society, and Fisheries and Oceans Canada. http://www.cwhc-rcsf.ca/docs/technical_reports/NARW_Incident_Report-%2020180405%20MD.pdf.

Davie, Emma. 2020. "New Population Estimate Suggests Only 356 North Atlantic Right Whales Left." *CBC Nova Scotia*. https://www.cbc.ca/news/canada/nova-scotia/356-north-atlantic-right-whales-left-2020-population-1.5779931.

Devet, Robert. 2019. "Documentary Tackles Effect of Deafening Seismic Blasts on Whales and Other Sea Life in Nova Scotia's Offshore." *The Nova Scotia Advocate*, 14 May. https://nsadvocate.org/2019/05/14/documentary-tackles-effect-of-deafening-seismic-blasts-on-whales-and-other-sea-life-in-nova-scotias-offshore/.

Diamond, Cora. 2008. "The Difficulty of Reality and the Difficulty of Philosophy." In *Philosophy and Animal Life*, edited by Stanley Cavell, Cora Diamond, John McDowell, Ian Hacking, and Cary Wolfe, 43–89. New York: Columbia University Press.

Europétrole. 2020. "PGS Completes 2020 Canada Acquisition and Prepares First Data." *Europétrole*, 23 September. https://www.euro-petrole.com/pgs-completes-2020-canada-acquisition-and-prepares-first-data-n-i-21121.

ExxonMobil. n.d. "Exploration Technology." *ExxonMobil*. https://corporate.exxonmobil.com/energy-and-environment/tools-and-processes/exploration-technology.

Fears, Darryl and Dino Grandoni. 2020. "Trump Plan to Allow Seismic Blasts in Atlantic Search for Oil Appears Dead." *The Washington Post*, 1 October. https://www.washingtonpost.com/climate-environment/2020/10/01/offshore-drilling-atlantic/.

Fitzpatrick, Ashley. 2018. "Province Tells Nalcor to Spend on Seismic Data." *The Telegram*, 18 April. https://www.thetelegram.com/business/province-tells-nalcor-to-spend-on-seismic-data-203104/.

Fisheries and Oceans Canada. 2018. "Review of the Effectiveness of Recovery Activities for North Atlantic Right Whales." https://waves-vagues.dfo-mpo.gc.ca/Library/4067986x.pdf

Ghosh, Amitav. 2016. *The Great Derangement: Climate Change and the Unthinkable.* Chicago: University of Chicago Press.

Giggs, Rebecca. 2015. "Whale Fall." *Granta,* 18 November. https://granta.com/whale-fall/.

Gulf of Maine Aquarium. 1995. "Comparing Oceans: Exerpted from Space Available, Aquatic Applications of Satellite Imagery. Gulf of Maine Aquarium. Portland, ME. 1995." *Gulf of Maine Aquarium.* http://www.gma.org/comparing_oceans.html.

Ha, Tu Thanh. 2017. "Starboard's Story: The Life and Death of a Right Whale." *The Globe and Mail,* 6 October. https://www.theglobeandmail.com/news/national/starboards-story-life-and-death-of-a-whale/article36520577/.

Hatch, Leila T., Christopher W. Clark, Sofie M. Van Parijs, Adam S. Frankel, and Dimitri W. Ponirakis. 2012. "Quantifying Loss of Acoustic Communication Space for Right Whales In and Around a US National Marine Sanctuary." *Conservation Biology* 26 (6), 983–994.

Haraway, Donna. 2003. *The Companion Species Manifesto: Dogs, People, and Significant Otherness.* Chicago: University of Chicago Press.

———. 2008. *When Species Meet.* Minneapolis: University of Minnesota Press.

———. 2012. "Awash in Urine: DES and Premarin® in Multispecies Response-ability." *Women's Studies Quarterly* 40 (1/2): 301–316.

Helble, Tyler A., Regina A. Guazzo, Cameron R. Martin, Ian N. Durbach, Gabriela C. Alongi, Stephen W. Martin, John K. Boyle, and E. Elizabeth Henderson. 2020. "Lombard Effect: Minke Whale Boing Call Source Levels Vary with Natural Variations in Ocean Noise." *Journal of the Acoustical Society of America* 147: 698–712. doi:10.1121/10.0000596.

Hildebrand, J. A. 2005. "Impacts of Anthropogenic Sound." In *Marine Mammal Research: Conservation beyond Crisis,* edited by John E. Reynolds III, William F. Perrin, Randall R. Reeves, Suzanne Montgomery, and Timothy J. Ragen, 101–124. Baltimore: Johns Hopkins University Press.

Hof, Patrick R., and Estel Van der Gucht. 2007. "Structure of the Cerebral Cortex of the Humpback Whale, *Megaptera novaeangliae* (Cetacea, Mysticeti, Balaenopteridae)." *The Anatomical Record: Advances in Integrative Anatomy and Evolutionary Biology* 290 (1): 1–31.

International Fund for Animal Welfare. 2013. "2013-09-26-IFAW_Madagascar_5830sonar.jpg." In "Mass Whale Stranding Proven to be Caused by Sonar." *HuffPost,* 26 September. http://images.huffingtonpost.com/2013-09-26-IFAW_Madagascar_5830sonar.jpg.

IUCN – SSC Cetacean Specialist Group. 2013. "WCS_AntsohihyCatolog2.jpg." In "A 2008 Whale Mass Stranding in Madagascar Linked to Sonar Mapping for the First Time." International Union for the Conservation of Nature, 2 October. https://iucn-csg.org/a-2008-whale-mass-stranding-in-madagascar-linked-to-sonar-mapping-for-the-first-time/.

Kershaw, Francine. 2019. "The Cruel Reality of Right Whale Entanglements." NRDC. https://www.nrdc.org/experts/francine-kershaw/cruel-reality-right-whale-entanglements.

Ketten, Darlene R. 1994. "Functional Analyses of Whale Ears: Adaptations for Underwater Hearing." In *Proceedings of OCEANS'94* (1), I/264–I/270. Brest, France: Institute of Electrical and Electronics Engineers. doi:10.1109/OCEANS.1994.363871.

———. 2000. "Cetacean Ears." In *Hearing by Whales and Dolphins,* edited by Whitlow W. L. Au and Richard R. Fay, 43–108. New York: Springer.

Kohn, Eduardo. 2013. *How Forests Think: Toward an Anthropology beyond the Human.* Berkeley, CA: University of California Press.

Koopman, Heather N., Suzanne M. Budge, Darlene R. Ketten, and Sara J. Iverson. 2006. "Topographical Distribution of Lipids inside the Mandibular Fat Bodies of Odontocetes: Remarkable Complexity and Consistency." *IEEE Journal of Oceanic Engineering* 31 (1), 95–106. doi:10.1109/JOE.2006.872205.

Latour, Bruno. 2004. *Politics of Nature: How to Bring the Sciences into Democracy.* Translated by Catherine Porter. Cambridge, MA: Harvard University Press.

———. 2005. *Reassembling the Social: An Introduction to Actor-Network-Theory.* Oxford: Oxford University Press.

———. 2014. "Agency at the Time of the Anthropocene." *New Literary History* 45 (1), 1–18.

Lee, Jane J. 2015. "Can You Hear Me Now? What Whale Ears Have That Ours Don't." *National Geographic,* March 11. https://www.nationalgeographic.com/news/2015/03/150311-whales-hearing-underwater-ears-echolocation-ocean-animals-science/.

Lévi-Strauss, Claude. 1962. *Totemism.* Translated by Rodney Needham. Presses Universitaires de France.

LGL Limited. 2017. "Environmental Assessment Update (2017) of the MKI Southern Grand Banks Seismic Program, 2014–2018." Prepared for Multi Klient Invest AS & TGS_NOPEC Geophysical Company ASA. LGL Report No. FA0119. https://www.cnlopb.ca/assessments/mkisgbss/.

———. 2020. "Environmental Assessment Update (2020) of the MKI Southern Grand Banks Seismic Program, 2018-2023. Prepared for Multi Klient Invest AS. LGL Report No. FA0204. https://www.cnlopb,ca/wp-content/uploads/mkiasseis/2020eaupdate.pdf

Lilly, John. 1961. *Man & Dolphin.* New York: Doubleday.

Macdonald, Graeme. 2013. "Research Note: The Resources of Fiction." *Reviews in Cultural Theory* 4 (2), 1–24.

Madrigal, Alexis. 2013. "You're Eye-to-Eye with a Whale in the Ocean—What Does It See?" *TheAtlantic,* 28 March. https://www.theatlantic.com/technology/archive/2013/03/youre-eye-to-eye-with-a-whale-in-the-ocean-what-does-it-see/274448/.

Mascarelli, Amanda. 2009. "Dead Whales Make for an Underwater Feast." *Audobon News,* November-December. https://www.audubon.org/magazine/november-december-2009/dead-whales-make-underwater-feast.

Mazzariol, Sandro, Giovanni Di Guardo, Antonio Petrella, Letizia Marsili, Cristina M. Fossi, Claudio Leonzio, Nicola Zizzo, Salvatrice Vizzini, Stefania Gaspari, Gianni Pavan, Michela Podestà, Fulvio Garibaldi, Margherita Ferrante, Chiara Copat, Donato Traversa, Federica Marcer, Sabina Airoldi, Alexandros Frantzis, Yara De Bernaldo Quirós, Bruno Cozzi, and Antonio Fernández. 2011. "Sometimes Sperm Whales (*Physeter macrocephalus*) Cannot Find their Way Back to the High Seas: A Multidisciplinary Study on a Mass Stranding." *PLoS ONE* 6 (5): e19417. doi:10.1371/journal.pone.0019417.

McCauley, Robert D., Ryan D. Day, Kerrie M. Swadling, Quinn P. Fitzgibbon, Reg A. Watson, and Jayson M. Semmens. 2017. "Widely Used Marine Seismic Survey Air Gun Operations Negatively Impact Zooplankton." *Nature Ecology & Evolution* 1 (7): 0195.

McKenna, Megan Frances. 2011. "Blue Whale Response to Underwater Noise from Commercial Ships." PhD dissertation, UC San Diego. http://cetus.ucsd.edu/Publications/Dissertations/McKennaPhD2011.pdf.

Migiro, Geoffrey. 2018. "The Foggiest Places on Earth." *World Atlas.* https://www.worldatlas.com/articles/foggiest-places-on-earth.html.

Monsarrat, Sophie, M. Grazia Pennino, Tim D. Smith, Randall R. Reeves, Christine N. Meynard, David M. Kaplan, and Ana SL Rodrigues. 2016. "A Spatially Explicit Estimate of the Prewhaling Abundance of the Endangered North Atlantic Right Whale." *Conservation Biology* 30 (4): 783–791. doi: 10.1111/cobi.12664.

Nalcor Energy Oil and Gas. 2011. "Seismic Survey to Further Exploration Efforts in Offshore Newfoundland and Labrador." https://oilandgas.nalcorenergy.com/press-release/seismic-survey-to-further-exploration-efforts-in-offshore-newfoundland-and-labrador/.

National Oceanic and Atmospheric Administration (NOAA). 2018. "Incidental Harassment Authorization." United States Department of Commerce, 30 November. https://www.fisheries.noaa.gov/webdam/download/84146686.

National Oceanic and Atmospheric Administration (NOAA) Fisheries. 2018. "Incidental Take Authorization: Oil and Gas Industry Geophysical Survey Activity in the Atlantic Ocean: Notice | New England/Mid-Atlantic, Southeast." https://www.fisheries.noaa.gov/action/incidental-take-authorization-oil-and-gas-industry-geophysical-survey-activity-atlantic.

———. 2020. "2017–2020 North Atlantic Right Whale Unusual Mortality Event." https://www.fisheries.noaa.gov/national/marine-life-distress/2017-2020-north-atlantic-right-whale-unusual-mortality-event.

Nixon, Rob. 2011. *Slow Violence and the Environmentalism of the Poor.* Cambridge, MA: Harvard University Press.

Norris, Kenneth S. 1980. "Peripheral Sound Processing in Odontocetes." In *Animal Sonar Systems*, edited by René-Guy Busnel & James F. Fish, 495–509. Boston: Springer. doi:10.1007/978-1-4684-7254-7_21.

Parks, Susan E., Mark Johnson, Douglas Nowacek, and Peter L. Tyack. 2011. "Individual Right Whales Call Louder in Increased Environmental Noise." *Biology Letters* 7, 33–35. doi:10.1098/rsbl.2010.0451.

Peach, Aaron. 2006. "E. J. Pratt (1882–1964)." Heritage Newfoundland and Labrador. https://www.heritage.nf.ca/articles/arts/e-j-pratt.php.

PGS. 2020a. "Eliminating the Ghost." https://www.pgs.com/marine-acquisition/tools-and-techniques/geostreamer/technology/eliminating-the-ghost/.

———. 2020b. "Ramform Titan." https://www.pgs.com/marine-acquisition/tools-and-techniques/the-fleet/active-fleet/ramform-titan/.

Pratt, E. J. 1936. "Silences." *Canadian Poetry Online.* https://canpoetry.library.utoronto.ca/pratt/poem4.htm.

Raffles, Hugh. 2010. *Insectopedia.* New York: Vintage.

Rendell, Luke, and Hal Whitehead. 2001. "Culture in Whales and Dolphins." *Behavioral and Brain Sciences* 24 (2), 309–324.

Roberts, Terry. 2017. "Super-Sized Seismic Activity Planned for Newfoundland's Offshore this Year." *CBC*, 6 June. https://www.cbc.ca/news/canada/newfoundland-labrador/seismic-offshore-newfoundland-1.4132538.

Rolland, Rosalind M., Susan E. Parks, Kathleen E. Hunt, Manuel Castellote, Peter J. Corkeron, Douglas P. Nowacek, Samuel K. Wasser and Scott D. Kraus. 2012. "Evidence that Ship Noise Increases Stress in Right Whales." *Proceedings of the Royal Society B* 279: 2363–2368. doi:10.1098/rspb.2011.2429.

Rutledge, A. K., and D.S. Leonard (ExxonMobil Exploration Company). 2001. "Role of Multibeam Sonar in Oil and Gas Exploration and Development." Offshore Technology Conference, Houston, TX. doi: 10.4043/12956-MS. https://www.onepetro.org/conference-paper/OTC-12956-MS.

Siebert, Charles. 2009. "Watching Whales Watching Us." *New York Times Magazine*, 8 July. https://www.nytimes.com/2009/07/12/magazine/12whales-t.html.

Southall, Brandon L., Teri Rowles, Frances Gulland, Robin W. Baird, and Paul D. Jepson. 2013. "Final Report of the Independent Scientific Review Panel Investigating Potential Contributing Factors to a 2008 Mass Stranding of Melon-headed Whales (*Peponocephala electra*) in Antsohihy, Madagascar." http://www.cascadiaresearch.org/Hawaii/Madagascar_ISRP_Final_report.pdf.

Stevenson. Lisa. 2014. *Life Beside Itself: Imagining Care in the Canadian Arctic.* Berkeley: University of California Press.

Stevenson, Lisa and Eduardo Kohn. 2015. "*Leviathan:* An Ethnographic Dream." *Visual Anthropology Review* 31 (1), 49–53. doi: 10.1111/var.12062.

Stewart, Kathleen. 2011. "Atmospheric Attunements." *Environment & Planning D* 29, 444–453.

Szeman, Imre, Jennifer Wenzel, and Patricia Yaeger. 2017. *Fueling Culture: 101 Words for Energy and Environment.* New York: Fordham University Press.

TallBear, Kim. 2011. "Why Interspecies Thinking Needs Indigenous Standpoints." *Fieldsights,* November 18. https://culanth.org/fieldsights/why-interspecies-thinking-needs-indigenous-standpoints.

Taussig, Michael. 2011. *I Swear I Saw This: Drawings in Fieldwork Notebooks, Namely My Own.* Chicago: University of Chicago Press.

Todd, Zoe. 2018. "Refracting the State through Human-Fish Relations: Fishing, Indigenous Legal Orders and Colonialism in North/Western Canada." *Decolonization: Indigeneity, Education & Society* 7 (1), 60–75.

Tokarczuk, Olga. 2017. *Flights.* Translated by Jennifer Croft. New York: Riverhead.

Trumble, Stephen J., Stephanie A. Norman, Danielle D. Crain, Farzaneh Mansouri, Zach C. Winfield, Richard Sabin, Charles W. Potter, Christine M. Gabriele, and Sascha Usenko. 2018. "Baleen Whale Cortisol Levels Reveal a Physiological Response to 20th Century Whaling." *Nature Communications* 9 (1): 1–8.

Tsing, Anna. 2015. *The Mushroom at the End of the World: On the Possibility of Life in Capitalist Ruins.* Princeton: Princeton University Press.

Van Dooren, Thom. 2013. "Life at the Edge of Extinction: Spectral Crows, Haunted Landscapes and the Environmental Humanities." Humanities Australia. http://www.humanities.org.au/wp-content/uploads/2017/04/AAH-Hancock-Lect-Van-Dooren-2013.pdf.

Wagner, D. R. 1968. "Occasional Poem #1: The Whales Dream." *Wormwood Review* 8 (4), 3. https://ddd.uab.cat/pub/wormwood/wormwood_a1968v8n4i32/wormwood_a1968v8n4i32p3.pdf.

Wartzok, Douglas, and Darlene R. Ketten. 1999. "Marine Mammal Sensory Systems." In *Biology of Marine Mammals*, edited by John E. Reynolds III & Sentiel A. Rommel, 117–175. Washington: Smithsonian Institution Press.

Watts, Vanessa. 2013. "Indigenous Place-Thought and Agency amongst Humans and Non-humans First Woman and Sky Woman go on a European Tour!" *Decolonization, Indigeneity, Education and Society* 2 (1), 20–34.

Weilgart, Lindy. 2013. "A Review of the Impacts of Seismic Airgun Surveys on Marine Life." Submitted to the CBD Expert Workshop on Underwater Noise and its Impacts on Marine and Coastal Biodiversity, 25–27 February 2014, London, UK. http://www.cbd.int/doc/?meeting=MCBEM-2014-01.

———. 2018. "The Impact of Ocean Noise Pollution on Fish and Invertebrates." Report for OceanCare, Switzerland. https://www. oceancare. org/wp-content/uploads/2017/10/OceanNoise_FishInvertebrates_May2018.pdf.

Weir, Caroline R. 2008. "Overt Responses of Humpback Whales (*Megaptera novaeangliae*), Sperm Whales (*Physeter macrocephalus*), and Atlantic Spotted Dolphins (*Stenella frontalis*) to Seismic Exploration off Angola." *Aquatic Mammals* 34 (1): 71–83. doi: 10.1578/AM.34.1.2008.71.

Wright, Andrew J., Natacha Aguilar Soto, Ann Linda Baldwin, Melissa Bateson, Colin M. Beale, Charlotte Clark, Terrence Deak, Elizabeth F. Edwards, Antonio Fernández, Ana Godinho, Leila T. Hatch, Antje Kakuschke, David Lusseau, Daniel Martineau, L. Michael Romero, Linda S. Weilgart, Brendan A. Wintle, Giuseppe Notarbartolo-di-Sciara, and Vidal Martin. 2007. "Do Marine Mammals Experience Stress Related to Anthropogenic Noise?" *International Journal of Comparative Psychology* 20: 274–316.

INDEX